现代数学基础

非线性微分方程的同伦分析方法（下卷）

■ 廖世俊　著
■ 崔继峰　刘　曾　杨小岩　译

中国教育出版传媒集团

高等教育出版社·北京

图书在版编目（CIP）数据

非线性微分方程的同伦分析方法. 下卷 / 廖世俊著；崔继峰，刘曾，杨小岩译. -- 北京：高等教育出版社，2025.1. -- ISBN 978-7-04-062903-3

I. O175

中国国家版本馆 CIP 数据核字第 202415UG48 号

非线性微分方程的同伦分析方法
Feixianxing Weifen Fangcheng de Tonglun Fenxi Fangfa

| 策划编辑 | 吴晓丽 | 责任编辑 | 吴晓丽 | 封面设计 | 张 楠 | 版式设计 | 杜微言 |
| 责任校对 | 马鑫蕊 | 责任印制 | 高 峰 | | | | |

出版发行	高等教育出版社	网 址	http://www.hep.edu.cn
社 址	北京市西城区德外大街 4 号		http://www.hep.com.cn
邮政编码	100120	网上订购	http://www.hepmall.com.cn
印 刷	固安县铭成印刷有限公司		http://www.hepmall.com
开 本	787mm×1092mm 1/16		http://www.hepmall.cn
印 张	22.75		
字 数	430 千字	版 次	2025 年 1 月第 1 版
购书热线	010-58581118	印 次	2025 年 1 月第 1 次印刷
咨询电话	400-810-0598	定 价	79.00 元

本书如有缺页、倒页、脱页等质量问题，请到所购图书销售部门联系调换
版权所有 侵权必究
物 料 号 62903-00

中文版序

从 1989 年本人初步形成"同伦分析方法"的雏形,到 2022 年独立发表一篇论文指出小分母问题在"同伦分析方法"框架下可以完全避免,已历时 33 个春秋;本人从一个 26 岁青年,步入花甲之年. 岁月如梭,令人感叹!

三十多年来,"同伦分析方法"在理论上不断完善和发展,得到国内、国际学术界和工程界越来越多的应用,SCI 他引上万次. 1991 年本人在美国伍兹霍尔 (Woods Hole) 海洋研究所举行的国际学术会议上首次报告这个方法时,称该方法为 "Process Analysis Method",但在 1992 年本人的博士论文中正式改为 "homotopy analysis method" (同伦分析方法),因为一位杂志的匿名评审人指出: 所谓的"连续变化过程"就是拓扑理论中的同伦. "同伦"这个对力学专业来说陌生的名词,让不少初学者望而生畏,也让不少同仁误以为我是数学专业毕业的. 但实际上,"同伦分析方法"并不需要拓扑理论的其他知识,很容易被理工科学生理解和掌握.

早期的同伦分析方法建立在传统的"同伦"概念之上,但并不能确保强非线性问题级数解收敛. 思考了多年之后,本人 1997 年通过引入没有任何物理意义的"收敛控制参数",扩展了传统的同伦概念之内涵,提出"广义同伦"这个全新的概念,并在此新概念基础上重构了同伦分析方法,使其能确保强非线性问题级数解收敛,极大地完善了同伦分析方法,从理论上克服了摄动理论几乎所有的局限性. 此后,同伦分析方法被成功应用于科学、工程、金融等领域各种类型的强非线性问题. 因此,广义同伦概念的提出,是同伦分析方法发展过程中的关键,具有重要的理论意义.

本人毕业于上海交通大学船舶与海洋工程系的潜艇专业，之所以对强非线性问题感兴趣，源于在硕士时期旁听了本系刘应中教授关于摄动理论的一门研究生课程，了解到摄动理论的一些局限性。在德国留学时，本人非常幸运地有选择研究方向的自由，想起了刘先生在课堂上指出的摄动理论之局限性，希望自己能克服它们。现在回想起来，自己当时真是年少无知，根本不知道有多少人曾经在这个研究方向尝试、失败过。但或许正是因为无知，所以无畏：年轻人没有什么害怕失去的。在此，本人要特别感谢刘应中教授。他在授课时不仅传授知识，更批判地指出现有理论和方法的局限性，给年轻人指出可能的前进方向。这种批判比简单地传授知识对学生的影响或许更大。也要特别感谢本人的两位博士生导师：无人深潜器专家朱继懋教授和船舶水动力学专家 H. Soeding 教授，给我探索的自由，宽容我的失败。即使在三十多年后的今天，自由探索和宽容失败，都是非常奢侈的事情。本人确实非常幸运。

今天，同伦分析方法已被成功应用于求解科学、工程、金融等领域的各种类型的非线性问题。这些成功的、广泛的应用，验证了同伦分析方法的有效性和一般性。局限于本人狭窄的研究范围，这种有效性和一般性的验证，是无法由本人和学生完成的。因此，特别感谢国内外不同领域的研究人员不断扩展同伦分析方法的应用范围。

在逻辑上，同伦分析方法包含大多数求解非线性方程的传统解析近似方法，如人工小参数法、δ 展开法、Adomain 分解法，甚至著名的欧拉变换等。除此之外，在同伦分析方法后面提出的一些方法，如 1998 年提出的"同伦摄动方法"(homotopy perturbation method)、2008 年提出的"最优同伦渐近方法"(optimal homotopy asymptotic method) 等，都是同伦分析方法的特例：应用这些方法求解的所有问题，在同伦分析方法框架下都可以获得完全一样的结果 (详见本书第六章)。因此，同伦分析方法理论上更具一般性。

2022 年，本人应用同伦分析方法证明：小分母问题是完全可以避免的 (详见本人发表的论文："Avoiding small denominator problems by means of the homotopy analysis method". Advances in Applied Mathematics and Mechanics, 15(2): 267–299, 2023)。众所周知，小分母问题是非线性力学领域的一个著名难题，且与混沌、三体问题等密切相关，因此具有重要的理论意义。本人 1989 年开始尝试克服摄动理论之局限性时，根本不敢奢想有一天它能求解小分母难题！这再一次显示了同伦分析方法求解非线性难题的能力。实际上，许多弱非线性问题用摄动理论就足够了，没有必要用同伦分析方法求解。同伦分析方法应该被更多地应用于求解其他解析近似方法不能解决的强非线性问题，特别是那些著名的非线性难题.

2002 年本人出版了第一本专著《超越摄动——同伦分析方法导论》(Beyond Perturbation), 介绍同伦分析方法的基本思想和一些应用. 该专著对同伦分析方法的推广和传播发挥了较大作用, 至今已被 SCI 他引四千余次. 2012 年, 为了系统地总结同伦分析方法理论上的最新研究,进展、介绍其一些重要的应用, 本人出版了英文专著 Homotopy Analysis Method in Nonlinear Differential equations, 至今已被 SCI 他引上千次. 衷心感谢崔继峰博士、刘曾博士、杨小岩博士将其翻译成中文. 这对进一步在中国学术界特别是本科生、研究生和青年学者中推广和普及同伦分析方法, 并应用它攻克一些遗留的非线性难题颇有益处.

衷心感谢三十多年来老师、同仁、学生的关心、支持和交流, 衷心感谢家人的关爱、理解和包容, 衷心感谢上海交通大学给予的自由和宽容, 以及德国学术交流中心、国家自然科学基金委员会等机构科研经费的支持.

最后, 致敬让青春自由翱翔的时代.

廖世俊

2023 年春, 上海松江

前言

众所周知，当非线性变强时，非线性问题的摄动解和渐近近似往往失效. 因此，摄动理论和渐近方法一般仅适用于弱非线性方程.

本人1992年提出的同伦分析方法 (Homotopy Analysis Method, 简称 HAM)，是一种适用于强非线性问题的解析近似方法. 首先，不同于摄动理论，同伦分析方法的有效性与是否存在物理小 (大) 参数毫无关系：在同伦分析方法框架内，无论是否存在物理小 (大) 参数，总能将一个非线性问题转化为无穷多个线性子问题. 其次，与其他所有解析近似方法不同，即使是非线性非常强的问题，同伦分析方法也能提供一个方便的途径确保级数的收敛. 而且，基于拓扑理论的同伦概念，同伦分析方法提供了极大的自由以选择初始猜测解、基函数、线性子问题方程类型等，以简便地求解复杂的非线性问题. 最后，同伦分析方法逻辑上包含其他传统的解析近似方法，如 Lyapunov 人工小参数法，Adomian 分解法，δ 展开法，欧拉变换等，更具一般性. 因此，同伦分析方法为求解科学、工程、金融等领域的强非线性问题提供了一个有效的工具.

本专著包含三个部分. 第一部分简述同伦分析方法的基本思想，特别是其理论上的完善和发展，包括最优收敛控制参数的选取，控制和加速收敛的方法，与形变方程、同伦导数算子等相关的定理，与欧拉变换之联系，等等.

第二部分，有感于同伦分析方法在众多不同领域中的成功应用，以及软件 Mathematica 和 Maple 强大的 "基于函数而非数的计算" 的能力，本人提供了一个同伦分析方法求解非线性边值问题的 Mathematica 软件包 BVPh 1.0，并用多个具体实例展示该软件包在求解有限 (或者无限) 区间内具有奇点、多解、多边界条件的强非线性常微分方程之有效性. 应用该软件包, 甚至可以求解一些非线性偏微分方程. 作为开源代码, 读者可在网上免费下载 BVPh 1.0 且本书为读者提供了使用指南.

第三部分, 举例说明应用同伦分析方法可以成功求解一些复杂的强非线性偏微分方程, 从而丰富和加深对这些非线性问题的理解和认识. 例如, 应用同伦分析方法, 成功获得美式认沽期权最优行权边界的一个显式解析近似公式, 其有效时间可长达几十年, 而相应的摄动或者渐近公式, 仅在十几天或者数周内有效. 基于该公式, 本书提供了一个 Mathematica 程序, 可在数秒内获得美式认沽期权的最优行权价格. 再例如, 应用同伦分析方法, 首次发现了任意数量重力行进波的波浪共振条件, 其在逻辑上包含了 Phillips 提出的小振幅条件下的四波共振条件. 所有这些都显示了同伦分析方法在求解科学、工程、金融等领域强非线性问题的原创性、有效性和普遍性.

本书适合应用数学、物理、金融、工程等领域对非线性方程解析近似解感兴趣的科研人员和研究生阅读. 书中所有 Mathematica 程序及其相关输入数据, 都在附录中给出.

感谢多位合作者的有益讨论和交流, 以及多位研究生的辛勤汗水和不懈努力. 感谢国家自然科学基金委员会对该研究提供的经费支持. 最后, 衷心感谢父母、妻子和女儿在过去二十多年来的鼓励和鼎力支持.

廖世俊

2011 年 3 月, 上海

目录

上 卷

第一章 绪论
1.1 研究目的与动机
1.2 同伦分析方法的特点
1.3 本书大纲
参考文献

第二章 同伦分析方法的基本思想
2.1 同伦的概念
2.2 例 2.1: 广义牛顿迭代公式
2.3 例 2.2: 非线性振动
    2.3.1 解的特征分析
    2.3.2 简明数学公式
    2.3.3 同伦级数解的收敛性
    2.3.4 收敛控制参数 c_0 的本质
    2.3.5 利用同伦–帕德方法加速收敛
    2.3.6 通过最优初始近似加速收敛
    2.3.7 通过迭代加速收敛
    2.3.8 辅助线性算子选取的灵活性

2.4　本章小结及讨论

致谢

问题

参考文献

附录 2.1　公式 (2.57) 中 δ_n 的推导

附录 2.2　通过第二种方法推导 (2.55)

附录 2.3　定理 2.3 的证明

附录 2.4　例 2.2 的 Mathematica 程序 (无迭代)

附录 2.5　例 2.2 的 Mathematica 程序 (迭代)

第三章　最优同伦分析方法

3.1　绪论

3.2　说明性描述

  3.2.1　基本思想

  3.2.2　不同类型的优化方法

3.3　系统性描述

3.4　本章小结及讨论

问题

参考文献

附录 3.1　Blasius 流动的 Mathematica 程序

第四章　系统性描述与相关定理

4.1　同伦分析方法的简要框架

4.2　同伦导数的性质

4.3　形变方程

  4.3.1　简史

  4.3.2　高阶形变方程

  4.3.3　例子

4.4　收敛定理

4.5　解表达

  4.5.1　初始近似的选取

  4.5.2　辅助线性算子的选取

4.6　收敛控制与加速

  4.6.1　最优收敛控制参数

4.6.2　最优初始近似

　　　4.6.3　同伦迭代方法

　　　4.6.4　同伦–帕德方法

4.7　值得讨论和有待解决的问题

参考文献

第五章　与欧拉变换的关系

5.1　绪论

5.2　广义泰勒级数

5.3　同伦变换

5.4　同伦分析方法与欧拉变换的关系

5.5　本章小结

致谢

参考文献

第六章　基于同伦分析方法的一些方法

6.1　同伦分析方法简史

6.2　同伦摄动方法

6.3　最优同伦渐近方法

6.4　谱同伦分析方法

6.5　广义边界元方法

6.6　广义比例边界有限元方法

6.7　预测同伦分析方法

参考文献

下　卷

第七章　Mathematica 软件包 BVPh ····································· 1

7.1　绪论 ··· 1

　　　7.1.1　适用范围 ·· 3

　　　7.1.2　简明数学公式 ·· 4

　　　7.1.3　基函数和初始猜测解的选取 ··························· 8

　　　7.1.4　辅助线性算子的选取 ··································· 11

　　　7.1.5　辅助函数的选取 ··· 13

　　　7.1.6　收敛控制参数 c_0 的选取 ····························· 14

7.2 解的近似和迭代 15
7.2.1 多项式 16
7.2.2 三角函数 17
7.2.3 混合基函数 18
7.3 BVPh 1.0 使用指南 21
7.3.1 关键模块 21
7.3.2 控制参数 22
7.3.3 输入 23
7.3.4 输出 24
7.3.5 全局变量 24
参考文献 25
附录 7.1 Mathematica 软件包 BVPh 1.0 31

第八章 具有多解的非线性边值问题 50
8.1 绪论 50
8.2 简明数学公式 51
8.3 例子 54
8.3.1 非线性扩散反应模型 54
8.3.2 非线性三点边值问题 61
8.3.3 具有多解的通道流动问题 68
8.4 本章小结 72
问题 73
参考文献 74
附录 8.1 例 8.3.1 的 BVPh 输入数据 76
附录 8.2 例 8.3.2 的 BVPh 输入数据 78
附录 8.3 例 8.3.3 的 BVPh 输入数据 80

第九章 具有变系数的非线性特征方程 82
9.1 绪论 82
9.2 简明数学公式 84
9.3 例子 89
9.3.1 轴向载荷作用下的非均匀梁 89
9.3.2 Gelfand 方程 101
9.3.3 具有奇点和变系数的方程 105
9.3.4 具有多解的多点边值问题 110

9.3.5 具有复系数的流动稳定性 Orr-Sommerfeld 方程 ················ 115

9.4 本章小结 ·· 119

问题 ··· 121

参考文献 ·· 122

附录 9.1 例 9.3.1 的 BVPh 输入数据 ··································· 124

附录 9.2 例 9.3.2 的 BVPh 输入数据 ··································· 125

附录 9.3 例 9.3.3 的 BVPh 输入数据 ··································· 127

附录 9.4 例 9.3.4 的 BVPh 输入数据 ··································· 128

附录 9.5 例 9.3.5 的 BVPh 输入数据 ··································· 130

第十章　具有无穷多解的边界层流动 ······························· **132**

10.1 绪论 ·· 132

10.2 呈指数衰减的解 ·· 134

10.3 呈代数衰减的解 ·· 139

10.4 本章小结 ··· 146

参考文献 ··· 147

附录 10.1 呈指数衰减的解的 BVPh 输入数据 ······················ 149

附录 10.2 呈代数衰减的解的 BVPh 输入数据 ······················ 150

第十一章　非相似边界层流动 ··· **153**

11.1 绪论 ·· 153

11.2 简明数学公式 ··· 157

11.3 同伦级数解 ·· 161

11.4 本章小结 ··· 166

参考文献 ··· 167

附录 11.1 BVPh 的输入数据 ·· 170

第十二章　非定常边界层流动 ··· **173**

12.1 绪论 ·· 173

12.2 摄动近似 ··· 176

12.3 同伦级数解 ·· 178

　12.3.1 简明数学公式 ··· 178

　12.3.2 同伦近似 ··· 182

  12.4 本章小结 ·················· 188

  参考文献 ·················· 189

  附录 12.1 BVPh 的输入数据 ·················· 190

第十三章 金融领域的应用：美式认沽期权 ·················· **193**

  13.1 数学模型 ·················· 193

  13.2 简明数学公式 ·················· 196

  13.3 显式同伦近似的有效性 ·················· 204

  13.4 可供商业使用的实用程序 ·················· 211

  13.5 本章小结 ·················· 213

  参考文献 ·················· 214

  附录 13.1 $f_n(\tau)$ 和 $g_n(\tau)$ 的详细推导 ·················· 217

  附录 13.2 美式认沽期权的 Mathematica 程序 ·················· 219

  附录 13.3 可供商人使用的 Mathematica 程序 `APOh` ·················· 227

第十四章 二维和三维 Gelfand 方程 ·················· **231**

  14.1 绪论 ·················· 231

  14.2 二维 Gelfand 方程的同伦近似 ·················· 233

    14.2.1 简明数学公式 ·················· 233

    14.2.2 同伦近似 ·················· 239

  14.3 三维 Gelfand 方程的同伦近似 ·················· 246

  14.4 本章小结 ·················· 252

  参考文献 ·················· 253

  附录 14.1 二维 Gelfand 方程的 Mathematica 程序 ·················· 255

  附录 14.2 三维 Gelfand 方程的 Mathematica 程序 ·················· 260

第十五章 非线性水波与非均匀来流的相互作用 ·················· **266**

  15.1 绪论 ·················· 266

  15.2 数学模型 ·················· 267

    15.2.1 原始边值方程 ·················· 267

    15.2.2 Dubreil-Jacotin 变换 ·················· 268

  15.3 简明数学公式 ·················· 270

    15.3.1 解表达 ·················· 270

    15.3.2 零阶形变方程 ·················· 271

| 15.3.3 高阶形变方程 ………………………………………… 273
| 15.3.4 连续求解过程 ………………………………………… 275
| 15.4 同伦近似 …………………………………………………………… 277
| 15.5 本章小结 …………………………………………………………… 289
| 参考文献 …………………………………………………………………… 290
| 附录 15.1 波流相互作用的 Mathematica 程序 …………………………… 292

第十六章　任意数量周期性行进波之间的共振 …………………… 300
  16.1 绪论 ………………………………………………………………… 300
  16.2 两个小幅值基波的共振条件 ………………………………………… 302
      16.2.1 简明数学公式 ………………………………………………… 302
      16.2.2 非共振波 ……………………………………………………… 310
      16.2.3 共振波 ………………………………………………………… 315
  16.3 任意数量基波的共振条件 …………………………………………… 325
      16.3.1 小幅值波的共振条件 ………………………………………… 325
      16.3.2 大幅值波的共振条件 ………………………………………… 328
  16.4 本章小结 …………………………………………………………… 331
  参考文献 …………………………………………………………………… 333
  附录 16.1 高阶形变方程的详细推导 …………………………………… 335

第七章 Mathematica 软件包 BVPh

BVPh 1.0 是为求解具有奇点和 (或) 多边界条件的强非线性边值 (特征值) 问题而开发的 Mathematica 软件包. 该软件包结合了同伦分析方法 (HAM) 和科学计算软件 Mathematica, 为求解非线性常微分方程 (ODEs) 甚至一些非线性偏微分方程 (PDEs) 提供了一个方便的解析工具. 本章简要描述了其适用范围、简明数学公式以及基函数、初始猜测解和辅助线性算子的选取等, 并附有使用指南. 开源代码 BVPh 1.0 在本章的附录中给出.

7.1 绪论

借助高性能计算机和数值方法 (比如 Runge-Kutta 法), 我们可以方便地求得除混沌系统外 [31,49] 的大多数非线性初值问题 (IVPs) 的高精度数值近似解. 然而, 求解边值问题 (BVPs) 却比较困难, 特别是当边值问题具有强非线性、多解、奇点和无限区间时.

BVP4c 是基于 MATLAB 语言 [25,62], 为求解多点边值问题而开发的著名软件包. BVP4c 可以求解许多线性常微分方程 (ODEs). 它采用打靶法 [62], 先添加一些猜测的初始条件将边值问题 (BVP) 转化为初值问题 (IVP), 然后逐步修正, 以满足所有原始边界条件. 因此, BVP4c 本质上是数值方法. BVP4c 通常利用诸如牛顿迭代的传统迭代法求解非线性问题. 遗憾的是, 这些传统的迭代法并不能保证迭代的收敛性, 尤其是当非线性很强时. 除此之外, 利用 BVP4c 寻找具有多解的边值问题的所有解通常是困难的, 特别是当这些解非常接近, 和 (或) 对打靶法的猜测初始条件十分敏感时. 此外, 基于数值计算, BVP4c 很难解决控制方程和 (或) 边界条件中的奇点, 例如当 $z = 0$ 时的 $\sin(\pi z)/z$, 尽管 $\sin(\pi z)/z$ 当 $z \to 0$ 时的极限为 π. 再者, BVP4c 将无限区间视为一种奇点, 并用有限区间替代: 这将给解平添不确定性和不精确性.

Chebfun [17, 70] 是基于 MATLAB 语言 [25, 62], 并由牛津大学的 Nick Trefethen 和 Zachary Battles 自 2002 年起开发的一系列算法和软件系统, 随后特拉华大学的 Toby Driscoll 于 2008 年也参与进来. 如需详细了解 Chebfun, 请参考 [17, 70]. 正如 Chebfun 开发团队所提到的, Chebfun 是基于 Chebyshev 展开、快速 Fourier 变换和重心插值等数值方法开发的一款强有力的 "数值" 求解工具. 不同于 BVP4c, Chebfun 给出的微分方程的解由诸如 Chebyshev 多项式这样的光滑基函数组合而成, 其解可在任意给定点上进行任意次精确微分. 尽管 Chebfun 和 BVP4c 都是基于 MATLAB 语言开发的, 但是两者在本质上截然不同. 特别地, 使用这类基函数, 我们可以自然且相对容易地解决控制方程和 (或) 边界条件中的奇点, 并能精确求解无限区间上的微分方程. 所以, "基于函数而非数的计算" [70] 的确是一个了不起的想法.

虽然 Chebfun 可以方便地求解线性微分方程, 但是只有少许关于非线性微分方程求解的例子. 这主要是因为, Chebfun 同 BVP4c 一样利用牛顿迭代法来求解非线性问题, 但众所周知的是, 牛顿迭代法的收敛性强烈依赖于初始猜测解, 换句话说, 解的收敛性得不到保证. 除此之外, Chebfun 选取不同的猜测近似解来搜索非线性常微分方程的多解. 然而, 如何获得这些猜测近似解尚不明晰. 因此, Chebfun 4.0 求解具有多解的强非线性微分方程似乎很困难.

有感于同伦分析方法 (HAM) [28, 33–42, 44–46, 86] 在众多领域求解强非线性问题 [1–8, 10–16, 19–24, 26, 27, 29–32, 47, 48, 50–52, 54–61, 64–69, 71–85, 87–95] 的普遍有效性以及 Mathematica [9] 和 Maple 等科学计算软件 "基于函数而非数的计算" [70] 的能力, 作者基于 Mathematica 语言开发了求解具有多解和奇点的强非线性微分方程的软件包 BVPh 1.0. 我们的目标是为尽可能多的非线性边值问题开发一种解析工具, 以便能够方便地找到强非线性边值问题的多解, 并且能够很容易地解决控制方程和 (或) 边界条件的无限区间和奇点.

如本书上卷所述, 同伦分析方法比传统的解析近似方法具有更明显的优势. 首先, 同伦分析方法基于拓扑中的同伦概念, 不依赖于物理小 (大) 参数. 因此, 即使非线性问题不包含任何摄动量, 同伦分析方法也是有效的. 其次, 廖世俊教授于 1997 年引入同伦控制参数 c_0, 为保证级数解的收敛性提供了一个简便的途径. 因此, 与其他解析近似方法不同, 同伦分析方法对强非线性问题的求解仍然有效. 再次, 同伦分析方法还为我们选取初始猜测解和辅助线性算子 $\mathcal{L}$ 提供了极大的自由, 从而可以很容易地找到非线性问题的多解. 最后, "基于函数而非数的计算" [70] 的思想, 借助 Mathematica、Maple 等科学计算软件可以简单自然地解决控制方程和 (或) 边界条件的无限区间和奇点. 因此, 在同伦分析方法的框架下, 我们为求解非线性边值问题开发的 Mathematica 软件包具有如下特点:

• **保证收敛**: 在同伦分析方法的框架下, 通过选取适当的收敛控制参数 c_0 来保证级数解的收敛性;

• **多解**: 利用同伦分析方法在选取不同猜测近似解和辅助线性算子上的自由度, 找出非线性问题的多解;

• **奇点**: 运用科学计算软件 Mathematica "基于函数而非数的计算" 的能力来解析解决控制方程和 (或) 边界条件中的奇点;

• **无限区间**: 通过选取无限区间上的适当基函数来解析求解无限区间上的方程.

Mathematica 软件包 BVPh 1.0 将在本章的附录中给出. 这里, 我们简要介绍了它的适用范围、简明数学公式以及基函数、初始猜测解、辅助线性算子 $\mathcal{L}$、辅助函数和收敛控制参数 c_0 的选取, 并给出了使用指南. 本章使用 12 个例子来证明 BVPh 1.0 对某些类型的非线性常微分方程和偏微分方程的有效性, 相应的输入数据文件在附录中给出.

7.1.1 适用范围

BVPh 1.0 为我们提供了求解某些非线性常微分方程 (ODEs) 和偏微分方程 (PDEs) 的解析工具, 主要涉及如下问题:

1. 一类定义在有限区间 $z \in [0, a]$ 上的非线性边值方程 $\mathcal{F}[z, u] = 0$;
2. 一类定义在无限区间 $z \in [b, +\infty)$ 上的非线性边值方程 $\mathcal{F}[z, u] = 0$;
3. 一类定义在有限区间 $z \in [0, a]$ 上的非线性特征方程 $\mathcal{F}[z, u, \lambda] = 0$;
4. 一类与非相似性和 (或) 非定常边界层流动有关的非线性偏微分方程,

其中 $\mathcal{F}$ 为非线性算子, $u(z)$ 为未知函数, λ 为未知特征值, $a > 0$ 和 $b \geqslant 0$ 均为已知常数. 这些问题的边界条件是线性的, 可以在包括两个端点在内的多个点上定义.

后续章节中给出了一些例子来验证 BVPh 1.0 对于上述问题的有效性. 第八章阐述了如何利用不同的初始猜测解和基函数来找到非线性常微分方程的多

解. 第九章用五个例子来说明如何使用 BVPh 1.0 求解具有奇点和 (或) 实 (或复) 变系数和 (或) 多边界条件的强非线性特征值问题. 第十章描述了如何用 BVPh 1.0 求解无限区间上的非线性常微分方程, 其解在无穷远处呈指数或代数衰减. 第十一章和第十二章论述了 BVPh 1.0 甚至可以求解一些非线性偏微分方程, 如与非相似性和 (或) 非定常边界层流动有关的非线性偏微分方程. 这些例子都证明了 BVPh 1.0 对于某些具有多解和奇点的强非线性常微分方程和偏微分方程的有效性.

需要注意的是, BVPh 1.0 只是我们尝试的开始. 未来将发布针对更多类型的, 具有更复杂和更强奇异性的非线性常微分方程和偏微分方程的改进版本. 事实上, 开发一个适用于尽可能多的非线性常微分方程和偏微分方程的 Mathematica 软件包是非常困难的. 正因为如此, 我们的开源代码放在网上供大家下载, 以便不同国家和不同时代的科研人员为共同完成这项艰巨的工作做出贡献, 因为科学是属于全人类的.

7.1.2 简明数学公式

BVPh 1.0 是基于同伦分析方法 (HAM) 开发的. 这里, 我们将简要描述数学公式.

7.1.2.1 有限区间上的边值问题

考虑有限区间上 n 阶非线性常微分方程的非线性边值问题

$$\mathcal{F}[z,u] = 0, \quad z \in [0,a], \tag{7.1}$$

满足 n 个线性边界条件

$$\mathcal{B}_k[z,u] = \gamma_k, \quad 1 \leqslant k \leqslant n, \tag{7.2}$$

其中 $\mathcal{F}$ 为 n 阶非线性算子, $\mathcal{B}_k$ 为线性算子, $u(z)$ 为光滑函数, $a > 0$ 和 γ_k 均为常数. 边界条件可以在包括两个端点在内的多个点上定义. 假设该边值问题至少存在一个解, 并且所有解都是光滑的.

令 $q \in [0,1]$ 表示嵌入变量, $u_0(z)$ 为 $u(z)$ 的初始猜测解. 在同伦分析方法的框架中, 我们构造了这样一种连续的变化 (或形变) $\phi(z;q)$, 即当 q 从 0 增加到 1 时, $\phi(z;q)$ 从初始猜测解 $u_0(z)$ 连续变化到 (7.1) 和 (7.2) 的解 $u(z)$. 这种连续的变化满足零阶形变方程

$$(1-q)\mathcal{L}\left[\phi(z;q) - u_0(z)\right] = c_0 q H(z)\mathcal{F}[z,\phi(z;q)], z \in [0,a], q \in [0,1] \tag{7.3}$$

和边界条件
$$\mathcal{B}_k[z,\phi(z;q)] = \gamma_k, \quad 1 \leqslant k \leqslant n, \tag{7.4}$$

其中 $\mathcal{L}$ 为辅助线性算子, c_0 为收敛控制参数, $H(z)$ 为辅助函数.

如第一部分所述, 同伦分析方法为我们选取辅助线性算子 $\mathcal{L}$、收敛控制参数 c_0 和辅助函数 $H(z)$ 提供了极大的自由度. 假设它们都选取适当, 则同伦-Maclaurin 级数

$$\phi(z;q) = u_0(z) + \sum_{m=1}^{+\infty} u_m(z)q^m \tag{7.5}$$

在 $q=1$ 时绝对收敛, 其中

$$u_m(z) = \mathcal{D}_m[\phi(z;q)] = \frac{1}{m!}\left.\frac{\partial^m \phi(z;q)}{\partial q^m}\right|_{q=0},$$

这里 $\mathcal{D}_m$ 称为 m 阶同伦导数算子. 然后, 我们得到同伦级数解

$$u(z) = u_0(z) + \sum_{m=1}^{+\infty} u_m(z), \tag{7.6}$$

其中 $u_m(z)$ 满足 m 阶形变方程

$$\mathcal{L}[u_m(z) - \chi_m u_{m-1}(z)] = c_0 H(z)\delta_{m-1}(z), \quad z \in [0,a] \tag{7.7}$$

和 n 个线性边界条件

$$\mathcal{B}_k[z, u_m(z)] = 0, \quad 1 \leqslant k \leqslant n, \tag{7.8}$$

其中

$$\chi_m = \begin{cases} 0, & m \leqslant 1, \\ 1, & m > 1 \end{cases} \tag{7.9}$$

和

$$\delta_k(z) = \mathcal{D}_k\{\mathcal{F}[z,\phi(z;q)]\} = \left\{\frac{1}{k!}\frac{\partial^k \mathcal{F}[z,\phi(z;q)]}{\partial q^k}\right\}\bigg|_{q=0} \tag{7.10}$$

由上卷第四章中证明的定理很容易得到.

需要注意的是, m 阶形变方程 (7.7) 是线性的, 且满足 n 个线性边界条件 (7.8), 因此在理论上很容易求解, 特别是利用 Mathematica 这样的科学计算软

件. 特别地, 收敛控制参数 c_0 为保证同伦级数解 (7.6) 的收敛性提供了一个简便的途径, 其最优值由控制方程 (7.1) 在 $u(z)$ 的高阶近似下的平方残差的最小值确定. 详情请参考第八章.

还要注意的是, 对于有限区间 $z \in [0, a]$ 上的边值问题, 我们应该为 BVPh 1.0 设置控制参数 TypeEQ=1.

7.1.2.2 无限区间上的边值问题

考虑无限区间上 n 阶非线性常微分方程的边值问题

$$\mathcal{F}[z, u] = 0, \quad z \in [b, +\infty), \tag{7.11}$$

满足 n 个线性边界条件

$$\mathcal{B}_k[z, u] = \gamma_k, \quad 1 \leqslant k \leqslant n, \tag{7.12}$$

其中 $\mathcal{F}$ 为 n 阶非线性算子, $\mathcal{B}_k$ 为线性算子, $u(z)$ 为光滑函数, $b \geqslant 0$ 和 γ_k 均为常数. 边界条件可以在包括两个端点在内的多个点上定义. 假设该边值问题至少存在一个解, 并且所有解都是光滑的.

除有限区间 $z \in [0, a]$ 被替换为无限区间 $z \in [b, +\infty)$ 外, 所有相关公式均与本小节给出的公式相同. 然而, 它们的解是由完全不同的基函数表示的, 所以这两种类型的边值问题使用了完全不同的初始猜测解 $u_0(z)$ 和辅助线性算子 $\mathcal{L}$. 详情请参考第十章.

当使用 BVPh 1.0 求解无限区间 $z \in [b, +\infty)$ 上的边值问题时, 我们应该设置 TypeEQ=1, zR=infinity.

7.1.2.3 有限区间上的特征值问题

考虑有限区间上 n 阶非线性常微分方程的特征值问题

$$\mathcal{F}[z, u, \lambda] = 0, \quad z \in [0, a], \tag{7.13}$$

满足 n 个线性边界条件

$$\mathcal{B}_k[z, u] = \gamma_k, \quad 1 \leqslant k \leqslant n, \tag{7.14}$$

其中 $\mathcal{F}$ 为 n 阶非线性算子, $\mathcal{B}_k$ 为线性算子, $u(z)$ 为光滑的特征函数, $a > 0$ 和 γ_k 均为常数. 边界条件可以在包括两个端点在内的多个点上定义. 假设该特征值问题至少存在一个特征函数和一个特征值, 并且所有特征函数都是光滑的.

特征值问题通常有无穷多个特征函数和特征值. 为了区分不同的特征函数和特征值, 我们添加附加边界条件

$$\mathcal{B}_0[z, u] = \gamma_0, \tag{7.15}$$

其中 $\mathcal{B}_0$ 为线性算子, γ_0 为常数.

令 $q \in [0,1]$ 表示嵌入变量, $u_0(z)$ 为特征函数 $u(z)$ 的初始猜测解, λ_0 为特征值 λ 的初始猜测值. 在同伦分析方法的框架中, 我们构造了 $\phi(z;q)$ 和 $\Lambda(q)$ 这两个连续变化 (或形变), 即当 q 从 0 增加到 1 时, $\phi(z;q)$ 从初始猜测解 $u_0(z)$ 连续变化到 (7.13) 和 (7.14) 的特征函数 $u(z)$, $\Lambda(q)$ 从初始猜测值 λ_0 连续变化到特征值 λ. 这两个连续变化满足零阶形变方程

$$(1-q)\mathcal{L}\left[\phi(z;q) - u_0(z)\right] = c_0 q H(z) \mathcal{F}[z, \phi(z;q), \Lambda(q)], \quad z \in [0, a] \tag{7.16}$$

和 n 个边界条件

$$\mathcal{B}_k[z, \phi(z;q)] = \gamma_k, \quad 1 \leqslant k \leqslant n, \tag{7.17}$$

以及附加边界条件

$$\mathcal{B}_0[z, \phi(z;q)] = \gamma_0, \tag{7.18}$$

其中 $\mathcal{L}$ 为辅助线性算子, c_0 为所谓的收敛控制参数, $H(z)$ 为辅助函数.

同伦分析方法为我们选取辅助线性算子 $\mathcal{L}$、收敛控制参数 c_0 和辅助函数 $H(z)$ 提供了极大的自由度. 假设它们都选取适当, 则同伦–Maclaurin 级数

$$\phi(z;q) = u_0(z) + \sum_{m=1}^{+\infty} u_m(z) q^m, \quad \Lambda(q) = \lambda_0 + \sum_{m=1}^{+\infty} \lambda_m q^m \tag{7.19}$$

在 $q = 1$ 时绝对收敛. 然后, 我们得到同伦级数解

$$u(z) = u_0(z) + \sum_{m=1}^{+\infty} u_m(z), \quad \lambda = \lambda_0 + \sum_{m=1}^{+\infty} \lambda_m, \tag{7.20}$$

其中未知的 $u_m(z)$ 满足 m 阶形变方程

$$\mathcal{L}[u_m(z) - \chi_m u_{m-1}(z)] = c_0 H(z) \delta_{m-1}(z), \quad z \in [0, a] \tag{7.21}$$

和 n 个线性边界条件

$$\mathcal{B}_k[z, u_m(z)] = 0, \quad 1 \leqslant k \leqslant n, \tag{7.22}$$

其中

$$\delta_k(z) = \mathcal{D}_k\{\mathcal{F}[z, \phi(z;q), \Lambda(q)]\} = \left\{\frac{1}{k!}\frac{\partial^k \mathcal{F}[z, \phi(z;q), \Lambda(q)]}{\partial q^k}\right\}\bigg|_{q=0} \quad (7.23)$$

由第四章中证明的定理很容易得到. 这里, $\mathcal{D}_k$ 是 k 阶同伦导数算子. 注意, (7.21) 中的 $\delta_{m-1}(z)$ 包含

$$\lambda_0, \lambda_1, \lambda_2, \cdots, \lambda_{m-1}.$$

另外, 未知项 λ_{m-1} 由附加边界条件

$$\mathcal{B}_0[z, u_m(z)] = 0 \quad (7.24)$$

确定. 详情请参考第九章.

事实上, BVPh 1.0 适用于更广泛的边界条件

$$\mathcal{B}_k[z, u, \lambda] = 0, \quad 1 \leqslant k \leqslant n, \quad (7.25)$$

它包含未知的特征值 λ.

需要注意的是, 当使用 BVPh 1.0 求解有限区间 $z \in [0, a]$ 上的特征值问题时, 我们应该设置 `TypeEQ=2`.

7.1.3 基函数和初始猜测解的选取

在同伦分析方法的框架中, 我们先应选取一组基函数

$$\{e_0(z), e_1(z), e_2(z), \cdots\},$$

它足以近似非线性边值问题的未知解 $u(z)$. 换句话说, 我们应该选取这样一组基函数, 使得 $u(z)$ 可以表示为

$$u(z) = \sum_{m=0}^{+\infty} a_m e_m(z), \quad (7.26)$$

其中 a_m 是系数. 上述表达式称为 $u(z)$ 的解表达, 它在同伦分析方法的框架中对辅助线性算子 $\mathcal{L}$、辅助函数 $H(z)$ 和初始猜测解 $u_0(z)$ 的选取起着重要作用.

对于有限区间 $z \in [0, a]$ 上的边值 (特征值) 问题, 其解可以用不同的基函数表示, 如幂级数

$$u(z) = \sum_{m=0}^{+\infty} A_m z^m, \quad (7.27)$$

或 Chebyshev 级数 [18,53]

$$u(z) = \sum_{m=0}^{+\infty} B_m T_m(z), \tag{7.28}$$

其中 $T_m(z)$ 为 m 阶第一类 Chebyshev 多项式, A_m 和 B_m 均为系数. 此外, 众所周知, 有限区间 $z \in [0, a]$ 上的光滑函数 $u(z)$ 可以用 Fourier 级数

$$u(z) = \sum_{m=0}^{+\infty} \left(\bar{A}_m \cos \frac{m\pi z}{a} + \bar{B}_m \sin \frac{m\pi z}{a} \right) \tag{7.29}$$

表示, 其中 $\bar{A}_m$, $\bar{B}_m$ 为系数. 此外, $u(z)$ 可以用混合基更有效地表示, 即多项式和三角函数的一种组合, 如 §7.2.3 所述. 因此, 多项式、三角函数及其组合提供了一组完整的基函数来近似有限区间 $z \in [0, 1]$ 上的光滑解 $u(z)$.

此外, 在求解给定的边值问题之前, 通常可以得到解的性质, 这对于我们给出更精确的解表达很有价值. 例如, 如果有限区间 $z \in [0, a]$ 上边值问题的解是奇函数解, 则相应的解表达用幂级数表示为

$$u(z) = \sum_{m=0}^{+\infty} A_{2m+1} z^{2m+1}, \tag{7.30}$$

用 Fourier 级数表示为

$$u(z) = \sum_{m=0}^{+\infty} \bar{B}_m \sin \frac{m\pi z}{a}, \tag{7.31}$$

类似地, 对于 $z \in [0, a]$ 上边值问题的偶函数解, 则相应的解表达用幂级数表示为

$$u(z) = \sum_{m=0}^{+\infty} A_{2m} z^{2m}, \tag{7.32}$$

用 Fourier 级数表示为

$$u(z) = \sum_{m=0}^{+\infty} \bar{A}_m \cos \frac{m\pi z}{a}. \tag{7.33}$$

对于无限区间 $z \in [b, +\infty)$ 上的边值 (特征值) 问题, 其中 $b \geqslant 0$ 是有界常数, 对于呈指数衰减的解 (在无穷远处), $u(z)$ 可以表示为

$$u(z) = \sum_{k=0}^{+\infty} \sum_{m=0}^{+\infty} \alpha_{k,m} z^k \exp(-m\gamma z), \tag{7.34}$$

或者，对于呈代数衰减的解 (在无穷远处)，$u(z)$ 可以表示为

$$u(z) = \sum_{m=0}^{+\infty} \frac{\beta_m}{(1+\gamma z)^m}, \tag{7.35}$$

其中 $\gamma > 0$ 为参数，$\alpha_{k,m}$ 和 β_m 均为系数.

初始猜测解 $u_0(z)$ 应符合解表达 (7.26)，并且尽可能满足所有边界条件. 因此，初始猜测解 $u_0(z)$ 通常表示为

$$u_0(z) = \sum_{m=1}^{K} b_m e_m(z), \quad K \geqslant n, \tag{7.36}$$

其中 n 为所有边界条件的个数，b_m 为系数，$e_m(z)$ 表示基函数，$K \geqslant n$ 为正整数. 在 $K = n$ 的情况下，初始猜测解由 n 个边界条件确定. 然而，当 $\mu = K - n > 0$ 时，这种初始猜测解 $u_0(z)$ 为我们提供了 μ 个额外的自由度. 这种未知参数被称为多解控制参数，为我们寻找多解和 (或) 保证同伦级数的收敛性提供了一个简便的途径，如第八章和第九章所述.

需要注意的是，对于上述边值 (特征值) 问题，基函数的选取主要由区间和解的性质决定，而不由控制方程决定. 因此，有限区间 $z \in [0, a]$ 上的光滑函数应该用 (7.27)，(7.28) 和 (7.29) 近似，但对于无限区间 $z \in [b, +\infty)$ 上的光滑函数，其中 $b > 0$，呈指数衰减的解应该用 (7.34) 近似，呈代数衰减的解应该用 (7.35) 近似. 给定一个非线性边值问题，在求解前通过分析控制方程和 (或) 边界条件，一般不难得出解的某些性质，如解在无穷远处的渐近性质、奇偶性和对称性等. 所有这些解的性质都为我们选取足够好的基函数提供了有价值的信息.

应该强调的是，与以物理小 (大) 参数为出发点的摄动方法不同，我们将解的基函数作为同伦分析方法的出发点. 这主要是因为基函数是解析近似解的关键：它与数值近似的数字一样重要. "基于函数而非数的计算" 意味着基函数对 BVPh 1.0 非常重要.

还要注意的是，使用基函数而非数字，并且通过科学计算软件，许多奇异项，如当 $z \to 0$ 时的 $\sin(\pi z)/z$ 可以很容易解决，因为当 $z \to 0$ 时，$\sin(\pi z)/z$ 的分子 $\sin(\pi z)$ 和分母 z 都被视为函数，而非数字 0[①]，除此之外，用科学计算软件 Mathematica 很容易求得极限

$$\lim_{z \to 0} \frac{\sin(\pi z)}{z} = \pi.$$

[①] 注意，0/0 型未定式对数值计算没有意义，这会导致许多数值工具 (如 BVP4c) 出现奇点.

此外，某些类型的基函数在无限区间上定义良好，因此它们可以方便地用于高精度近似无限区间上非线性边值问题的解. 这样，BVPh 1.0 的许多类型的奇点都可以通过科学计算软件 Mathematica 提供的"基于函数而非数的计算"能力轻松解决.

由于有限区间 $z \in [0, a]$ 上的所有光滑函数都可以用 Chebyshev 级数 (7.28) 和 (或) Fourier 级数 (7.29) 来表示，因此有限区间 $z \in [0, a]$ 上完全不同的边值方程在同伦分析方法的框架中可能具有相同的解表达. 这表明开发一个适用于一般非线性边值问题的 Mathematica 软件包是有可能的.

7.1.4 辅助线性算子的选取

辅助线性算子 $\mathcal{L}$ 的选取主要由解 $u(z)$ 的基函数决定. 原则上，应该适当选取辅助线性算子 $\mathcal{L}$，使得

- 所有解表达必须满足;
- 所有高阶形变方程都有唯一解，并可通过 Mathematica 等科学计算软件轻松求解;
- 所有同伦级数的收敛性可通过选取适当的收敛控制参数来保证.

对于有限区间 $z \in [0, a]$ 上的 n 阶边值 (特征值) 方程，当 $u(z)$ 用幂级数 (7.27) 或 Chebyshev 级数 (7.28) 表示时，我们选取辅助线性算子

$$\mathcal{L}(u) = \frac{d^n u(z)}{dz^n}. \tag{7.37}$$

在这种情况下，我们应该为 BVPh 1.0 设置 `TypeL = 1`.

但是，当 $u(z)$ 用 Fourier 级数 (7.29) 或 §7.2.3 中所述的混合基函数来表示时，我们通常会选取如下辅助线性算子

$$\begin{aligned}
\mathcal{L}(u) &= u'' + \omega_1^2 u, & n &= 2, \\
\mathcal{L}(u) &= u''' + \omega_1^2 u', & n &= 3, \\
\mathcal{L}(u) &= u'''' + (\omega_1^2 + \omega_2^2) u'' + \omega_1^2 \omega_2^2 u, & n &= 4, \\
\mathcal{L}(u) &= u''''' + (\omega_1^2 + \omega_2^2) u''' + \omega_1^2 \omega_2^2 u', & n &= 5, \\
&\cdots\cdots\cdots\cdots
\end{aligned}$$

对于有限区间 $z \in [0, a]$ 上的 n 阶边值方程，其中 $'$ 表示对 z 求导，$\omega_i > 0$ 为频率. 根据不同的边界条件，$\omega_i > 0$ 可能互不相同，如

$$\omega_i = i\left(\frac{\kappa \pi}{a}\right), \tag{7.38}$$

也可能相同，如

$$\omega_i = \frac{\kappa\pi}{a}, \tag{7.39}$$

其中 $\kappa \geqslant 1$ 和 $i \geqslant 1$ 是正整数. 上述辅助线性算子可以表示为

$$\mathcal{L}(u) = \left[\prod_{i=1}^{m}\left(\frac{d^2}{dx^2} + \omega_i^2\right)\right]u, \quad n = 2m \tag{7.40}$$

或

$$\mathcal{L}(u) = \left[\prod_{i=1}^{m}\left(\frac{d^2}{dx^2} + \omega_i^2\right)\right]u', \quad n = 2m+1. \tag{7.41}$$

需要注意的是，它们都具有性质

$$\mathcal{L}[\cos(\omega_i z)] = \mathcal{L}[\sin(\omega_i z)] = 0,$$

其中 $\omega_i > 0$ 为上述提及的频率. 在 (7.38) 或 (7.39) 中, 具有不同 κ 值的辅助线性算子 (7.40) 或 (7.41) 为我们提供了一个简便的途径来找到许多非线性边值 (特征值) 方程的多解, 如 §9.3.1、§9.3.2 和 §9.3.3 所述. 注意, 这种辅助线性算子对应于 BVPh 1.0 的 `TypeL = 2`.

需要注意的是, 第八章和第九章中有限区间 $z \in [0, a]$ 上的边值或特征值问题的所有示例都可通过 (7.37) 或 (7.40) 或 (7.41) 来求解. 因此, 对于有限区间 $z \in [0, a]$ 上的非线性边值问题或特征值问题, 我们强烈建议先尝试使用这两种辅助线性算子进行计算.

对于无限区间 $z \in [b, +\infty)$ 上具有呈指数衰减解的 n 阶边值问题, 其中 $b \geqslant 0$ 为有界常数, 我们通常 (但并不总是) 选取辅助线性算子

$$\mathcal{L}(u) = \frac{d^n u}{dz^n} + \sum_{m=0}^{n-1} \check{a}_m \frac{d^m u}{dz^m}, \tag{7.42}$$

使得

$$\mathcal{L}[\exp(m\gamma z)] = 0, \quad 1 + n' - n \leqslant m \leqslant n' \tag{7.43}$$

成立, 其中 n' 是无穷远处边界条件数, $\gamma > 0$ 是待定参数, $\check{a}_m$ 是系数. (7.42) 中的 n 个未知系数 $\check{a}_m$ 由 (7.43) 给出的 n 个线性代数方程唯一确定.

对于无限区间 $z \in [b, +\infty)$ 上具有呈代数衰减的解 $u \sim z^\beta$ $(z \to +\infty)$ 的 n 阶边值问题, 其中 $b > 0$ 和 β 为有界常数, 我们通常 (但并不总是) 选取辅助线

性算子

$$\mathcal{L}(u) = z^n \frac{d^n u}{dz^n} + \sum_{m=0}^{n-1} \check{b}_m z^m \frac{d^m u}{dz^m}, \quad (7.44)$$

使得

$$\mathcal{L}\left(z^{\beta+m}\right) = 0, \quad 1-n \leqslant m \leqslant 0 \quad (7.45)$$

成立，其中 $\check{b}_m$ 为未知常数. (7.44) 中的 n 个未知系数由 (7.45) 给出的 n 个线性代数方程唯一确定. 对于无限区间 $z \in [0, +\infty)$ 上的边值问题，我们应该先利用变换

$$\xi = 1 + \gamma z,$$

使得 $\xi \in [1, +\infty)$，其中 $\gamma > 0$ 为参数，然后使用上述的辅助线性算子进行计算. 这样，我们也可以把 γ 看作一种收敛控制参数，如 §10.3 所述. 注意，对于无限区间 $z \in [b, +\infty)$ 上的边值问题，我们必须为 BVPh 1.0 设置 `zR=infinity`.

在同伦分析方法的框架中，借助 Mathematica 等科学计算软件提供的"基于函数而非数的计算"能力，无限区间 $z \in [b, +\infty)$ 与有限区间 $z \in [0, a]$ 在本质上没有区别：只是使用的基函数、辅助线性算子和初始猜测解不同. BVPh 1.0 认为无限区间 $z \in [b, +\infty)$ 上的两个端点相同且无根本区别：将两个端点的所有边界条件视为一种极限，即 $z \to b$ 或 $z \to +\infty$. 这样，我们可以很容易地解决边值问题的控制方程和 (或) 边界条件中的无限区间及多种类型的奇点.

由于同伦分析方法为选取辅助线性算子 $\mathcal{L}$ 提供了极大的自由，我们可以在必要时选取其他形式的辅助算子 $\mathcal{L}$. 在这种情况下，BVPh 1.0 的输入数据文件中必须明确定义辅助线性算子.

需要注意的是，对于边值问题，辅助线性算子的选取主要取决于基函数. 如上所述，基函数的选取主要取决于解的区间和渐近性质. 因此，辅助线性算子的选取主要取决于解的区间和渐近性质.

7.1.5 辅助函数的选取

本质上，辅助函数 $H(z)$ 的选取可以被视为辅助线性算子 $\mathcal{L}$ 选取的一部分，因为在 (7.3) 这样的零阶形变方程中，辅助线性算子 $\mathcal{L}$ 和辅助函数 $H(z)$ 可以进行组合，即

$$\check{\mathcal{L}}(u) = \frac{\mathcal{L}(u)}{H(z)}.$$

因此，与辅助线性算子 $\mathcal{L}$ 的选取一样，辅助函数 $H(z)$ 的选取也主要由区间和解的性质决定. 原则上, 辅助函数 $H(z)$ 应满足以下条件:
- 解表达必须符合;
- 高阶形变方程的解存在且唯一;
- 同伦级数解的收敛性可通过选取适当的收敛控制参数来保证.

在大多数情况下，我们可以简单地选取辅助函数 $H(z) = 1$，特别是有限区间 $z \in [0, a]$ 上的边值问题. 对于无限区间 $[b, +\infty)$ 上的边值问题，我们有时可以选取辅助线性算子 $H(z) = \exp(\kappa\gamma z)$ 得到如下呈指数衰减的解

$$u(z) = \sum_{m=0}^{+\infty} a_m \exp(-m\gamma z),$$

或者选取 $H(z) = z^\kappa$ 得到如下呈代数衰减的解

$$u(z) = \sum_{m=\mu}^{+\infty} \frac{b_m}{z^m},$$

其中 $\gamma > 0$ 和 $\mu \geqslant 0$ 为参数，κ 是由 $u(z)$ 在无穷远处的渐近性质和本书上卷中的 "系数遍历原则" (见上卷 §2.3.4 [38]) 确定的参数，即随着 $m \to +\infty$，上述解表达中的每个系数 a_m 和 b_m 都可以被修正.

7.1.6 收敛控制参数 c_0 的选取

与其他所有解析近似方法不同，同伦分析方法在零阶形变方程 (如 (7.3) 和 (7.16)) 中引入收敛控制参数 c_0，为我们提供了一种简便的方法来保证级数解的收敛性. 事实上, 收敛控制参数 c_0 的引入正是同伦分析方法与其他解析方法的本质区别，如上卷所述.

如第三章所述，m 阶同伦近似下的最优收敛控制参数由控制方程的平方残差 E_m 的最小值确定，对应于

$$\frac{dE_m}{dc_0} = 0, \tag{7.46}$$

其中

$$E_m = \int_\Omega \mathcal{F}\left[z, \sum_{k=0}^{m} u_k(z)\right]^2 dz \tag{7.47}$$

是有限区间 $\Omega : z \in [0, a]$ 或无限区间 $\Omega : z \in [b, +\infty)$ 上边值问题的平方残差,

$$E_m = \int_\Omega \mathcal{F}\left[z, \sum_{k=0}^{m} u_k(z), \sum_{k=0}^{m-1} \lambda_m\right]^2 dz \tag{7.48}$$

是有限区间 $z \in [0, a]$ 上特征值问题的平方残差,其中 $\mathcal{F}[z, u] = 0$ 和 $\mathcal{F}[z, u, \lambda] = 0$ 分别表示边值问题和特征值问题的控制方程. 为了提高计算效率,我们通过一些离散点数值计算这些积分 (默认 `Nintegral=50`).

在大多数情况下,平方残差 E_m 仅取决于收敛控制参数 c_0. 然而,为了在同伦分析方法的框架中寻找多解,我们通常在初始猜测解 $u_0(z)$ 中引入多解控制参数 σ,这为我们提供了更大的自由度来获得多解. 在这种情况下,最优收敛控制参数 c_0 和最优多解控制参数 σ 由公式

$$\frac{\partial E_m}{\partial c_0} = 0, \quad \frac{\partial E_m}{\partial \sigma} = 0 \tag{7.49}$$

确定. 相关示例,请参考第八章和第九章.

7.2 解的近似和迭代

虽然高阶形变方程 (7.7) 和 (7.21) 总是线性的,但右端项 $\delta_{m-1}(z)$ 可能相当复杂,所以在一般情况下仍不容易求解. 例如,当选取辅助线性算子 (7.40) 或 (7.41) 时,线性微分方程

$$\mathcal{L}(u) = z^i \left(\cos \frac{j\kappa\pi z}{a} + \sin \frac{j\kappa\pi z}{a} \right) \tag{7.50}$$

对于任意整数 $i \geq 0$ 和 $j \geq 1$,有一个简短的封闭解. 然而,项 $\delta_0(z) = \mathcal{F}[z, u_0(z)]$ 或 $\delta_0(z) = \mathcal{F}[z, u_0(z), \lambda_0]$ 可能非常复杂,因为非线性算子 $\mathcal{F}$ 相当普遍.

例如,在 $\mathcal{F}[z, u, \lambda] = u'' + \lambda \sin u$ 的情况下,选取辅助线性算子 (7.40) 和初始猜测解 $u_0 = \cos \frac{\kappa\pi z}{a}$,我们求解线性微分方程

$$u'' + \left(\frac{\kappa\pi z}{a} \right)^2 u = \sin \left(\cos \frac{\kappa\pi z}{a} \right),$$

该方程没有封闭解,因此右端项必须展开成无穷级数

$$\sin \left(\cos \frac{\kappa\pi z}{a} \right) = \cos \frac{\kappa\pi z}{a} - \frac{1}{3!} \cos^3 \frac{\kappa\pi z}{a} + \cdots .$$

在这种情况下,$u_m(z)$ 的长度是无限的,因此很难获得高阶近似解. 为了避免这种情况,我们选取合理数量的基函数来近似 $\delta_{m-1}(z)$,以便以足够高的精度来有效地求解 m 阶形变方程 (7.7) 或 (7.21).

更重要的是,如第二章所述,迭代可以大大加快同伦级数的收敛速度. 然而,在同伦分析方法的框架中使用迭代法的必要条件是通过合理数量的基函数来高

精度近似高阶形变方程的解，即

$$u_m(z) \approx \sum_{m=0}^{N_t} a_m e_m(z),$$

其中 $e_m(z)$ 为基函数，N_t+1 为截断项的个数，a_m 为系数. 如上所述，$\delta_{m-1}(z)$ 可能相当复杂. 因此，为了得到上述形式的 $u_m(z)$，我们将 $\delta_{m-1}(z)$ 近似为如下形式

$$\delta_{m-1}(z) \approx \sum_{m=0}^{N_t} b_m e_m(z),$$

其中系数 b_m 由 $\delta_{m-1}(z)$ 和基函数 $e_m(z)$ 唯一确定.

如第二章所述，同伦分析方法为我们选取初始猜测解 $u_0(z)$ 提供了极大的自由. 由于 $u_m(z)$ 具有固定的长度，因此，令 M 阶近似解作为新的初始猜测解 $u_0^*(z)$ 进行迭代相当方便，即

$$u_0(z) + \sum_{m=1}^{M} u_m(z) \to u_0^*(z). \tag{7.51}$$

上述表达式为我们提供了所谓的 M 阶迭代公式. 这样，同伦级数的收敛速度可以大大加快，如第八章和第九章所述.

BVPh 1.0 的迭代法目前仅适用于有限区间 $z \in [0,a]$ 上的非线性边值 (特征值) 问题. 因此，我们在这里仅考虑有限区间 $z \in [0,a]$ 上光滑函数 $f(z)$ 的近似.

7.2.1 多项式

众所周知，有限区间 $z \in [0,a]$ 上的光滑解 $u(z)$ 可以由如下多项式很好地近似

$$u(z) \approx \sum_{k=0}^{N_t} a_k z^k, \tag{7.52}$$

其中 a_k 是系数，N_t+1 是截断项的个数. 此外，我们有最优近似

$$u(z) \approx \frac{b_0}{2} + \sum_{k=1}^{N_t} b_k T_k(z), \tag{7.53}$$

其中 $T_k(z)$ 是 k 阶第一类 Chebyshev 多项式，N_t 是 Chebyshev 多项式的个数，且

$$b_k = \frac{2}{\pi} \int_0^\pi u\left[\frac{a}{2}(1+\cos\theta)\right] \cos(k\theta) d\theta.$$

BVPh 1.0 利用多项式求解有限区间 $z \in [0, a]$ 上的边值 (特征值) 问题时, 应该设置

$$\texttt{TypeBase} = 0, \quad \texttt{TypeL} = 1,$$

其中多项式 (7.52) 设置 `ApproxQ=0`, Chebyshev 多项式 (7.53) 设置 `ApproxQ=1`.

7.2.2 三角函数

众所周知, 连续函数 $f(z)$ 在 $z \in (-a, a)$ 上的 Fourier 级数

$$\frac{a_0}{2} + \sum_{n=1}^{+\infty} \left(b_n \sin \frac{n\pi z}{a} + a_n \cos \frac{n\pi z}{a} \right)$$

收敛于 $f(z)$, 其中

$$a_n = \frac{1}{a} \int_{-a}^{a} f(t) \cos \frac{n\pi t}{a} dt, \quad b_n = \frac{1}{a} \int_{-a}^{a} f(t) \sin \frac{n\pi t}{a} dt.$$

对于 $[0, a]$ 上的连续函数 $f(z)$, 我们可以在 $z \in (0, a)$ 上定义 $f(-z) = f(z)$, 则偶函数的 Fourier 级数为

$$f(z) = \frac{a_0}{2} + \sum_{n=1}^{+\infty} a_n \cos \frac{n\pi z}{a}. \tag{7.54}$$

或者在 $z \in (0, a)$ 上定义 $f(-z) = -f(z)$, 则奇函数的 Fourier 级数为

$$f(z) = \sum_{n=1}^{+\infty} b_n \sin \frac{n\pi z}{a}. \tag{7.55}$$

在实际计算过程中, 我们有相应的近似

$$f(z) = \frac{a_0}{2} + \sum_{n=1}^{N_t} a_n \cos \frac{n\pi z}{a} \tag{7.56}$$

或

$$f(z) = \sum_{n=1}^{N_t} b_n \sin \frac{n\pi z}{a}, \tag{7.57}$$

其中 N_t 表示截断项的个数.

BVPh 1.0 利用三角函数求解有限区间 $z \in [0, a]$ 上的边值 (特征值) 问题时, 我们应该设置

$$\texttt{ApproxQ} = 1, \quad \texttt{TypeL} = 2,$$

其中, 奇函数的 Fourier 级数 (7.57) 设置 `TypeBase=1`, 偶函数的 Fourier 级数 (7.56) 设置 `TypeBase=2`.

7.2.3 混合基函数

需要注意的是, 偶函数的 Fourier 级数 (7.54) 的一阶导数在两个端点 $z = 0$ 和 $z = a$ 处等于零, 但原始函数 $f(z)$ 的 $f'(0)$ 和 $f'(a)$ 可以为任意值. 因此, 在 $f'(0) \neq 0$ 和 (或) $f'(a) \neq 0$ 的情形下, 我们必须使用偶函数的 Fourier 级数 (7.54) 的许多项来获得在两个端点附近的高精度近似值. 为了克服这个缺点, 我们首先将 $f(z)$ 表示为

$$f(z) \approx Y(z) + w(z), \tag{7.58}$$

其中

$$Y(z) = \left\{ f'(0)z - \frac{[f'(0) + f'(a)]}{2a}z^2 \right\} \cos\frac{\pi z}{a}. \tag{7.59}$$

然后, 利用偶函数的 Fourier 级数

$$w(z) \sim \frac{\bar{a}_0}{2} + \sum_{n=1}^{N_t} \bar{a}_n \cos\frac{n\pi z}{a} \tag{7.60}$$

来近似 $w(z) = f(z) - Y(z)$. 其中 Fourier 系数

$$\bar{a}_n = \frac{2}{a} \int_0^a [f(t) - Y(t)] \cos\frac{n\pi t}{a} dt,$$

这里 $N_t + 1$ 是截断项的个数. 注意, $Y(z)$ 满足条件

$$Y'(0) = f'(0), \quad Y'(a) = f'(a),$$

使得 $w'(0) = w'(a) = 0$. 因此, 我们经常需要用偶函数的 Fourier 级数的几项来高精度近似 $w(z)$. 还需要注意的是, 在 (7.58) 中, 三角函数和多项式被用来近似 $f(z)$. 研究发现, 相比于传统的 Fourier 级数, 这种基于混合基函数对 $[0, a]$ 内给定的光滑函数 $f(z)$ 近似所需的项通常要少得多. 例如, $z \in [0, \pi]$ 上 $f(z) = 1/(1 + z)$ 的 15 项混合基近似比 50 项传统的 Fourier 近似要好得多, 特别是在 $z = 0$ 附近, 如图 7.1 所示.

另外, 对于 $[0, a]$ 上的连续函数 $f(z)$, 我们选取

$$Y(z) = f(0) + \frac{[f(a) - f(0)]}{a}z \tag{7.61}$$

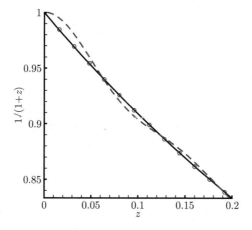

图 7.1 $z \in [0, \pi]$ 上的 $1/(1+z)$ 的不同近似值的比较. 实线: $1/(1+z)$; 空心圆: 15 项混合基近似; 虚线: 50 项传统的 Fourier 近似

或

$$Y(z) = \frac{[f(0) + f(a)]}{2} + \frac{[f(0) - f(a)]}{2} \cos \frac{\pi z}{a}, \quad (7.62)$$

并利用奇函数的 Fourier 级数

$$w(z) \sim \sum_{n=1}^{N_t} \bar{b}_n \sin \frac{n\pi z}{a} \quad (7.63)$$

来近似 $w(z)$, 其中

$$\bar{b}_n = \frac{2}{a} \int_0^a [f(t) - Y(t)] \sin \frac{n\pi t}{a} dt.$$

对于偶函数 $f(z)$, 建议使用满足 (7.59) 的混合基近似 (7.58), 对于奇函数 $f(z)$, 可以使用满足 (7.61) 或 (7.62) 的混合基近似 (7.58). 如果 $f(z)$ 既不是奇函数也不是偶函数, 则两者都有效.

BVPh 1.0 可以用上述混合基近似方法求解 m 阶形变方程 (7.7) 和 (7.21). 基于该方法, 我们选取适当数目的截断项来近似 $\delta_i(z)$, 例如

$$\delta_i(z) \approx Y_i(z) + w_i(z), \quad (7.64)$$

其中

$$Y_i(z) = \left\{ \delta_i'(0) z - \frac{[\delta_i'(0) + \delta_i'(a)]}{2a} z^2 \right\} \cos \frac{\pi z}{a} \quad (7.65)$$

和
$$w_i(z) \sim \frac{\check{a}_0}{2} + \sum_{n=1}^{N_t} \check{a}_n \cos \frac{n\pi z}{a}, \qquad (7.66)$$

其中 Fourier 系数
$$\check{a}_n = \frac{2}{a} \int_0^a [\delta_i(t) - Y_i(t)] \cos \frac{n\pi t}{a} dt,$$

或者 (特别是当解是奇函数时), 我们也可以选取
$$Y_i(z) \approx \delta_i(0) + \frac{[\delta_i(a) - \delta_i(0)]}{a} z \qquad (7.67)$$

或
$$Y_i(z) \approx \frac{[\delta_i(0) + \delta_i(a)]}{2} + \frac{[\delta_i(0) - \delta_i(a)]}{2} \cos \frac{\pi z}{a}, \qquad (7.68)$$

其中
$$w_i(z) \sim \sum_{n=1}^{N_t} \check{b}_n \sin \frac{n\pi z}{a} \qquad (7.69)$$

且
$$\check{b}_n = \frac{2}{a} \int_0^a [\delta_i(t) - Y_i(t)] \sin \frac{n\pi t}{a} dt.$$

这样, 只要 N_t 足够大, 原始高阶形变方程 (7.7) 和 (7.21) 就可以被高精度的
$$\mathcal{L}[u_m(z) - \chi_m u_{m-1}(z)] \approx c_0 H(z) [Y_{m-1}(z) + w_{m-1}(z)] \qquad (7.70)$$

替代, 这里我们选取辅助线性算子 (7.40) 或 (7.41). 然后, 上述线性方程可以分解为 (7.50) 形式的有限个线性微分方程, 这些方程通过科学计算软件 Mathematica 很容易求解 [9]. 更重要的是, 给定 N_t, 即截断项的个数, $u_m(z)$ 总是具有固定的长度, 即使近似阶数很高, 该长度也不会增加. 因此, 基于这种混合基近似方法, 无论非线性算子 $\mathcal{F}$ 有多复杂, 我们都能有效地求解高阶形变方程 (7.7) 和 (7.21).

当使用上述混合基近似时, 我们有更大的自由来选取初始猜测解 $u_0(z)$. 例如, 对于有限区间 $z \in [0, a]$ 上的二阶边值 (特征值) 问题, 我们可以选取初始猜测解
$$u_0(z) = B_0 + B_1 \cos \frac{\kappa \pi z}{a} + B_2 \sin \frac{\kappa \pi z}{a} \qquad (7.71)$$

或

$$u_0(z) = \left(B_0 + B_1 z + B_2 z^2\right) \cos\frac{\kappa\pi z}{a}, \tag{7.72}$$

其中 B_0, B_1, B_2 由线性边界条件确定.

BVPh 1.0 利用上述混合基近似方法求解有限区间 $z \in [0, a]$ 上的边值 (特征值) 问题时, 我们应该设置

$$\texttt{ApproxQ} = 1, \quad \texttt{TypeL} = 2,$$

其中奇函数的 Fourier 级数 (7.63) 设置 TypeBase=1, 偶函数的 Fourier 级数 (7.60) 设置 TypeBase=2.

7.3 BVPh 1.0 使用指南

7.3.1 关键模块

BVPh   模块 BVPh[k_,m_] 给出了上述定义的非线性边值问题 (TypeEQ=1) 或非线性特征值问题 (TypeEQ=2) 的 k 至 m 阶同伦近似. 它是基本模块. 例如, BVPh[1,10] 给出了 1 至 10 阶同伦近似, BVPh[11,15] 给出了 11 至 15 阶同伦近似.

iter   模块 iter[k_,m_,r_] 通过 r 阶迭代公式 (7.51) 为我们提供了第 k 至 m 次迭代的同伦近似. 它调用基本模块 BVPh. 例如, iter[1,10,3] 用 3 阶迭代公式给出了第 1 至 10 次迭代的同伦近似. 此外, iter[11,20,3] 给出了第 11 至 20 次迭代的同伦近似. 对于特征值问题, 特征值的初始猜测值由代数方程确定. 因此, 如果特征值有一个以上的初始猜测值, 则通过输入整数, 例如 1 或 2, 然后从中选取一个来对应特征值的第一个或第二个初始猜测解. 如果在迭代开始时收敛控制参数 c_0 未知, 我们在第 1 次迭代时给出控制方程在高达 3 阶近似下的平方残差与 c_0 的关系曲线, 以选取适当的 c_0 值, 该 c_0 值将在 Nupdate 次迭代后更新. 当控制方程的平方残差小于临界值 ErrReq (默认值为 10^{-20}) 时, 迭代停止.

GetErr   模块 GetErr[k_] 给出了控制方程在模块 BVPh 获得的 k 阶同伦近似或模块 iter 获得的第 k 次迭代同伦近似下的平方残差. 需要注意的是, error[k_] 提供了由 BVPh 得到的 k 阶同伦近似或由 iter 得到的第 k 次迭代同伦近似下控制方程的残差, Err[k_] 给出了由 BVPh 得到的控制方程在 k 阶同伦近似下的平均平方残差, ERR[k_] 给出了由 iter 得到的控制方程在第 k 次迭代同伦近似下的平均平方残差.

GetMin1D　模块 GetMin1D[f_,x_,a_,b_,Npoint_] 将区间 $[a,b]$ 划分为 Npoint 等份, 并在区间 $x \in [a,b]$ 上搜索实函数 $f(x)$ 的局部最小值. 如果 Npoint 足够大, 它会给出位置 x 的所有局部最小值. 一般建议设置 Npoint=20. 该模块通常用于寻找最优收敛控制参数 c_0, 或非线性边值问题的多解, 其调用模块 GetMin1D0[f_,x_,a_,b_,Npoint_].

GetMin2D　模块 GetMin2D[f_,x_,a_,b_,y_,c_,d_,Npoint_] 将区间 $a \leqslant x \leqslant b, c \leqslant y \leqslant d$ 划分为 Npoint×Npoint 等份来搜索实函数 $f(x,y)$ 在该区间上的局部最小值. 如果 Npoint 足够大, 那么它会给出位置 (x,y) 的所有局部最小值. 一般建议设置 Npoint=20. 该模块常用于搜索非线性边值问题的多解, 其调用模块 GetMin2D[f_,x_,a_,b_,y_,c_,d_,Npoint_].

hp　模块 hp[f_,m_,n_] 给出同伦近似 f 的 $[m,n]$ 阶同伦-帕德近似, 其中 f[0],f[1],f[2] 分别表示 f 的零阶, 一阶和二阶同伦近似. 同伦-帕德近似的相关信息, 请参见第二章.

7.3.2　控制参数

TypeEQ　控制方程类型的控制参数. TypeEQ = 1 对应于非线性边值方程, TypeEQ=2 对应于非线性特征值问题.

TypeL　基函数类型的控制参数. TypeL=1 对应于多项式 (7.52) 或 Chebyshev 多项式 (7.53), TypeL=2 对应于 §7.2.3 中描述的三角近似或混合基近似. 它只适用于有限区间 $z \in [0,a]$ 上的边值 (特征值) 问题, 其中 $a > 0$ 为常数.

ApproxQ　近似解的控制参数. 当 ApproxQ=1 时, 所有高阶形变方程的右端由 Chebyshev 多项式 (7.53) 或 §7.2.3 中描述的混合基近似. 当 ApproxQ=0 时, 这种近似不存在. 当采用模块 iter 时, 软件将自动赋值 ApproxQ=1. 对于 BVPh 1.0, 该参数仅适用于有限区间 $z \in [0,a]$ 上的边值 (特征值) 问题, 其中 $a > 0$ 为常数.

TypeBase　近似解的基函数类型的控制参数. TypeBase=0 对应于 Chebyshev 多项式 (7.53), TypeBase=1 对应于奇函数的 Fourier 近似 (7.63) 的混合基 (7.58), TypeBase=2 对应于偶函数的 Fourier 近似 (7.60) 的混合基 (7.58). 该参数仅适用于当 ApproxQ=1 时, 有限区间 $z \in [0,a]$ 上的边值 (特征值) 问题.

Ntruncated　正整数 $N_t > 0$ 的控制参数, 对应 (7.53), (7.60) 和 (7.63) 中截断项的个数, N_t 的值越大, 近似效果越好, 但 CPU 时间更长. 它仅适用于当 ApproxQ=1 时, 有限区间 $z \in [0,a]$ 上的边值 (特征值) 问题. 默认值为 10.

NtermMax　模块 truncated 中使用的正整数, 该模块忽略所有阶数高于 NtermMax 的多项式项. 默认值为 90.

ErrReq　用于终止模块 iter 迭代的控制方程平方残差的临界值. 默认值为 10^{-10}.

NgetErr　模块 BVPh 中使用的正整数. 当近似阶数除以 NgetErr 为整数时, 软件将计算控制方程的平方残差. 默认值为 2.

Nupdate　更新收敛控制参数 c_0 的迭代次数的临界值. 默认值为 10.

Nintegral　数值计算函数积分时等距离散点的个数, 被用于模块 int 中. 默认值为 50.

ComplexQ　复变量的控制参数. ComplexQ=1 对应于控制方程和 (或) 边界条件中存在复变量的情况. ComplexQ=0 对应于控制方程和 (或) 边界条件中不存在复变量的情况. 默认值为 0.

c0L　用于确定收敛控制参数 c_0 区间的实数, 以便在模块 iter 中绘制控制方程的平方残差与 c_0 的关系曲线. 默认值为 -2, 对应于区间 $-2 \leqslant c_0 \leqslant 0$. c0L 的值可为正, 例如 c0L=1, 对应于区间 $0 \leqslant c_0 \leqslant 1$.

7.3.3　输入

f[z_,u_,lambda_]　控制方程, 对应于有限区间 $z \in [0,a]$ 或无限区间 $z \in [b,+\infty)$ 上的非线性边值问题 $\mathcal{F}[z,u] = 0$ 或对应于有限区间 $z \in [0,a]$ 上的非线性特征值问题 $\mathcal{F}[z,u,\lambda] = 0$, 其中 $a > 0, b \geqslant 0$ 均为有界常数. 注意, lambda 表示非线性特征值问题的特征值, 但其对于非线性边值问题没有任何意义.

BC[k_,z_,u_,lambda_]　第 k 个边界条件, 对应于非线性边值问题的 $\mathcal{B}_k[z,u] = 0$ 或非线性特征值问题的 $\mathcal{B}_k[z,u,\lambda] = 0$, 其中 $0 \leqslant k \leqslant n$. 注意, lambda 表示非线性特征值问题的特征值, 但其对于非线性边值问题没有任何意义.

u[0]　初始猜测解 $u_0(z)$.

L[f_]　辅助线性算子. 对于有限区间 $z \in [0,a]$ 上的边值 (特征值) 问题, 当 TypeL=1 时, 辅助线性算子自动选取为 (7.37), 否则自动选取为辅助线性算子 (7.40) 或 (7.41). 对于无限区间 $z \in [b,+\infty)$ 上的边值问题, 呈指数衰减的解可以选取 (7.42), 呈代数衰减的解可以选取 (7.44), 其中 $b \geqslant 0$ 为有界常数. 在任何情况下, 辅助线性算子都必须被明确定义和适当选取.

H[z_]　辅助函数. 默认 H[z_]:=1.

OrderEQ　边值方程 $\mathcal{F}[z,u]=0$ 或特征方程 $\mathcal{F}[z,u,\lambda]=0$ 的阶数.

zL　区间 $z\in[a,b]$ 的左端点, 对应于 $z=a$. 例如, zL=1 对应于 $a=1$. 对于有限区间 $z\in[0,a]$ 上的边值 (特征值) 问题, 自动赋值 zL=0. 默认值为 0.

zR　区间 $z\in[a,b]$ 的右端点, 对应于 $z=b$. 例如, zR=1 对应于 $b=1$. 对于无限区间上的边值问题, BVPh 1.0 中必须设置 zR=infinity.

7.3.4 输出

U[k]　基本模块 BVPh 给出的 $u(z)$ 的 k 阶同伦近似.

V[k]　迭代模块 iter 给出的 $u(z)$ 的第 k 次迭代的同伦近似.

Uz[k]　基本模块 BVPh 给出的 $u'(z)$ 的 k 阶同伦近似.

Vz[k]　迭代模块 iter 给出的 $u'(z)$ 的第 k 次迭代的同伦近似.

Lambda[k]　基本模块 BVPh 给出的特征值 λ 的 k 阶同伦近似.

LAMBDA[k]　迭代模块 iter 给出的特征值 λ 的第 k 次迭代的同伦近似.

error[k]　k 阶同伦近似 (由基本模块 BVPh 得到) 或第 k 次迭代的同伦近似 (由迭代模块 iter 得到) 给出的控制方程的残差.

Err[k]　k 阶同伦近似 (由基本模块 BVPh 得到) 给出的控制方程的平均平方残差.

ERR[k]　k 次迭代的同伦近似 (由迭代模块 iter 得到) 给出的控制方程的平均平方残差.

7.3.5 全局变量

上述所有控制参数和输出变量均为全局变量. 除此之外, 以下变量和参数也是全局的.

z　自变量 z.

u[k]　k 阶形变方程的解 $u_k(z)$.

lambda[k]　常变量, 即 λ_k.

delta[k]　依赖于 z 的函数, 即高阶形变方程的右端项 $\delta_k(z)$.

L　辅助线性算子 $\mathcal{L}$.

Linv　$\mathcal{L}$ 的逆算子, 即 $\mathcal{L}^{-1}$.

nIter　迭代次数, 用于模块 iter.

sNum　正整数, 当 λ_0 存在多解时, 用于选取特征值的初始猜测值 λ_0.

参考文献

[1] Abbas, Z., Wang, Y., Hayat, T., Oberlack, M.: Hydromagnetic flow in a viscoelstic fluid due to the oscillatory stretching surface. Int. J. Nonlin. Mech. **43**, 783–793 (2008).

[2] Abbasbandy, S.: The application of the homotopy analysis method to nonlinear equations arising in heat transfer. Phys. Lett. A. **360**, 109–113 (2006).

[3] Abbasbandy, S.: The application of homotopy analysis method to solve a generalized HirotaSatsuma coupled KdV equation. Phys. Lett. A. **361**, 478–483 (2007).

[4] Abbasbandy, S.: Solitary wave equations to the Kuramoto-Sivashinsky equation by means of the homotopy analysis method. Nonlinear Dynam. **52**, 35–40 (2008).

[5] Abbasbandy, S., Magyari, E., Shivanian, E.: The homotopy analysis method for multiple solutions of nonlinear boundary value problems. Communications in Nonlinear Science and Numerical Simulation. **14**, 3530–3536 (2009).

[6] Abbasbandy, S., Parkes, E.J.: Solitary smooth hump solutions of the Camassa-Holm equation by means of the homotopy analysis method. Chaos Soliton. Fract. **36**, 581–591 (2008).

[7] Abbasbandy, S., Parkes, E.J.: Solitary-wave solutions of the DegasperisProcesi equation by means of the homotopy analysis method. Int. J. Comp. Math. **87**, 2303–2313 (2010).

[8] Abbasbandy, S., Shivanian, E.: Predictor homotopy analysis method and its application to some nonlinear problems. Commun. Nonlinear Sci. Numer. Simulat. **16**, 2456–2468 (2011).

[9] Abell, M.L., Braselton, J.P.: Mathematica by Example (3rd Edition). Elsevier Academic Press. Amsterdam (2004).

[10] Akyildiz, F.T., Vajravelu, K.: Magnetohydrodynamic flow of a viscoelastic fluid. Phys. Lett. A. **372**, 3380–3384 (2008).

[11] Akyildiz, F.T., Vajravelu, K., Mohapatra, R.N., Sweet, E., Van Gorder, R.A.: Implicit differential equation arising in the steady flow of a Sisko fluid. Applied Mathematics and Computation. **210**, 189–196 (2009).

[12] Alizadeh-Pahlavan, A., Aliakbar, V., Vakili-Farahani, F., Sadeghy, K.: MHD flows of UCM fluids above porous stretching sheets using two-auxiliary-parameter homotopy analysis method. Commun. Nonlinear Sci. Numer. Simulat. **14**, 473–488 (2009).

[13] Alizadeh-Pahlavan, A., Borjian-Boroujeni, S.: On the analytical solution of viscous fluid flow past a flat plate. Physics Letters A. **372**, 3678–3682 (2008).

[14] Allan, F.M.: Derivation of the Adomian decomposition method using the homotopy analysis method. Appl. Math. Comput. **190**, 6–14 (2007).

[15] Allan, F.M.: Construction of analytic solution to chaotic dynamical systems using the homotopy analysis method. Chaos, Solitons and Fractals. **39**, 1744–1752 (2009).

[16] Allan, F.M., Syam, M.I.: On the analytic solutions of the nonhomogeneous Blasius problem. J. Comp. Appl. Math. **182**, 362–371 (2005).

[17] Battles, Z., Trefethen, L.N.: An extension of Matlab to continuous functions and operators. SIAM J. Sci. Comp. **25**, 1743–1770 (2004).

[18] Boyd, J.P.: Chebyshev and Fourier Spectral Methods. DOVER Publications, Inc. New York (2000).

[19] Cai, .W.H.: Nonlinear Dynamics of Thermal-Hydraulic Networks. PhD dissertation, University of Notre Dame (2006).

[20] Cheng, J.: Application of the Homotopy Analysis Method in Nonlinear Mechanics and Finance. PhD dissertation, Shanghai Jiao Tong University (2008).

[21] Gao, L.M.: Analysis of the Propagation of Surface Acoustic Waves in Functionally Graded Material Plate. PhD dissertation, Tongji University (2007).

[22] Hayat, T., Sajid, M.: On analytic solution for thin film flow of a fourth grade fluid down a vertical cylinder. Phys. Lett. A. **361**, 316–322 (2007).

[23] Jiao, X.Y.: Approximate Similarity Reduction and Approximate Homotopy Similarity Reduction of Several Nonlinear Problems. PhD dissertation, Shanghai Jiao Tong University (2009).

[24] Jiao, X.Y., Gao, Y., Lou, S.Y.: Approximate homotopy symmetry method-Homotopy series solutions to the sixth-order Boussinesq equation. Science in China (G). **52**, 1169–1178 (2009).

[25] Kierzenka, J., Shampine, L.F.: A BVP solver based on residual control and the MATLAB PSE. ACM TOMS. **27**, 299–316 (2001).

[26] Kumari, M., Nath, G.: Unsteady MHD mixed convection flow over an impulsively stretched permeable vertical surface in a quiescent fluid. Int. J. Non-Linear Mech. **45**, 310–319 (2010).

[27] Kumari, M., Pop, I., Nath, G.: Transient MHD stagnation flow of a non-Newtonian fluid due to impulsive motion from rest. Int. J. Non-Linear Mech. **45**, 463–473 (2010).

[28] Li, Y.J., Nohara, B.T., Liao, S.J.: Series solutions of coupled van der Pol equation by means of homotopy analysis method. J. Mathematical Physics **51**, 063517 (2010). doi:10.1063/1.3445770.

[29] Liang, S.X.: Symbolic Methods for Analyzing Polynomial and Differential Systems. PhD dissertation, University of Western Ontario (2010).

[30] Liang, S.X., Jeffrey, D.J.: Comparison of homotopy analysis method and homotopy perturbation method through an evelution equation. Commun. Nonlinear Sci. Numer. Simulat. **14**, 4057–4064 (2009a).

[31] Liang, S.X., Jeffrey, D.J.: An efficient analytical approach for solving fourth order boundary value problems. Computer Physics Communications. **180**, 2034–2040 (2009b).

[32] Liang, S.X., Jeffrey, D.J.: Approximate solutions to a parameterized sixth order boundary value problem. Computers and Mathematics with Applications. **59**, 247–253 (2010).

[33] Liao, S.J.: The Proposed Homotopy Analysis Technique for the Solution of Nonlinear Problems. PhD dissertation, Shanghai Jiao Tong University (1992).

[34] Liao, S.J.: A kind of approximate solution technique which does not depend upon small parameters——(II) An application in fluid mechanics. Int. J. Nonlin. Mech. **32**, 815–822 (1997).

[35] Liao, S.J.: An explicit, totally analytic approximation of Blasius viscous flow problems. Int. J. Nonlin. Mech. **34**, 759–778 (1999a).

[36] Liao, S.J.: A uniformly valid analytic solution of 2D viscous flow past a semi-infinite flat plate. J. Fluid Mech. **385**, 101–128 (1999b).

[37] Liao, S.J.: On the analytic solution of magnetohydrodynamic flows of non-Newtonian fluids over a stretching sheet. J. Fluid Mech. **488**, 189–212 (2003a).

[38] Liao, S.J.: Beyond Perturbation-Introduction to the Homotopy Analysis Method. Chapman & Hall/ CRC Press, Boca Raton (2003b).

[39] Liao, S.J.: On the homotopy analysis method for nonlinear problems. Appl. Math. Comput. **147**, 499–513 (2004).

[40] Liao, S.J.: A new branch of solutions of boundary-layer flows over an impermeable stretched plate. Int. J. Heat Mass Tran. **48**, 2529–2539 (2005).

[41] Liao, S.J.: Series solutions of unsteady boundary-layer flows over a stretching flat plate. Stud. Appl. Math. **117**, 2529–2539 (2006).

[42] Liao, S.J.: Notes on the homotopy analysis method-Some definitions and theorems. Commun. Nonlinear Sci. Numer. Simulat. **14**, 983–997 (2009a).

[43] Liao, S.J.: On the reliability of computed chaotic solutions of non-linear differential equations. Tellus. **61A**, 550–564 (2009b).

[44] Liao, S.J.: An optimal homotopy-analysis approach for strongly nonlinear differential equations. Commun. Nonlinear Sci. Numer. Simulat. **15**, 2003–2016 (2010).

[45] Liao, S.J., Campo, A.: Analytic solutions of the temperature distribution in Blasius viscous flow problems. J. Fluid Mech. **453**, 411–425 (2002).

[46] Liao, S.J., Tan, Y.: A general approach to obtain series solutions of nonlinear differential equations. Stud. Appl. Math. **119**, 297–355 (2007).

[47] Liu, Y.P.: Study on Analytic and Approximate Solution of Differential Equations by Symbolic Computation. PhD Dissertation, East China Normal University (2008).

[48] Liu, Y.P., Li, Z.B.: The homotopy analysis method for approximating the solution of the modified Korteweg-de Vries equation. Chaos Soliton. Fract. **39**, 1–8 (2009).

[49] Lorenz, E. N.: Deterministic nonperiodic flow. J. Atmos. Sci. **20**, 130–141 (1963).

[50] Mahapatra, T. R., Nandy, S.K., Gupta, A.S.: Analytical solution of magnetohydrodynamic stagnation-point flow of a power-law fluid towards a stretching surface. Applied Mathematics and Computation. **215**, 1696–1710 (2009).

[51] Marinca, V., Herişanu, N.: Application of optimal homotopy asymptotic method for solving nonlinear equations arising in heat transfer. Int. Commun. Heat Mass. **35**, 710–715 (2008).

[52] Marinca, V., Herişanu, N.: An optimal homotopy asymptotic method applied to the steady flow of a fourth-grade fluid past a porous plat. Appl. Math. Lett. **22**, 245–251 (2009).

[53] Mason, J.C., Handscomb, D.C.: Chebyshev Polynomials. Chapman & Hall/CRC Press, Boca Raton (2003).

[54] Molabahrami, A., Khani, F.: The homotopy analysis method to solve the Burgers-Huxley equation. Nonlin. Anal. B. **10**, 589–600 (2009).

[55] Motsa, S.S., Sibanda, P., Shateyi, S.: A new spectral homotopy analysis method for solving a nonlinear second order BVP. Commun. Nonlinear Sci. Numer. Simulat. **15**, 2293–2302 (2010a).

[56] Motsa, S.S., Sibanda, P., Auad, F.G., Shateyi, S.: A new spectral homotopy analysis method for the MHD Jeffery-Hamel problem. Computer & Fluids. **39**, 1219–1225 (2010b).

[57] Niu, Z., Wang, C.: A one-step optimal homotopy analysis method for nonlinear differential equations. Commun. Nonlinear Sci. Numer. Simulat. **15**, 2026–2036 (2010).

[58] Pandey, R.K., Singh, O.P., Baranwal, V.K.: An analytic algorithm for the space-time fractional advection-dispersion equation. Computer Physics Communications. **182**, 1134–1144 (2011).

[59] Pirbodaghi, T., Ahmadian, M.T., Fesanghary, M.: On the homotopy analysis method for non-linear vibration of beams. Mechanics Research Communications. **36**, 143–148 (2009).

[60] Sajid, M.: Similar and Non-similar Analytic Solutions for Steady Flows of Differential Type Fluids. PhD dissertation, Quaid-I-Azam University (2006).

[61] Sajid, M., Hayat, T.: Comparison of HAM and HPM methods in nonlinear heat conduction and convection equations. Nonlinear Anal. B. **9**, 2296–2301 (2008).

[62] Shampine, L.F., Gladwell, I., Thompson, S.: Solving ODEs with MATLAB. Cambridge University Press. Cambridge (2003).

[63] Shampine, L.F., Reichelt, M.W., Kierzenka, J.: Solving boundary value problems for ordinary differential equations in MATLAB with BVP4c. Available at http://www.mathworks.com/bvp_tutorial. Accessed 15 April 2011.

[64] Shidfar, A., Babaei, A., Molabahrami, A.: Solving the inverse problem of identifying an unknown source term in a parabolic equation. Computers and Mathematics with Applications. **60**, 1209–1213 (2010).

[65] Shidfar, A., Molabahrami, A.: A weighted algorithm based on the homotopy analysis method - Application to inverse heat conduction problems. Commun. Nonlinear Sci. Numer. Simulat. **15**, 2908–2915 (2010).

[66] Siddheshwar, P.G.: A series solution for the Ginzburg-Landau equation with a timeperiodic coefficient. Applied Mathematics. **3**, 542–554 (2010). Online available at http://www.SciRP.org/journal/am. Accessed 15 April 2011.

[67] Singh, O.P., Pandey, R.K., Singh, V.K.: An analytic algorithm of LaneEmden type equations arising in astrophysics using modified homotopy analysis method. Computer Physics Communications. **180**, 1116–1124 (2009).

[68] Song, H., Tao, L.: Homotopy analysis of 1D unsteady, nonlinear groundwater flow through porous media. J. Coastal Res. **50**, 292–295 (2007).

[69] Tao, L., Song, H., Chakrabarti, S.: Nonlinear progressive waves in water of finite depth-An analytic approximation. Coastal Engineering. **54**, 825–834 (2007).

[70] Trefethen, L.N.: Computing numerically with functions instead of numbers. Math. in Comp. Sci. **1**, 9–19 (2007).

[71] Turkyilmazoglu, M.: Purely analytic solutions of the compressible boundary layer flow due to a porous rotating disk with heat transfer. Physics of Fluids. **21**, 106104 (2009).

[72] Turkyilmazoglu, M.: A note on the homotopy analysis method. Appl. Math. Lett. **23**, 1226–1230 (2010a).

[73] Turkyilmazoglu, M.: Series solution of nonlinear two-point singularly perturbed boundary layer problems. Computers and Mathematics with Applications. **60**, 2109–2114 (2010b).

[74] Turkyilmazoglu, M.: An optimal analytic approximate solution for the limit cycle of Duffing-van der Pol equation. ASME J. Appl. Mech. **78**, 021005 (2011a).

[75] Turkyilmazoglu, M.: Numerical and analytical solutions for the flow and heat transfer near the equator of an MHD boundary layer over a porous rotating sphere. Int. J. Thermal Sciences. **50**, 831–842 (2011b).

[76] Turkyilmazoglu, M.: An analytic shooting-like approach for the solution of nonlinear boundary value problems. Math. Comp. Modelling. **53**, 1748–1755 (2011c).

[77] Turkyilmazoglu, M.: Some issues on HPM and HAM methods-A convergence scheme. Math. Compu. Modelling. **53**, 1929–1936 (2011d).

[78] Van Gorder, R.A., Vajravelu, K.: Analytic and numerical solutions to the Lane-Emden equation. Phys. Lett. A. **372**, 6060–6065 (2008).

[79] Van Gorder, R.A., Vajravelu, K.: On the selection of auxiliary functions, operators, and convergence control parameters in the application of the homotopy analysis method to nonlinear differential equations-A general approach. Commun. Nonlinear Sci. Numer. Simulat. **14**, 4078–4089 (2009).

[80] Van Gorder, R.A., Sweet, E., Vajravelu, K.: Analytical solutions of a coupled nonlinear system arising in a flow between stretching disks. Applied Mathematics and Computation. **216**, 1513–1523 (2010a).

[81] Van Gorder, R.A., Sweet, E., Vajravelu, K.: Nano boundary layers over stretching surfaces. Commun. Nonlinear Sci. Numer. Simulat. **15**, 1494–1500 (2010b).

[82] Van Gorder, R.A., Vajravelu, K.: Convective heat transfer in a conducting fluid over a permeable stretching surface with suction and internal heat generation/absorption. Applied Mathematics and Computation. **217**, 5810–5821 (2011).

[83] Wu, Y.Y.: Analytic Solutions for Nonlinear Long Wave Propagation. PhD dissertation, University of Hawaii (2009).

[84] Wu, Y., Cheung, K.F.: Explicit solution to the exact Riemann problems and application in nonlinear shallow water equations. Int. J. Numer. Meth. Fl. **57**, 1649–1668 (2008).

[85] Wu, Y.Y., Cheung, K.F.: Homotopy solution for nonlinear differential equations in wave propagation problems.Wave Motion. **46**, 1–14 (2009).

[86] Xu, H., Lin, Z.L., Liao, S.J., Wu, J.Z., Majdalani, J.: Homotopy-based solutions of the Navier-Stokes equations for a porous channel with orthogonally moving walls. Physics of Fluids. **22**, 053601 (2010). doi:10.1063/1.3392770.

[87] Yabushita, K., Yamashita, M., Tsuboi, K.: An analytic solution of projectile motion with the quadratic resistance law using the homotopy analysis method. J. Phys. A-Math. Theor. **40**, 8403–8416 (2007).

[88] Zand, M.M., Ahmadian, M.T: Application of homotopy analysis method in studying dynamic pull-in instability of microsystems. Mechanics Research Communications. **36**, 851–858 (2009).

[89] Zand, M.M., Ahmadian, M.T., Rashidian, B.: Semi-analytic solutions to nonlinear vibrations of microbeams under suddenly applied voltages. J. Sound and Vibration. **325**, 382–396 (2009).

[90] Zhao, J.,Wong, H.Y.:A closed-form solution to American options under general diffusions (2008). Available at SSRN: http://ssrn.com/abstract=1158223. Accessed 15 April 2011.

[91] Zhu, J.: Linear and Non-linear Dynamical Analysis of Beams and Cables and Their Combinations. PhD dissertation, Zhejiang University (2008).

[92] Zhu, S.P.: A closed-form analytical solution for the valuation of convertible bonds with constant dividend yield. ANZIAM J. **47**, 477–494 (2006a).

[93] Zhu, S.P.: An exact and explicit solution for the valuation of American put options. Quant. Financ. **6**, 229–242 (2006b).

[94] Zou, L.: A Study of Some NonlinearWater Wave Problems Using Homotopy Analysis Method. PhD dissertation, Dalian University of Technology (2008).

[95] Zou, L., Zong, Z.,Wang, Z., He, L.: Solving the discrete KdV equation with homotopy analysis method. Phys. Lett. A. **370**, 287–294 (2007).

附录 7.1　Mathematica 软件包 BVPh 1.0

BVPh 1.0 是廖世俊于 2010 年在上海交通大学基于同伦分析方法开发的 Mathematica 软件包, 用于求解某些类型的 n 阶非线性边值 (特征值) 问题.

版 权 声 明

版权所有 ⓒ2011 上海交通大学和 BVPh 开发者. 保留所有权利.

如果满足以下条件, 则允许以源代码和二进制形式重新发布和使用, 无论是否修改:

- 源代码的重新发布必须保留上述版权声明、此条件列表和以下免责声明.
- 以二进制形式重新发布必须在随发布提供的文档和 (或) 其他材料中复制上述版权声明、此条件列表和以下免责声明.
- 未经 BVPh 开发者书面同意, 不得出于盈利目的以源代码和二进制形式重新发布或使用, 无论是否修改.

本软件由版权所有人和贡献者 "按原样" 提供, 不提供任何明示或暗示的保证. 在任何情况下, 版权所有人或贡献者均不对任何损害承担责任.

Mathematica 软件包 BVPh 1.0

廖世俊
上海交通大学
2010 年 8 月

```
<<Calculus'Pade',
<<Graphics'Graphics';

(*********************************************)
(* Default of control parameters              *)
(*********************************************)
```

```
Ntruncated       = 10;
Nupdate          = 10;
NtermMax         = 90;
Nintegral        = 50;
ErrReq           = 10^(-20);
NgetErr          = 2;
ComplexQ         = 0;
kappa            = 1;
PRN              = 1;
cOL              = -2;
ZL               = 0;
c0               = .;


(****************************************************)
(* Default auxiliary function                       *)
(****************************************************)
H[z_] := 1;

U[0] = u[0];
(****************************************************)
(* Define Linv                                      *)
(* Find solution of linear equation: L[u ] = f      *)
(****************************************************)
Linv[f_] := Module[{temp, w},
temp = DSolve[ L[w[z]] == f,  w[z],  z];
temp[[1, 1, 2]] /. C[_]->0 // TrigReduce
];


(****************************************************)
(* Property of the inverse operator of L            *)
(* Linv[f_+g_] := Linv[f] + Linv[g]                 *)
(****************************************************)
Linv[p_Plus] := Map[Linv, p];
Linv[c_*f_] := c*Linv[f] /; FreeQ[c, z];


(****************************************************)
(* Define Getdelta[k]                               *)
(* Get delta[k] automatically based on              *)
(*     the definition of f[z, u, lambda]           *)
```

附录 7.1 Mathematica 软件包 BVPh 1.0

```
(****************************************************)
Getdelta[k_]:=Module[{temp, phi, LAMBDA, w,
                     lamb, q, eq, m, n, coeff, Coeff},
eq = f[z, phi[z, q], LAMBDA[q]];
temp[0]  =   D[eq, {q, k}]/k! // Expand;
temp[1]  =   temp[0]/.D[phi[z, q], {z, m_}, {q, n_}]
                  ->n!*diff[w[n], {z, m}];
temp[2]  =   temp[1]/.D[LAMBDA[q], {q, n_}]
                  ->n!*lamb[n];
temp[3]  =   temp[2]/.{phi[z, q]->w[0],
                  LAMBDA[q]->lamb[0]};
temp[4] = temp[3]/.diff[w[m_], {z, 0}]->w[m];
temp[5] = temp[4]/.{w->u, diff->D};
delta0[k] = temp[5] /.lamb->lambda;
If[ApproxQ == 0,
   temp[-1] = delta0[k]//Expand//TrigReduce
   ];
If[ApproxQ == 1 && TypeL == 1,
If[TypeEQ == 1,
temp[-1]=ChebyApproxA[delta0[k], z, 0, zR, Ntruncated],
For[m=0, m<=k, m++,
   coeff[m]=Coefficient[temp[5], lamb[m]]
   ];
coeff[-1] = temp[5]/. lamb[_]->0;
For[n=-1, n<=k, n++,
Coeff[n]=ChebyApproxA[coeff[n], z, 0, zR, Ntruncated];
   ];
lamb[-1] = 1;
temp[-1] = Sum[Coeff[n]*lamb[n], {n, -1, k}]
             /. lamb->lambda;
];
];
If[ApproxQ == 1 && TypeL == 2,
If[TypeEQ == 1,
temp[-1] = TrigApprox[delta0[k], z, zR,
                     Ntruncated, TypeBase],
For[m=0, m<=k, m++,
   coeff[m]=Coefficient[temp[5], lamb[m]]
   ];
```

```
coeff[-1] = temp[5]/. lamb[_]->0;
For[n=-1, n<=k, n++,
    Coeff[n] = TrigApprox[coeff[n], z, zR,
                         Ntruncated, TypeBase];
    ];
lamb[-1] = 1;
temp[-1] = Sum[Coeff[n]*lamb[n], {n, -1, k}]
           /. lamb->lambda;
];
];
delta[k] = temp[-1]//Expand//GetDigit;
];


(*****************************************************)
(* Define GetBC[k]                                   *)
(* Get boundary conditions automatically             *)
(*       based on the definition BC[n, z, u, lambda]*)
(*****************************************************)
GetBC[i_, k_] := Module[{temp, j, phi, q, LAMB},
phi = Sum[u[j]*q^j, {j, 0, k}];
LAMB = Sum[lambda[j]*q^j, {j, 0, k}];
temp[0] = BC[i, z, phi, LAMB];
temp[1] = Series[temp[0], {q, 0, k}]//Normal;
temp[2] = Coefficient[temp[1], q^k];
temp[2]//Expand//GetDigit
];


(*****************************************************)
(* Define functions chi[m                            *)
(*****************************************************)
chi[m_] := If[m <= 1, 0, 1];


(*****************************************************)
(* Define truncated[z^n_. ]                          *)
(*****************************************************)
truncated[a_] := a /; FreeQ[a, z];
truncated[a_*f_] := a*truncated[f] /; FreeQ[a, z];
truncated[p_Plus] := Map[truncated, p];
truncated[z^n_. ] := If[n > NtermMax, 0, z^n];
```

```
(******************************************)
(* Define GetReal[] and GetDigit[]         *)
(******************************************)
Naccu = 100;
GetReal[c_] := N[IntegerPart[c*10^Naccu]
                /10^Naccu, Naccu] /; NumberQ[c];
GetDigit[c_] := N[IntegerPart[c*10^Naccu]
                 /10^Naccu, Naccu] /; NumberQ[c];
Default[GetDigit, 1] = 0;
GetDigit[p_Plus] := Map[GetDigit, p];
GetDigit[c_. *f_] := GetReal[c]*f /; NumberQ[c];
GetDigit[c_. ] := 0 /; NumberQ[c]
                       && Abs[c] < 10^(-Naccu+1);


(*******************************************)
(* Define int[f, x, x0, x1, Nintegral]     *)
(* Integration of f in the interval [x0, x1] *)
(*      by Nintegral points                *)
(*******************************************)
int[f_, x_, x0_, x1_, Nintegral_]
:= Module[{temp, dx, s, t, i, M},
M = Nintegral;
dx = N[x1-x0, 100]/M;
temp[0] = Series[f, {x, 0, 2}]//Normal;
temp[0] = temp[0] /. x^_. ->0 ;
s = temp[0];
For[i=1, i<=M, i=i+1,
   t = x0 + i*dx;
   temp[i] = f /. x->t;
   s = s + temp[i]//Expand;
   ];
temp[M+1] = (temp[0]+temp[M])/2;
(s - temp[M+1])*dx//Expand
];


(********************************************)
(* Define ChebyApproxA[f, x, a, b, M]       *)
(* Approximate a function by Chebyshev polynomial *)
```

```
(***********************************************)
ChebyApproxA[f_, x_, a_, b_, Ntruncated_]
:= Module[{temp, n, z, t, F, A},
temp[0] = f/. x-> a + (b-a)*(z+1)/2;
F = temp[0] /. z -> Cos[t];
For[n=0, n<=Ntruncated, n++,
   A[n] = 2*int[F*Cos[n*t], t, 0, Pi, 100]/Pi
   ];
For[n = 0, n<=Ntruncated, n++,
temp[0] = A[n]//Expand;
A[n] = GetDigit[temp[0]];
];
temp[1] = A[0]/2 + Sum[A[n]*ChebyshevT[n, z],
                   {n, 1, Ntruncated}];
temp[1] /. z-> -1 + 2*(x-a)/(b-a)//Expand
];


(***********************************************)
(* Define TrigApprox[f, x, Ntruncated, TypeBase] *)
(*        Approximate f[x] in [0, xR] :          *)
(*            TypeBase = 0 : Fourier series      *)
(*            TypeBase = 1 : hybrid-base         *)
(*            Ntruncated : number of truncated terms *)
(***********************************************)
TrigApprox[f_, x_, xR_, Ntruncated_, TypeBase_]
:= Module[{temp, a, b, c, y, i, j, t},
If[TypeBase == 0, temp[0] = f/. x->t];
If[TypeBase == 1,
   temp[0] = Series[f, {x, 0, 2}]//Normal;
   temp[1] = temp[0] /. x^_. -> 0;
   temp[2] = f /. x->xR;
   temp[3] = temp[1]+(temp[2]-temp[1])/xR*x//Expand;
   y = GetDigit[temp[3]];
   temp[0] = f - y /. x->t;
   ];
If[TypeBase == 2,
   temp[0] = D[f, x]//Expand;
   temp[1] = Series[temp[0], {x, 0, 2}]//Normal;
   temp[1] = temp[1] /. x^_. -> 0;
```

```
    temp[2] = temp[0] /.  x->xR;
    a = temp[1];
    b = -(temp[1] + temp[2])/2/xR;
    temp[3] = (a*x+b*x^2)*Cos[Pi*x/xR]//Expand;
    y = GetDigit[temp[3]];
    temp[0] = f - y /.  x-> t ;
    ];
If[TypeBase == 0 || TypeBase == 2,
    For[i = 0,  i <= Ntruncated,  i++,
        c[i] = 2*int[temp[0]*Cos[i*t*Pi/xR],
                    t,0,xR,Nintegral]/xR;
        ];
    temp[4] = c[0]/2
        + Sum[c[j]*Cos[j*x*Pi/xR], {j, 1, Ntruncated}];
    ];
If[TypeBase == 1,
    For[i = 1,  i <= Ntruncated,  i++,
        c[i] = 2*int[temp[0]*Sin[i*t*Pi/xR],
                    t,0,xR,Nintegral]/xR;
        ];
    temp[4] = Sum[c[j]*Sin[j*x*Pi/xR],
                    {j, 1, Ntruncated}];
    ];
If[TypeBase == 0,
    temp[5] = GetDigit[temp[4]],
    temp[5] = GetDigit[temp[4] + y]
    ];
temp[5]
];


(*******************************************************)
(* Define GetMin1D[f, x, a, b, Npoint]              *)
(*******************************************************)
GetMin1D[f_, x_, x0_, x1_, Npoint_] := Module[
{temp, fmin, Xmin, xmin0, xmin1, dx, Num, i, j, X, s},
Num = 0;
dx = (x1-x0)/Npoint //GetDigit;
For[i = 0,  i <= Npoint,  i++,
    X[i] = x0 + i*dx //GetDigit;
```

```
          temp[i] = f /. x -> X[i];
          ];
    For[i = 1, i <= Npoint-1, i++,
          If[temp[i] < temp[i-1] && temp[i] < temp[i+1],
              Num = Num + 1;
              fmin[Num] = temp[i];
              Xmin[Num] = X[i];
              ];
          ];
    For[i = 1, i <= Num, i++,
          j = 0;
          xmin0 = Xmin[i];
          Label[100];
          j = j + 1;
          temp[0] = xmin0 - dx/5^(j-1) //GetDigit;
          temp[1] = xmin0 + dx/5^(j-1) //GetDigit;
          s = GetMin1D0[f, x, temp[0], temp[1], 10];
          xmin1 = s[[2]];
          If[Abs[xmin1 - xmin0] < 10^(-20),
              Xmin[i] = xmin1;
              fmin[i] = f /. x-> xmin1;
              Goto[200];
              ];
          xmin0 = xmin1;
          Goto[100];
          Label[200];
          Print[" Minimum = ", fmin[i]//N, " at ",
                   x, " = ", N[Xmin[i], 20]];
          ];
    ];


GetMin1D0[f_, x_, x0_, x1_, Npoint_]
  := Module[{temp, fmin, Xmin, dx, i, z, F},
  dx = (x1-x0)/Npoint // GetDigit;
  fmin = 10^100;
  Xmin = 0;
  For[i = 1, i <= Npoint, i++,
      z = x0 + i*dx // GetDigit;
      F = f /. x->z // GetDigit;
```

```
        If[F < fmin,
            fmin = F;
            Xmin = z;
            ];
        ];
{fmin, Xmin}
];


(***********************************************)
(* Define GetMin2D[f, x, x0, x1, y, y0, y1, Npoint]    *)
(***********************************************)
GetMin2D[f_, x_, x0_, x1_, y_, y0_, y1_, Npoint_]
    :=Module[{temp, fmin, Xmin, Ymin, xmin0, xmin1,
              ymin0, ymin1, dx, dy, Num, i, j, X, Y, s},
Num = 0;
dx = (x1-x0)/Npoint //GetDigit;
dy = (y1-y0)/Npoint //GetDigit;
For[i = 0,  i <= Npoint,  i++,
    X[i] = x0 + i*dx //GetDigit;
    For[j = 0,  j <= Npoint,  j++,
        Y[j] = y0 + j*dy //GetDigit;
        temp[i, j] = f /. {x -> X[i], y->Y[j]};
        ];
    ];
For[i = 1, i <= Npoint-1, i++,
    For[j = 1, j <= Npoint-1, j++,
        If[temp[i, j] < temp[i-1, j]
            && temp[i, j] < temp[i+1, j]
            && temp[i, j] < temp[i, j-1]
            && temp[i, j] < temp[i, j+1],
            Num = Num + 1;
            fmin[Num] = temp[i, j];
            Xmin[Num] = X[i];
            Ymin[Num] = Y[j];
            ];
        ];
    ];
For[i = 1,  i <= Num,  i++,
    j = 0;
```

```
      xmin0 = Xmin[i];
      ymin0 = Ymin[i];
      Label[100];
      j = j + 1;
      X[0] = xmin0 - dx/5^(j-1) //GetDigit;
      X[1] = xmin0 + dx/5^(j-1) //GetDigit;
      Y[0] = ymin0 - dy/5^(j-1) //GetDigit;
      Y[1] = ymin0 + dy/5^(j-1) //GetDigit;
      s = GetMin2D0[f, x, X[0], X[1], y, Y[0], Y[1], 10];
      xmin1 = s[[2]];
      ymin1 = s[[3]];
      If[Abs[xmin1 - xmin0] < 10^(-20)
         && Abs[ymin1 - ymin0] < 10^(-20),
         Xmin[i] = xmin1;
         Ymin[i] = ymin1;
         fmin[i] = f /. {x-> xmin1, y->ymin1};
         Goto[200];
         ];
      xmin0 = xmin1;
      ymin0 = ymin1;
      Goto[100];
      Label[200];
      Print[" Minimum = ", fmin[i]//N];
      Print[" at ",  x,  " = ", N[Xmin[i], 20]];
      Print[" ",  y,  " = ", N[Ymin[i], 20]];
      ];
];

GetMin2D0[f_, x_, x0_, x1_, y_, y0_, y1_, Npoint_]
:= Module[{temp, fmin, Xmin, Ymin, dx, dy, i, X, Y, F},
dx = (x1-x0)/Npoint // GetDigit;
dy = (y1-y0)/Npoint // GetDigit;
fmin = 10^10;
Xmin = 10^10;
Ymin = 10^10;
For[i = 1,  i <= Npoint,  i++,
    X = x0 + i*dx // GetDigit;
    For[j = 1,  j <= Npoint,  j++,
       Y = y0 + j*dy // GetDigit;
```

```
            F = f /.  {x->X,  y->Y} // GetDigit;
         If[F < fmin,
               fmin = F;
               Xmin = X;
               Ymin = Y
               ];
            ];
      ];
{fmin, Xmin, Ymin}
];


(*****************************************************)
(* Define GetErr[m]                                 *)
(*     This module gives averaged squared residual  *)
(*****************************************************)
GetErr[k_] := Module[{temp},
error[k] = f[z, U[k], Lambda[k-1]];
If[ComplexQ == 0,  temp[0] = error[k]^2 ] ;
If[ComplexQ == 1,
   temp[1] = Re[error[k]];
   temp[2] = Im[error[k]];
   temp[0] = temp[1]^2 + temp[2]^2;
];
If[!NumberQ[zRintegral],
   If[FreeQ[zR, infinity] && zR < 100,
      zRintegral = zR,
      Print["Squared residual is integrated
                in the interval [", zL, ", b]"];
      Print[" The value of b = ? "];
      zRintegral = Input[];
      ];
   ];
temp[1] = Abs[zRintegral-zL];
Err[k] = int[temp[0], z, zL, zRintegral, Nintegral]
          /temp[1];
];


(***************************************************)
(* Define hp[f_, m_, n_]                          *)
```

```
(***************************************************)
hp[f_, m_, n_]:=Block[{k, i, df, res, q},
df[0] = f[0];
For[k=1, k<=m+n, k++, df[k]=f[k]-f[k-1]//Expand];
res = df[0] + Sum[df[i]*q^i, {i, 1, m+n}];
Pade[res, {q, 0, m, n}]/. q->1
];


(***************************************************)
(* Main Code                                      *)
(* HAM approach without iteration                 *)
(***************************************************)
BVPh[begin_, end_]:=Block[
{uSpecial, CC, temp, ss, w, i, j, EQ, Unknown, phi, LAM,
   sLeng, sMin, sMinI, RHS, base, a},
time[0] = SessionTime[];
For[k=begin, k<=end, k=k+1,
   If[PRN == 1,  Print["k = ", k]];
   If[k == 1 && NumberQ[c0],  c0 = GetDigit[c0]];
   If[k == 1,  Clear[lambda]];

(* Get special solution *)
   Getdelta[k-1];
   RHS = H[z]*delta[k-1]//Expand;
   temp[1] = Linv[RHS];
   uSpecial = c0*temp[1] + chi[k]*u[k-1]//Expand;

(* Get lambda[k-1] and u[k] *)
   temp[1] = DSolve[L[w[z]]==0, w[z], z];
   temp[2] = temp[1][[1, 1, 2]];
   Unknown = {};
   For[i = 1,  i <= OrderEQ,  i++,
      base[i] = Coefficient[temp[2], C[i]];
      If[FreeQ[base[i], Exp[n_ . *z]],
         Unknown = Union[Unknown, {CC[i, k]}],
         If[( Abs[base[i]] /.  z -> 1. ) < 1,
            Unknown = Union[Unknown, {CC[i, k]}],
            CC[i,k] = 0;
            base[i] = 0;
```

附录 7.1 Mathematica 软件包 BVPh 1.0

```
                    ];
                ];
            ];

ss = Sum[base[i]*CC[i, k], {i, 1, OrderEQ}];
u[k] = uSpecial + ss // Expand;
    For[i = 1,  i <= OrderEQ,  i++,
        eq[i] = GetBC[i, k]//Expand//GetDigit
        ];
    EQ = {};
    For[i=1,  i <= OrderEQ,  i++,
        If[FreeQ[BC[i, z, w[z], a], infinity],
            EQ = Union[EQ, {eq[i] == 0}];
            ];
        ];

If [TypeEQ != 1,
    eq[0] = GetBC[0, k]//Expand//GetDigit;
    EQ = Union[EQ, {eq[0] == 0}];
    Unknown = Union[Unknown, {lambda[k-1]}];
    ];

ss = Solve[EQ, Unknown];
sLeng = Length[ss];
If[sLeng == 1,  sNum = 1];
If[k == 1 && !NumberQ[sNum],
    For [j = 1,  j <= sLeng, j++,
    Temp [1] = lambda[0] /. ss[[j]];
    Print[j, " th solution of lambda[0] = ",
            temp[1]//N];
        ];
Print ["which solution is chosen ? "];
sNum = Max[1, Input[]];
];

If[k == 1 && TypeEQ != 1 && NumberQ[sNum]
            && NumberQ[LAMBDA[nIter-1]],
    sMin = 10^10;
    For[j = 1,  j <= sLeng,  j++,
```

```
               temp[1] = lambda[0] /.  ss[[j]];
               temp[2] = Abs[temp[1]-LAMBDA[nIter-1]];
               If[ temp[2] < sMin,
                  sNum = j;
                  sMin = temp[2];
                  ];
            ];
         ];

   For[i = 1,  i <= OrderEQ,  i++,
         temp[1] = CC[i, k] /.  ss[[sNum]]//Expand;
         CC[i, k] = temp[1]//GetDigit;
      ];
   u[k] = u[k]//Expand//GetDigit;
   U[k] = Expand[U[k-1] + u[k]];
   Uz[k] = D[U[k], z];

   If[TypeEQ != 1,
      temp[1] = lambda[k-1] /.  ss[[sNum]]//Expand;
      lambda[k-1] = temp[1]//GetDigit;
      Lambda[k-1] = Sum[lambda[i], {i, 0, k-1}]
                   //Expand//GetDigit;
      If[PRN == 1,
         Print[k-1, "th-order approx. of eigenvalue
                   = ", Lambda[k-1]//N ];
         ];
      ];


   (* Print results *)
      If[PRN == 1 && NumberQ[c0] && TypeEQ == 1,
         output[z, U, k]
         ];
      If[IntegerQ[k/NgetErr] && PRN == 1,
         GetErr[k];
         If[NumberQ[Err[k]],
            Print["Squared residual = ", Err[k]//N];
            ];
         If [Err[k] < ErrReq,
            Print["Congratulation: the required
```

```
                    squared residual is satisfied !"];
            Goto[end]
            ];
        ];

    If[PRN == 1,
        time[k] = SessionTime[];
        temp[0] = time[k]-time[0];
        Print["Used CPU time = ", temp[0],
            " (seconds) "];
        ];
    ];

If [PRN == 1, Print[" Successful !"]];
Label[end];
];


(*****************************************************)
(*               Main Code                          *)
(*     HAM approach with iteration                  *)
(*       begin : number of starting iteration       *)
(*       end : number of stopping iteration         *)
(*       OrderIter : order of iteration formula     *)
(*****************************************************)
iter [begin_, end_, OrderIter_]:=Block[{temp, time},
PRN = 0;
If[TypeL != 1, ApproxQ = 1];
For[nIter=Max[1, begin], nIter<=end, nIter=nIter+1,
    If[nIter == 1,
        sNum = . ;
        If[NumberQ[c0], c0 = GetDigit[c0]];
        Time[0] = SessionTime[];
        V[0] = u[0]//Expand//GetDigit;
        Vz[0] = D[V[0], z];
        ];
    If[nIter > 1,
        u[0] = V[nIter-1];
        U[0] = u[0];
        ];
```

```
Print[ nIter, " th iteration:"];

If [NumberQ[c0],  BVPh[1, OrderIter]];
If [!NumberQ[c0],
    BVPh[1, Max[3, OrderIter]];
    GetErr[1];
    GetErr[2];
    GetErr[3];
    Print[" Squared Residual of Governing EQ:"];
    Print[" Red line : 1st-order approx. "];
    Print[" Green line : 2nd-order approx. "];
    Print[" Blue line : 3rd-order approx. "];
    LogPlot[{Abs[Err[1]], Abs[Err[2]], Abs[Err[3]]},
       {c0, c0L, 0}, PlotStyle->{RGBColor[1, 0, 0],
          RGBColor[0, 1, 0], RGBColor[0, 0, 1]}];
    Print[" Adjust the interval of c0 by choosing
         a better end-point "];
    Print[" New end-point on left (c0 < 0)
          or right (c0 > 0) = ?
          (Input 0 to skip it) "];
    c0L = Input[];
    If[c0L != 0,
    LogPlot[{Abs[Err[1]], Abs[Err[2]], Abs[Err[3]]},
       {c0,c0L,0}, PlotStyle->{RGBColor[1,0,0],
          RGBColor[0,1,0],RGBColor[0,0,1]}];
      ];
    Print[" Choose the value of
           convergence-control parameter c0"];
    Print[" c0 = ? (Input 0 to stop the code)"];
    temp[0] = Input[];
    c0 = GetDigit[temp[0]];
    Print[" c0 = ", c0//N];
    If[c0==0, c0 =. ;Print["Stop!"];Goto[stop]];
    ];

 temp[0] = U[OrderIter]//Expand;
 If [PolynomialQ[temp[0], z],
    temp[0] = truncated[temp[0]]
    ];
```

```
        V[nIter] = GetDigit[temp[0]];
        Vz[nIter] = D[V[nIter], z];
        If[TypeEQ == 1,
            output[z, V, nIter],
            LAMBDA[nIter]=Lambda[OrderIter-1]//Expand;
            Print [" Eigenvalue = ", LAMBDA[nIter]//N ];
            ];

        GetErr [OrderIter];
        ERR[nIter] = Err[OrderIter];
        Print[" Squared Residual = ", ERR[nIter]//N];
        If[ERR[nIter] < ErrReq,
            Print["STOP:Required accuracy satisfied!"];
            Goto[stop];
           ];
        If[nIter > 1 && ERR[nIter] > ERR[nIter-1]
            && Err[nIter] < 10^(-7),
            Print["Squared residual does NOT decrease!"];
            Goto[stop];
            ];
        Time[nIter] = SessionTime[];
        Temp[-1] = Time[nIter]-Time[0];
        Print[" Used CPU time = ",  Temp[-1],
              " (seconds) "];
        If[IntegerQ[nIter/Nupdate]&&Abs[c0]< 1/2, c0 =. ];
        ];
Print[" End of iteration ! "];
Label[stop];
PRN = 1;
];


(****************************************************)
(* Print equation, boundary conditions          *)
(*    and control parameters                    *)
(****************************************************)
PrintInput[s_] := Module[{},
Print["----------------------------------------"];
Print["The values of control parameters: "];
Print[" Nupdate = ",  Nupdate];
```

```
Print[" Ntruncated = ",  Ntruncated];
Print[" NtermMax = ",  NtermMax];
Print[" Nintegral = ",  Nintegral];
Print[" ErrReq = ",  ErrReq//N];
Print[" NgetErr = ",  NgetErr];
Print[" ComplexQ = ",  ComplexQ];
Print[" PRN = ",  PRN];
Print[" cOL = ",  cOL];
Print[" zL = ",  zL];
Print["----------------------------------------"];
Print[" Governing Equation : ",
        f[z,s,lambda],  " = 0 "];
Print[" Interval of z : ",
        zL,  " < z < ",  zR];
If[TypeEQ != 1,
        Print[" ", 0, "th Boundary Condition : ",
              BC[0,  z,  s,  lambda], " = 0"];
   ];
For[i=1,  i<= OrderEQ,  i++,
   Print[" ", i, "th Boundary Condition : ",
         BC[i,  z,  s,  lambda], " = 0"];
   ];
Print["----------------------------------------"];
If[ApproxQ == 0,
   Print[" Delta[k] is NOT approximated
           by other base functions ! "],
   If[TypeL == 1,
      Print [" Base function : Chebyshev "],
      If[TypeBase == 1,
         Print[" Base function : hybrid-base
               with sine and polynomial"],
         Print[" Base function : hybrid-base
               with cosine and polynomial "]
         ];
      ];
   ];
Print[" Auxiliary linear operator L[u] = ", L[s]];
If [TypeL != 1,
   Print["                kapp         a = ", kappa]
```

```
    ];
Print[" Initial guess u[0] = ", u[0]];
Print[" Auxiliary function H[z] = ", H[z]];
If [NumberQ[c0],
    Print[" Convergence-control parameter c0 = ", c0]
    ];
Print["----------------------------------------"];
]
```

第八章 具有多解的非线性边值问题

本章使用三种不同类型的具有多解的非线性边值方程,验证了基于同伦分析方法的 Mathematica 软件包 BVPh 1.0 对于求解定义在有限区间 $0 \leqslant z \leqslant a$ 上,且满足 n 个线性边界条件 $\mathcal{B}_k[z,u] = \gamma_k$ $(1 \leqslant k \leqslant n)$ 的 n 阶非线性边值方程 $\mathcal{F}[z,u] = 0$ 的有效性,其中 $\mathcal{F}$ 是 n 阶非线性算子,$\mathcal{B}_k$ 是线性算子,γ_k 是常数. 特别地,我们在初始猜测解中引入多解控制参数来寻找多解,以 BVPh 1.0 为工具,通过这种多解控制参数可以找到某些非线性边值方程的多解.

8.1 绪论

本章阐述了 Mathematica 软件包 BVPh 1.0 对于具有多解的非线性边值问题的普遍有效性. 该问题由有限区间上的 n 阶非线性微分方程

$$\mathcal{F}[z,u] = 0, \quad z \in [0,a] \tag{8.1}$$

控制,且在两个端点和区间 $z \in (0,a)$ 上的一些分离点处满足 n 个线性多边界条件

$$\mathcal{B}_k[z,u] = \gamma_k, \tag{8.2}$$

其中 $\mathcal{F}$ 是 n 阶非线性算子,$\mathcal{B}_k$ 是线性算子,z 是自变量,$u(z)$ 是未知函数,γ_k 是常数. 假定此类非线性边值问题至少存在一个光滑解.

非线性边值方程一般很难通过数值技术找到多解,特别是当它们彼此接近和 (或) 数值不稳定时. 基于同伦分析方法 (HAM) [100–114, 117],我们提出了

一些解析方法以获得非线性边值问题的多解. 2005 年, 廖世俊应用同伦分析方法成功地获得了边界层方程

$$F''' + \frac{1}{2}FF'' - \beta F'^2 = 0, \quad F(0) = 0, F'(0) = 1, F'(+\infty) = 0$$

的解的两个分支, 其中 $-1 < \beta < +\infty$ 是常数. 引入额外的未知量 $\delta = F(+\infty)$, 通过求解与该未知量 δ 相关的非线性代数方程, 廖世俊 [108] 在 $\beta > 1$ 时发现了解的一个新分支, 该解从来没有被其他解析方法甚至数值方法得到过, 这主要是因为解的两个分支的 $F'''(0)$ 值之间的差异非常小, 很难区分它们. 这说明了同伦分析方法对于具有多解的非线性问题的巨大潜力. 此外, 本章证明了通过适当地引入未知参数可以找到一些非线性问题的多解.

2009 年, Abbasbandy 等人 [96] 应用同伦分析方法求解关于收敛控制参数的非线性代数方程, 获得了二阶反应扩散方程的多解. 2010 年, 徐航等人 [117] 应用同伦分析方法, 通过类似打靶的技术获得了两个移动平行板间黏性边界层流动问题的多解.

本章说明了 BVPh 1.0 可以通过在初始猜测解中引入多解控制参数来找到非线性边值方程的多解. 本章将用三个例子来验证这一方法的有效性, BVPh 1.0 的输入数据文件在本章的附录中给出.

8.2 简明数学公式

令 $u_0(z)$ 表示 (8.1) 和 (8.2) 的解 $u(z)$ 的初始猜测解, $q \in [0, 1]$ 为嵌入变量. 在同伦分析方法的框架中, 我们构造这样一个连续形变 (同伦) $\phi(z; q)$, 即当 q 从 0 增加到 1 时, $\phi(z; q)$ 从初始猜测解 $u_0(z)$ 变化到方程 (8.1) 和 (8.2) 的解 $u(z)$. 这种连续形变满足零阶形变方程

$$(1-q)\mathcal{L}\left[\phi(z;q) - u_0(z)\right] = c_0 q \mathcal{F}[z, \phi(z;q)] \tag{8.3}$$

和 n 个线性边界条件

$$\mathcal{B}_k[z, \phi(z;q)] = \gamma_k, \tag{8.4}$$

其中 $\mathcal{L}$ 是具有性质 $\mathcal{L}(0) = 0$ 的辅助线性算子, 其导数的最高阶数为 n, $c_0 \neq 0$ 为收敛控制参数. 显然, 当 $q = 0$ 和 $q = 1$ 时, 分别有

$$\phi(z;0) = u_0(z), \quad \phi(z;1) = u(z). \tag{8.5}$$

因此，如果辅助线性算子 $\mathcal{L}$ 和收敛控制参数 c_0 选取合适，使得同伦 – Maclaurin 级数

$$\phi(z;q) = u_0(z) + \sum_{m=1}^{+\infty} u_m(z) q^m \tag{8.6}$$

收敛至 $\phi(z;1)$，则我们有同伦级数解

$$u(z) = u_0(z) + \sum_{m=1}^{+\infty} u_m(z), \tag{8.7}$$

其中 $u_m(z)$ 满足 m 阶形变方程

$$\mathcal{L}\left[u_m(z) - \chi_m u_{m-1}(z)\right] = c_0 \delta_{m-1}(z) \tag{8.8}$$

和 n 个线性边界条件

$$\mathcal{B}_k[z, u_m] = 0, \tag{8.9}$$

其中

$$\delta_k(z) = \mathcal{D}_k\{\mathcal{F}[z, \phi(z;q)]\}$$

且

$$\mathcal{D}_k[\phi] = \frac{1}{k!} \left.\frac{\partial^k \phi}{\partial q^k}\right|_{q=0}, \quad \chi_k = \begin{cases} 0, & k \leqslant 1, \\ 1, & k > 1. \end{cases}$$

这里，$\mathcal{D}_k$ 是 k 阶同伦导数算子. 方程 (8.8) 由定理 4.15 给出，详情请参考第四章. 注意，$\delta_k(z)$ 只依赖于非线性算子 $\mathcal{F}[z,u]$，并且可以通过第四章中证明的定理轻松推导. 事实上，对于非线性常微分方程 (ODEs)，我们不需要通过 BVPh 1.0 推导出 $\delta_k(z)$ 项.

在 m 阶同伦近似

$$\tilde{u}(z) \approx u_0(z) + \sum_{k=1}^{m} u_k(z) \tag{8.10}$$

下，控制方程 (8.1) 的离散平方残差 E_m 定义为

$$E_m = \frac{1}{N_p + 1} \sum_{k=0}^{N_p} \{\mathcal{F}[z_k, \tilde{u}(z_k)]\}^2, \tag{8.11}$$

其中 $z_k = k\left(\frac{a}{N_p}\right)$ 表示 $N_p + 1$ 个不同点的坐标. 最优收敛控制参数 c_0 由平方残差 E_m 的最小值确定.

在同伦分析方法的框架中, 我们有极大的自由度来选取辅助线性算子 $\mathcal{L}$ 和初始猜测解 $u_0(z)$. 众所周知, 有限区间 $z \in [0, a]$ 上的连续函数 $f(z)$ 可以用 Chebyshev 级数或 Fourier 级数很好地近似. 如果利用幂级数或 Chebyshev 多项式来近似 $u(z)$, 我们通常选取辅助线性算子

$$\mathcal{L}(u) = \frac{d^n u}{dz^n}, \tag{8.12}$$

其中 n 是 (8.1) 的导数的最高阶数. 如果采用 §7.2.3 中描述的混合基近似方法来近似解 $u(z)$, 我们选取如下辅助算子

$$\begin{aligned}
\mathcal{L}(u) &= u'' + \omega_1^2 u, & n &= 2, \\
\mathcal{L}(u) &= u''' + \omega_1^2 u', & n &= 3, \\
\mathcal{L}(u) &= u'''' + (\omega_1^2 + \omega_2^2) u'' + \omega_1^2 \omega_2^2 u, & n &= 4, \\
\mathcal{L}(u) &= u''''' + (\omega_1^2 + \omega_2^2) u''' + \omega_1^2 \omega_2^2 u', & n &= 5, \\
&\cdots\cdots\cdots
\end{aligned}$$

其中 $\omega_k > 0$ 是基函数 $\cos(\omega_k z)$ 和 $\sin(\omega_k z)$ 的频率, 即

$$\mathcal{L}[\cos(\omega_k z)] = \mathcal{L}[\sin(\omega_k z)] = 0.$$

上述辅助线性算子可以表示为

$$\mathcal{L}(u) = \left[\prod_{i=1}^{m} \left(\frac{d^2}{dx^2} + \omega_i^2\right)\right] u, \quad n = 2m, \tag{8.13}$$

或

$$\mathcal{L}(u) = \left[\prod_{i=1}^{m} \left(\frac{d^2}{dx^2} + \omega_i^2\right)\right] u', \quad n = 2m + 1, \tag{8.14}$$

其中频率 ω_i 可能互不相同, 例如

$$\omega_i = i\left(\frac{\kappa \pi}{a}\right), \tag{8.15}$$

也可能相同, 例如

$$\omega_i = \frac{\kappa \pi}{a}, \tag{8.16}$$

其取决于边界条件 (8.2). 这里, $\kappa \geqslant 1$ 为正整数.

需要注意的是, 初始猜测解 $u_0(z)$ 应至少满足 n 个线性边界条件 (8.2), 并且应由多项式或三角函数或混合基函数表示.

关于辅助线性算子 $\mathcal{L}$ 和初始猜测解 $u_0(z)$ 的选取, 请参考第七章.

8.3 例子

许多非线性边值方程的多解通常很难通过数值方法得到. 本节使用三个具有多解的非线性边值问题来验证 BVPh 1.0 的有效性, BVPh 1.0 已在附录 7.1 中给出. 此外, 所有这些例子的 BVPh 1.0 输入数据文件均在本章的附录中给出.

8.3.1 非线性扩散反应模型

我们先考虑多孔催化剂中的非线性扩散反应模型

$$uu'' - \gamma^2 = 0, \quad u'(0) = 0, u(1) = 1, \tag{8.17}$$

其中 γ 为给定的参数. 上述方程具有隐式封闭解

$$z = \frac{u(0)}{i\gamma}\sqrt{\frac{\pi}{2}}\,\mathrm{Erf}\left[i\sqrt{\ln\frac{u(z)}{u(0)}}\right], \tag{8.18}$$

其中 $i = \sqrt{-1}$, $\mathrm{Erf}(z)$ 表示误差函数, $u(0)$ 由

$$1 = \frac{u(0)}{i\gamma}\sqrt{\frac{\pi}{2}}\,\mathrm{Erf}\left[i\sqrt{\ln\frac{1}{u(0)}}\right] \tag{8.19}$$

确定.

该例子是 (8.1) 的一个特例, 即

$$\mathcal{F}[z,u] = uu'' - \gamma^2.$$

在这种情况下, m 阶同伦近似由 (8.10) 给出, 其中 $u_k(z)$ 由 k 阶形变方程

$$\mathcal{L}[u_k(z) - \chi_k u_{k-1}(z)] = c_0 \delta_{k-1}(z) \tag{8.20}$$

控制, 满足边界条件

$$u'_k(0) = 0, \quad u_k(1) = 0, \tag{8.21}$$

其中由定理 4.1 可得

$$\delta_n(z) = \sum_{j=0}^{n} u_j u''_{n-j} - (1 - \chi_{n+1}) \gamma^2. \tag{8.22}$$

正如 Magyari [115] 指出的那样，当 $0 \leqslant \gamma \leqslant 0.775152$ 时，方程 (8.17) 有两个解．Abbasbandy 等人 [96] 使用以多项式为基函数的同伦分析方法，并通过求解关于收敛控制参数 c_0 的非线性代数方程得到了这两个解．本章提出了一种新的最优同伦分析方法，通过在初始猜测解中引入一个未知参数，即多解控制参数来找到多解．

不失一般性，我们先考虑 $\gamma = 3/5$ 的情况．当 $\gamma = 3/5$ 时，非线性代数方程 (8.19) 有 $u(0) = 0.101046$ 和 $u(0) = 0.779034$ 两个解，分别对应于隐式形式 (8.18) 的两个精确解．

我们应用 §7.2.3 中提到的混合基近似方法来求解具有多解的非线性边值问题．根据 (8.13)，选取辅助线性算子

$$\mathcal{L}(u) = u'' + \pi^2 u. \tag{8.23}$$

需要注意的是，如果 $u(z)$ 满足 (8.17)，那么 $u(-z) = u(z)$ 也是它的解．因此，$u(z)$ 是偶函数．这个结论对于我们选取初始猜测解 $u_0(z)$ 是有用的．因此，选取初始猜测解

$$u_0(z) = \frac{1}{2}(\sigma + 1) + \frac{1}{2}(\sigma - 1)\cos(\pi z), \tag{8.24}$$

其中 $\sigma = u_0(0)$ 为未知参数．注意，$u_0(z)$ 依赖于未知参数 σ，因此，不同取值的 σ 会得到不同的初始猜测解．此处引入未知参数 σ 以寻找多解，如下所述．

选取辅助线性算子 (8.23) 和初始猜测解 (8.24)，我们通过 BVPh 1.0 获得了 10 阶同伦近似．注意，该近似包含两个未知参数：收敛控制参数 c_0 和 (8.24) 中的未知参数 σ．然而，给定 c_0，平方残差 E_m 仅取决于 σ．例如，在 $c_0 = -1$ 的情况下，E_m 关于 σ 的曲线如图 8.1 所示，这清楚地表明，在大范围 $-0.8 < \sigma < 0.5$ 内，平方残差 E_m 随着近似阶数的增加而变小，并且 σ 的最优值 ($c_0 = -1$) 大约为 -0.2，对应于收敛最快的级数．确实如此，我们发现，当 $c_0 = -1$ 时，3 阶迭代法由 $\sigma = -1/2, -1/5, 1/2$ 得到收敛结果 $u(0) = 0.101046$，并且在 $\sigma = -1/5$ 时收敛最快，分别如表 8.1 和表 8.2 所示．此外，相应的 5 阶近似值与下分支精确解非常吻合，如图 8.3 所示．因此，当 $c_0 = -1$ 时，存在区间 $\Omega_1 \supset [-1/2, 1/2]$，对于任意 $\sigma \in \Omega_1$，对应的同伦级数收敛至 (8.17) 的下分支解．注意，如果 $\sigma \notin \Omega_1$，则无法获得下分支解的高精度近似值．在这个意义上，与 c_0 一样，初始猜测解 (8.24) 中的未知参数 σ 也可以视为一种收敛控制参数．

根据图 8.1, 存在另一个区间 $\Omega_2 \subset (0.5, 1)$, 使得平方残差 E_m 随着近似阶数的增加而变小. 然而, 当 $c_0 = -1$ 时, 该间隔相当小, 即 $\Omega_2 \subset (0.7276, 0.7292)$, 如图 8.2 所示. 研究发现, 随着近似阶数的增加, 这种区间越来越小, 以至于确定其精确边界非常耗时. 此外, 即使使用 20 阶近似下的最优值 $\sigma^* = 0.7284$ ($c_0 = -1$), 我们仍然无法获得收敛级数解. 若使用 c_0 的其他值, 例如 $c_0 = -1/4$, 这种区间 Ω_2 仍然非常小. 因此, 当 $c_0 = -1$ 时, 区间 Ω_2 很小, 以至于很难通过迭代法获得 (8.17) 上分支解的足够精确的近似值.

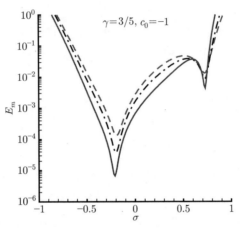

图 8.1 当 $\gamma = 3/5$ 和 $c_0 = -1$ 时, 选取辅助线性算子 (8.23) 和初始猜测解 (8.24), (8.17) 关于 $\sigma = u_0(0)$ 的平方残差 E_m. 虚线: 5 阶近似; 点划线: 6 阶近似; 实线: 8 阶近似

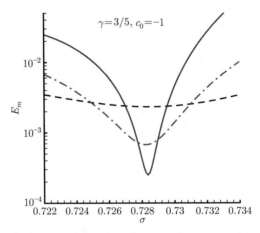

图 8.2 当 $\gamma = 3/5$ 和 $c_0 = -1$ 时, 选取辅助线性算子 (8.23) 和初始猜测解 (8.24), (8.17) 关于 $\sigma = u_0(0)$ 的平方残差 E_m. 虚线: 10 阶近似; 点划线: 15 阶近似; 实线: 20 阶近似

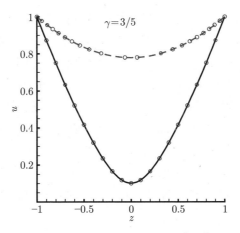

图 8.3 当 $\gamma = 3/5$ 时, 选取辅助线性算子 (8.23) 和初始猜测解 (8.24), (8.17) 的解 $u(z)$ 的两个分支的近似. 实线: 当 $c_0 = -1$ 和 $\sigma = -1/5$ 时, 下分支解的 5 阶同伦近似; 虚线: 当 $c_0 = -0.71848810, \sigma = 0.72834766$ 时, 上分支解的 10 阶同伦近似; 空心圆: 精确解

表 8.1 当 $N_t = 20, \gamma = 3/5, c_0 = -1$ 和 σ 取不同值时, 选取辅助线性算子 (8.23) 和初始猜测解 (8.24), m 次迭代下 $u(0)$ 的近似值

m	$\sigma = -1/2$	$\sigma = -1/5$	$\sigma = 1/2$
5	0.113536	0.101924	0.138348
10	0.100775	0.101008	0.099292
15	0.101057	0.101046	0.101032
20	0.101046	0.101045	0.101041
25	0.101046	0.101046	0.101045
30	0.101046	0.101046	0.101046

表 8.2 当 $N_t = 20, \gamma = 3/5, c_0 = -1$ 和 σ 取不同值时, 选取辅助线性算子 (8.23) 和初始猜测解 (8.24), m 次迭代下 (8.17) 的平方残差 E_m

m	$\sigma = -1/2$	$\sigma = -1/5$	$\sigma = 1/2$
5	4.6×10^{-4}	3.5×10^{-6}	3.2×10^{-3}
10	4.6×10^{-7}	1.5×10^{-8}	1.8×10^{-5}
15	1.5×10^{-8}	2.1×10^{-9}	1.5×10^{-7}
20	2.3×10^{-9}	1.9×10^{-9}	6.1×10^{-9}
25	1.9×10^{-9}	1.8×10^{-9}	2.0×10^{-9}
30	1.9×10^{-9}	1.8×10^{-9}	1.9×10^{-9}

为了得到 (8.17) 上分支解的足够精确的近似值,我们将 c_0 和 σ 均视为未知参数,并在给定的近似阶下寻找控制方程的平方残差的最小值. 通过 Mathematica 的命令 `Minimize`, 在 $\sigma > 1/2$ 的限制下 (否则,我们总是找到下分支解), 很容易获得平方残差的最小值以及相应的最优参数 c_0^* 和 σ^*, 如表 8.3 所示. 或者,可以使用软件包 BVPh 1.0 的模块 `GetMin2D` 来搜索平方残差的局部最小值. 在 10 阶近似下, 最优收敛控制参数 $c_0^* = -0.71848810$ 和 $\sigma^* = 0.72834766$, 对应于平方残差的最小值 $E_{10} = 2.68 \times 10^{-8}$. 注意, 相应的 $u(z)$ 的 10 阶同伦近似与上分支精确解非常吻合, 如图 8.3 所示. 此外, 它还给出了高精度近似 $u(0) = 0.779034$, 与精确值相等, 如表 8.3 所示.

表 8.3 当 $\gamma = 3/5$ 时, 选取辅助线性算子 (8.23) 和初始猜测解 (8.24), (8.17) 上分支解的 m 阶近似的最优参数 c_0^* 和 σ^*

m	c_0^*	σ^*	E_m	$u(0)$
4	-0.71489261	0.72844802	2.59×10^{-5}	0.778979
5	-0.71891704	0.72832456	1.51×10^{-6}	0.779088
6	-0.71353321	0.72835306	3.32×10^{-7}	0.779022
8	-0.71350498	0.72834803	3.20×10^{-8}	0.779032
10	-0.71848810	0.72834766	2.68×10^{-8}	0.779034

首先, 通过在初始猜测解 (8.24) 中引入未知参数 σ, 我们成功地得到了 (8.17) 的两个解. 我们称这种未知参数为多解控制参数. 正如收敛控制参数 c_0 为我们提供了一个保证级数解收敛的简便途径一样, 多解控制参数也为我们提供了一个寻找非线性微分方程多解的简便途径, 如下所述.

其次, 同伦分析方法为我们提供了极大的自由度来选取辅助线性算子和初始猜测解. 因此, 我们也可以选取辅助线性算子

$$\mathcal{L}(u) = u'' \tag{8.25}$$

以及初始猜测解

$$u_0(z) = \sigma + (1-\sigma)z^2, \tag{8.26}$$

以便借助多项式来近似 (8.17) 的解, 其中 $\sigma = u_0(0)$ 为未知参数, 此处用作多解控制参数.

BVPh 1.0 需要数秒的 CPU 时间来获得相应的 5 阶同伦近似. 通过 Mathematica 的命令 Minimize, 我们可以得到最优收敛控制参数 $c_0^* = -1.16236$ 和最优多解控制参数 $\sigma^* = 0.76500$, 对应于控制方程的平方残差的最小值 (5 阶同伦近似) 6.3×10^{-12}. 当 $c_0 = -1.16236$ 和 $\sigma = 0.76500$ 时, 对应的同伦级数相当快地收敛至上分支解: 即使是 8 阶同伦近似也给出了精确值 $u(0) = 0.779034$, 如表 8.4 所示. 令 $c_0 = -1$, 高达 5 阶近似的平方残差 E_m 与 σ 的曲线如图 8.4 所示. 这表明, 当 $c_0 = -1$ 时, 存在区间 Ω_2, 对于任意 $\sigma \in \Omega_2$, E_m 随着近似阶数的增加而变小. 研究发现, 当 $c_0 = -1$ 和 $0.6 \leqslant \sigma \leqslant 1$ 时, 同伦级数收敛. 因此, 当 $c_0 = -1$ 时, $\Omega_2 \supset [0.6, 1]$ 成立. 有趣的是, 通过具有任意多解控制参数 $\sigma \in [0.6, 1]$ 的 3 阶迭代法, 我们还得出了 (8.17) 上分支解的高精度近似.

表 8.4 当 $\gamma = 3/5$ 时, 选取 $c_0 = -1.16236$, $\sigma = 0.76500$ 的辅助线性算子 (8.25) 和初始猜测解 (8.26), m 阶近似下 (8.17) 的平方残差 E_m 和 $u(0)$

m	E_m	$u(0)$
4	1.5×10^{-9}	0.779002
8	2.4×10^{-13}	0.779034
12	1.1×10^{-17}	0.779034
16	8.2×10^{-23}	0.779034
20	3.4×10^{-25}	0.779034

根据图 8.4 可知, 在 $\sigma = 0$ 附近存在区间 Ω_1, 使得控制方程的平方残差 E_m 随着近似阶数的增加而变小. 不幸的是, 如图 8.5 所示, 该区间相当小, 以至于精确确定区间 Ω_1 的边界非常耗时. 为了避免这种情况, 我们使用 Mathematica 的命令 Minimize, 并限制 $\sigma < 1/10$ (否则, 我们总是会找到上分支解), 以获得最优同伦近似. 研究发现, 选取最优收敛控制参数 $c_0^* = -1.6951993251$ 和最优多解控制参数 $\sigma^* = 0.05122982326872$, 25 阶同伦近似下平方残差的最小值达到 2.7×10^{-9}, 且下分支解的高精度近似 $u(0) = 0.101046$, 如表 8.5 所示. 因此, 选取辅助线性算子 (8.25) 和初始猜测解 (8.26), 并通过多解控制参数 σ 得到了 (8.17) 的两个解.

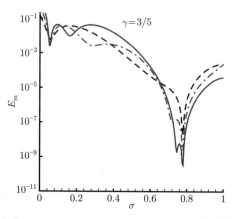

图 8.4 当 $\gamma = 3/5$ 和 $c_0 = -1$ 时，选取辅助线性算子 (8.25) 和初始猜测解 (8.26)，在区间 $\sigma \in [0,1]$ 上 (8.17) 关于 $\sigma = u_0(0)$ 的平方残差 E_m. 虚线: 3 阶近似; 点划线: 4 阶近似; 实线: 5 阶近似

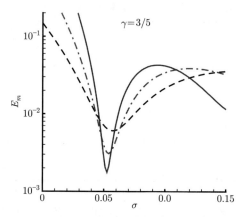

图 8.5 当 $\gamma = 3/5$ 和 $c_0 = -1$ 时，选取辅助线性算子 (8.25) 和初始猜测解 (8.26)，在局部区间 $\sigma \in [0, 0.15]$ 上 (8.17) 关于 $\sigma = u_0(0)$ 的平方残差 E_m. 虚线: 3 阶近似; 点划线: 4 阶近似; 实线: 5 阶近似

表 8.5 当 $\gamma = 3/5$ 时, 选取辅助线性算子 (8.25) 和初始猜测解 (8.26), (8.17) 下分支解的 m 阶近似的最优参数 c_0^* 和 σ^*

m	c_0^*	σ_0^*	E_m	$u(0)$
5	-1.52954	0.05137	6.3×10^{-4}	0.102379
10	-1.6299959583	0.05123009640906	2.3×10^{-5}	0.101165
15	-1.6648354484	0.05122982392509	1.1×10^{-6}	0.101060
20	-1.6833563121	0.05122982327004	5.2×10^{-8}	0.101047
25	-1.6951993251	0.05122982326872	2.7×10^{-9}	0.101046

所有这些表明, 在初始猜测解 (8.24) 和 (8.26) 中引入的多解控制参数 σ 确实为我们寻找非线性边值问题的多解提供了一个简便的途径. 对于 (8.17), 给定一个合理的收敛控制参数 c_0, 存在两个区间 Ω_1 和 Ω_2, 只要 $\sigma \in \Omega_1$ 或 $\sigma \in \Omega_2$, 相应的同伦级数就分别收敛于下分支解或上分支解. 有趣的是, 区间 Ω_1 和 Ω_2 不仅依赖于收敛控制参数 c_0, 还依赖于辅助线性算子和初始猜测解. 选取辅助线性算子 (8.23) 和初始猜测解 (8.24), 区间 Ω_1 比 Ω_2 大得多, 以至于获得下分支解比获得上分支解更容易. 然而, 选取辅助线性算子 (8.25) 和初始猜测解 (8.26), 区间 Ω_2 则比 Ω_1 大得多, 因此更易得到上分支解. 由于同伦分析方法为我们选取辅助线性算子和初始猜测解提供了极大的自由度, 我们可以选取辅助线性算子 (8.23) 和初始猜测解 (8.24) 来获得下分支解, 选取辅助线性算子 (8.25) 和初始猜测解 (8.26) 来获得上分支解. 通过此方法, 我们获得了 γ 可能取值的两个分支解[①]. 这也说明, BVPh 1.0 可以通过不同类型的辅助线性算子和包含未知参数 (称为多解控制参数) 的不同类型的初始猜测解来寻找非线性边值方程 (8.1) 的多解.

8.3.2 非线性三点边值问题

我们进一步考虑非线性三点边值问题

$$u'''' = \beta z \left(1 + u^2\right), \quad u(0) = u'(1) = u''(1) = 0, \ u''(0) - u''(\alpha) = 0, \quad (8.27)$$

其中 $\alpha \in (0,1)$ 和 β 是给定的常数. Graef 等人 [98, 99] 证明了当 $\alpha = 1/5$ 和 $\beta = 10$ 时, 上述方程至少存在两个正解.

该问题是 (8.1) 的一个特例, 即

$$\mathcal{F}[z, u] = u'''' - \beta z \left(1 + u^2\right).$$

类似地, m 阶同伦近似由 (8.10) 给出, 其中 $u_k(z)$ 由 k 阶形变方程

$$\mathcal{L}\left[u_k(z) - \chi_k u_{k-1}(z)\right] = c_0 \delta_{k-1}(z) \quad (8.28)$$

控制, 满足边界条件

$$u_k(0) = 0, \quad u'_k(1) = 0, \quad u''_k(1) = 0, \quad u''_k(0) - u''_k(\alpha) = 0, \quad (8.29)$$

由定理 4.1 可得

$$\delta_i(z) = u_i'''' - (1 - \chi_{i+1})\beta z - \beta z \sum_{j=0}^{i} u_j u_{i-j}. \quad (8.30)$$

[①] 由于当 $\gamma \to 0$ 时存在奇点, 则需要更多的截断项来获得小 γ 的精确下分支解.

需要注意的是, 控制方程 (8.27) 包含 z 项. 因此, 有限区间 $z \in [0,1]$ 上的解 $u(z)$ 可以用多项式表示. 选取辅助线性算子

$$\mathcal{L}(u) = u'''' \tag{8.31}$$

以及初始猜测解

$$u_0(z) = a_0 + \sum_{i=1}^{4} a_k z^k,$$

其中 a_k 是未知参数. 令 $\sigma = u(1)$ 表示未知的多解控制参数, 这为我们提供了附加条件

$$u(1) = \sigma. \tag{8.32}$$

然后, 利用四个边界条件 (8.27) 和附加边界条件 (8.32) 来确定初始猜测解 $u_0(z)$ 的五个未知常数 a_k $(0 \leqslant k \leqslant 4)$. 因此, 初始猜测解为

$$u_0(z) = \frac{\sigma}{2\alpha - 3} \left[2(3\alpha - 4)z + 6(1-\alpha)z^2 + 2\alpha z^3 - z^4 \right]. \tag{8.33}$$

需要注意的是, 上述初始猜测解取决于位置 $z = \alpha$, 其中存在多边界条件 $u''(0) = u''(\alpha)$.

多边界条件可以通过像 Mathematica 这样的科学计算软件轻松解决: 所有边界条件均以类似的方式得到满足. 令 $u_k^*(z)$ 表示 (8.28) 的一个特解. 根据 (8.31), 其通解为

$$u_k(z) = u_k^*(z) + C_0 + C_1 z + C_2 z^2 + C_3 z^3,$$

其中四个系数 C_0, C_1, C_2 和 C_3 由四个线性边界条件 (8.29) 确定. 这主要是因为像 Mathematica 这样的科学计算软件为我们提供了 "基于函数而非数的计算" 的能力 [116].

我们先考虑 $\alpha = 1/5$ 和 $\beta = 10$ 的情况. 将收敛控制参数 c_0 和多解控制参数 σ 均视为未知数, 控制方程在 m 阶同伦近似下的平方残差 E_m 取决于 c_0 和 σ. 研究发现, 5 阶同伦近似下平方残差的最小值为 1.6×10^{-20}, 对应于最优收敛控制参数 $c_0^* = -1.00661$ 和最优多解控制参数 $\sigma^* = 0.62018$. 当 $c_0 = -1$ 时, E_m 关于 σ 的曲线如图 8.6 所示, 这表明存在区间 Ω_1, 对于任意 $\sigma \in \Omega_1$, 平方残差 E_m 随着 m 的增加而变小. 研究发现, 选取 $c_0 = -1$, $\sigma = 0, 0.5, 1.1$, 我们可以得到相同的收敛级数解, 因此, $\Omega_1 \supset [0, 1.1]$. 特别地, 当 $c_0 = -1$ 和 $\sigma = 0.62$

时，通过 3 阶迭代法，我们可在数秒的 CPU 时间内获得 $u(1) = 0.627315$ 的精确结果，如表 8.6 所示.

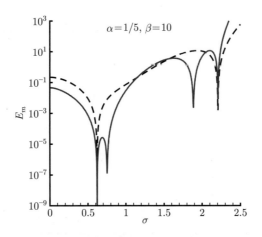

图 8.6 当 $c_0 = -1$, $\alpha = 1/5$ 和 $\beta = 10$ 时，选取辅助线性算子 (8.31) 和初始猜测解 (8.33), (8.27) 关于 $\sigma \in [0, 2.5]$ 的平方残差 E_m. 虚线: 3 阶近似; 实线: 5 阶近似

表 8.6 当 $\alpha = 1/5$ 和 $\beta = 10$ 时，选取 $c_0 = -1$, $\sigma = 0.62$ 的辅助线性算子 (8.31) 和初始猜测解 (8.33), m 次迭代近似下 (8.27) 的平方残差 E_m 和 $u(1)$

m	E_m	$u(1)$
1	1.7×10^{-8}	0.627299
2	1.2×10^{-10}	0.627313
3	8.7×10^{-13}	0.627315
5	4.4×10^{-17}	0.627315
10	8.2×10^{-28}	0.627315

根据图 8.6, $\sigma = 2$ 附近应该存在一个解. 令 Ω_2 表示这样一个区间, 对于任意 $\sigma \in \Omega_2$ 和 $c_0 = -1$, 我们可以得到第二个解的收敛同伦级数. 然而, 我们发现区间 Ω_2 非常小, 以至于精确确定它非常耗时. 为了避免这种情况, 我们通过在足够高的近似阶下平方残差的最小值来寻找第二个解. 利用 Mathematica 命令 `Minimize` 并限制 $\sigma > 2$ (否则我们总是得到第一个解), 发现 5 阶同伦近似下平方残差的最小值为 1.7×10^{-15}, 对应于最优收敛控制参数 $c_0^* = -1.0202828199$

和最优多解控制参数 $\sigma^* = 2.206223812447131$. 此外, 10 阶近似下平方残差的最小值为 7.4×10^{-31}, 对应于最优收敛控制参数 $c_0^* = -1.0166130822$ 和最优多解控制参数 $\sigma^* = 2.206223812318337$. 我们发现在 $z = 1$ 时, 5 至 10 阶同伦近似趋于相同的值 $u(1) = 2.2411770076$, 如表 8.7 所示. 因此, 当 $\alpha = 1/5$ 和 $\beta = 10$ 时, 我们确实得到了非线性三点边值方程 (8.27) 两个不同的正解, 如图 8.7 所示, 并在表 8.8 中列出.

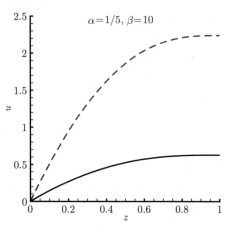

图 8.7 当 $\alpha = 1/5$ 和 $\beta = 10$ 时, 选取辅助线性算子 (8.31) 和初始猜测解 (8.33), (8.27) 的两个解 $u(z)$ 的近似. 实线: 当 $c_0 = -1$, $\sigma = 0.62$ 时, 第一个解的第 10 次同伦迭代近似; 虚线: 当最优参数 $c_0 = -1.0166130822$, $\sigma = 2.206223812318337$ 时, 第二个解的 10 阶同伦近似

表 8.7 当 $\alpha = 1/5$ 和 $\beta = 10$ 时, 选取辅助线性算子 (8.31) 和初始猜测解 (8.33), (8.27) 的第二个解在 m 阶近似下的最优收敛控制参数 c_0^* 和最优多解控制参数 σ^*

m	c_0^*	σ^*	E_m	$u(1)$
3	-1.0156751650	2.206225834958724	3.2×10^{-9}	2.2411852031
5	-1.0202828199	2.206223812447131	1.7×10^{-15}	2.2411770076
6	-1.0186293138	2.206223812311513	6.1×10^{-19}	2.2411770076
7	-1.0147412057	2.206223812318337	2.2×10^{-21}	2.2411770076
8	-1.0216642167	2.206223812318337	3.8×10^{-25}	2.2411770076
9	-1.0198947262	2.206223812318337	2.1×10^{-28}	2.2411770076
10	-1.0166130822	2.206223812318337	7.4×10^{-31}	2.2411770076

表 8.8 当 $\alpha = 1/5$ 和 $\beta = 10$ 时, 选取辅助线性算子 (8.31) 和初始猜测解 (8.33), (8.27) 的两个正解的收敛同伦近似

z	第一个解	第二个解
0	0	0
0.1	0.144853	0.508335
0.2	0.269514	0.948159
0.3	0.374042	1.319559
0.4	0.458715	1.623039
0.5	0.524156	1.859952
0.6	0.571473	2.033131
0.7	0.602414	2.147651
0.8	0.619519	2.211668
0.9	0.626286	2.237253
1.0	0.627315	2.241177

类似地, BVPh 1.0 可以找到 (8.27) 对于不同 α 和 β 值的多解. 例如, 当 $\alpha = 1/5$ 时, (8.27) 在区间 $\beta \in (0, 12.05]$ 上有两个正解, 如表 8.9 所示. 注意, 随着 β 趋于零, 上分支解的 $u(1)$ 迅速增大, 但下分支解的 $u(1)$ 趋于零, 如图 8.8 所示.

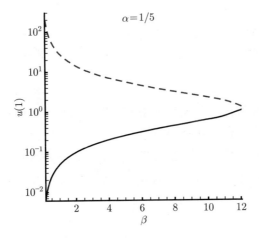

图 8.8 当 $\alpha = 1/5, \beta$ 取不同的值时, 选取辅助线性算子 (8.31) 和初始猜测解 (8.33), (8.27) 的两个正解 $u(1)$. 实线: 下分支解; 虚线: 上分支解

表 8.9 当 $\alpha = 1/5$, β 取不同的值时, 选取辅助线性算子 (8.31) 和初始猜测解 (8.33), (8.27) 的两个正解 $u(1)$

β	第一个解	第二个解
0.125	0.006111	230.072656
0.25	0.012224	115.027019
0.5	0.024455	57.494885
1	0.048974	28.710135
2	0.098462	14.279968
3	0.149017	9.435354
4	0.201271	6.985695
6	0.314218	4.474595
8	0.447555	3.141431
9	0.528490	2.660305
10	0.627315	2.241177
11	0.762509	1.843780
12	1.054690	1.332973
12.05	1.102400	1.275268

类似地, 当 $\beta = 10$ 时, 在边界条件

$$u''(0) - u''(\alpha) = 0$$

下, 我们找到了 (8.27) 对于不同 α 值的多解. 区间 $\alpha \in (0,1]$ 上至少存在两个正解, 如图 8.9 所示. 表 8.10 列出了 $u(1)$ 与 α 的相应值. 这些表明上述边界条件在分离点 $z = \alpha$ 处对非线性边值方程 (8.27) 解的影响.

需要注意的是, (8.27) 可以先转化为非线性特征方程, 再通过 BVPh 1.0 求解, 如例 9.3.4 所示.

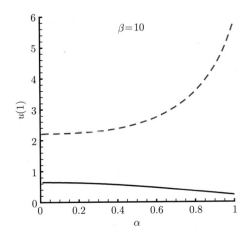

图 8.9 当 $\beta = 10$, α 取不同的值时，选取辅助线性算子 (8.31) 和初始猜测解 (8.33), (8.27) 的两个正解 $u(1)$. 实线: 下分支解; 虚线: 上分支解

表 8.10 当 $\beta = 10$, α 取不同的值时，选取辅助线性算子 (8.31) 和初始猜测解 (8.33), (8.27) 的两个正解 $u(1)$

α	第一个解	第二个解
0.01	0.648176	2.213774
0.05	0.646932	2.215093
0.1	0.643033	2.219512
0.2	0.627315	2.241177
0.3	0.601033	2.289274
0.4	0.564806	2.377541
0.5	0.520086	2.522182
0.6	0.468852	2.744448
0.7	0.413081	3.077674
0.8	0.354329	3.583159
0.9	0.293562	4.391860
0.95	0.262562	4.999196
1	0.231157	5.848855

该例子表明 BVPh 1.0 可以通过在初始猜测解中引入多解控制参数来获得非线性多点边值问题的多解.

8.3.3 具有多解的通道流动问题

黏性不可压缩流体在具有正交移动壁的二维多孔通道中的层流 [117] 由如下非线性边值微分方程描述

$$F'''' + \alpha(zF''' + 3F'') + (FF''' - F'F'') = 0, \quad 0 \leqslant z \leqslant 1, \tag{8.34}$$

满足边界条件

$$F(0) = 0, \quad F''(0) = 0, \quad F(1) = R, \quad F'(1) = 0, \tag{8.35}$$

其中 $'$ 表示对 z 求导, R 是横流雷诺数, α 是与壁膨胀比相关的物理常数. 这里, $F(z)$ 与流函数

$$\psi(x, y) = \frac{\nu x}{d} F(z), \quad z = y/d$$

有关, 其中 d 为两个平行板之间距离的一半, ν 为流体的运动黏度. 详情请参考 [97] 和 [117]. 令 $u(z) = F(z)/R$, 上述方程变为

$$u'''' + \alpha(zu''' + 3u'') + R(uu''' - u'u'') = 0, \quad 0 \leqslant z \leqslant 1, \tag{8.36}$$

满足边界条件

$$u(0) = 0, \quad u''(0) = 0, \quad u(1) = 1, \quad u'(1) = 0. \tag{8.37}$$

在同伦分析方法的框架中, 徐航等人 [117] 开发了一种解析方法, 以找到该问题的多解, 该方法原则上基于打靶法. 这里, BVPh 1.0 可以通过在初始猜测解中引入多解控制参数来找到它的多解.

该问题是 (8.1) 的一个特例, 即

$$\mathcal{F}[z, u] = F'''' + \alpha(zF''' + 3F'') + (FF''' - F'F'').$$

m 阶同伦近似由 (8.10) 给出, 其中 $u_k(z)$ 满足 k 阶形变方程

$$\mathcal{L}[u_k(z) - \chi_k u_{k-1}(z)] = c_0 \delta_{k-1}(z) \tag{8.38}$$

和边界条件

$$u_k(0) = 0, \quad u_k''(0) = 0, \quad u_k(1) = 0, \quad u_k'(1) = 0, \tag{8.39}$$

由定理 4.1 可得

$$\delta_i(z) = u_i'''' + \alpha\left(zu_i''' + 3u_i''\right) + R\sum_{j=0}^{i}\left(u_j u_{i-j}''' - u_j' u_{i-j}''\right). \tag{8.40}$$

通过分析控制方程 (8.36) 和边界条件 (8.37), 我们很容易发现解是奇函数. 此外, 我们注意到 (8.36) 包含 z 项, 可用 z 的多项式来表示 $u(z)$. 因此, 选取辅助线性算子

$$\mathcal{L}(u) = u'''' \tag{8.41}$$

以及初始猜测解

$$u_0(z) = a_1 z + a_2 z^3 + a_3 z^5,$$

其中 a_1, a_2, a_3 为常数. 注意, 该初始猜测解自动满足 $z=0$ 时的边界条件. 因此, 在 $z=1$ 处只剩下两个边界条件, 则三个未知常数 a_1, a_2, a_3 中有一个是不确定的. 这在初始猜测解中为我们提供了一个额外的自由度, 有助于找到多解, 如下所述. 为了寻找多解, 我们引入多解控制参数 $\sigma = u_0'(0)$. 那么, 三个未知常数由 $\sigma = u_0'(0)$ 和 $z=1$ 时的两个边界条件 (8.37) 唯一确定. 因此, 我们得到初始猜测解

$$u_0(z) = \sigma z + \frac{1}{2}(5-4\sigma)z^3 - \frac{1}{2}(3-2\sigma)z^5. \tag{8.42}$$

不失一般性, 我们考虑 $R=-11$ 和 $\alpha = 3/2$ 的情况. 将收敛控制参数 c_0 和多解控制参数 σ 均视为未知数, 则控制方程在 m 阶近似下的平方残差 E_m 取决于 c_0 和 σ. 显然, E_m 值越小, m 阶同伦近似效果越好. 因此, 最优近似由 E_m 的最小值给出. 此外, 任何一个很小的 E_m 值都对应于 $u(z)$ 的高精度近似. 在 $R=-11$ 和 $\alpha=3/2$ 的情况下, 为了获得关于 (8.36) 多解的相关信息, 我们先绘制了 $c_0=-1$ 时, E_6 和 E_{10} 与 σ 的关系曲线, 如图 8.10 所示. 注意, 这里存在三个局部最小值, 这表明三个解可能分别存在于区间 $\sigma<0$, $0<\sigma<1$ 和 $\sigma>1$ 上. 为了证实这一猜测, 我们使用 Mathematica 命令 `Minimize`, 并限制 $\sigma<0$ 来寻找 (8.36) 的解. 10 阶和 20 阶近似下平方残差的最小值分别减小至 1.8×10^{-7} 和 6.9×10^{-17}, $u'(0)$ 的值收敛到 -1.02377, 如表 8.11 所示. 因此, 区间 $\sigma<0$ 上确实存在一个解. 我们发现, 当 $c_0=-1$ 时, 存在区间 $\Omega_1 \supset [-2,-0.7]$, 对于任意 $\sigma \in \Omega_1$, 我们可由无迭代的 BVPh 1.0 得到相同的同伦近似且 $u'(0) = -1.02377$. 特别地, 选取接近其最优值的 $\sigma=-1$ 和 $c_0=-1$, 笔记本电脑可在数秒的 CPU 时间内获得收敛解.

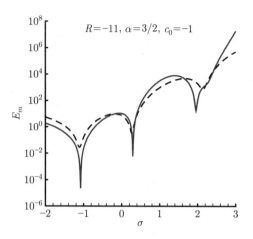

图 8.10 当 $R = -11$, $\alpha = 3/2$ 时, 选取 $c_0 = -1$ 的辅助线性算子 (8.41) 和初始猜测解 (8.42), (8.36) 关于 σ 的平方残差 E_m. 实线: E_{10}; 虚线: E_6

类似地, 使用 Mathematica 命令 `Minimize`, 限制 $0 < \sigma < 1$, 我们借助 BVPh 1.0 找出 $R = -11$ 和 $\alpha = 3/2$ 时 (8.36) 的第二个解. 20 阶近似下平方残差的最小值减小至 1.8×10^{-10}, 且选取最优收敛控制参数 c_0^* 和最优多解控制参数 σ^* 时的最优同伦近似给出了收敛值 $u'(0) = 0.16935$, 如表 8.12 所示. 令 Ω_2 表示这样一个区间, 对于任意 $\sigma \in \Omega_2$, 相应的同伦级数收敛至第二个解. 然而, 我们发现区间 Ω_2 非常小, 以至于精确找到它的边界非常耗时. 因此, 无迭代的 BVPh 1.0 可以更方便地获得第二个解.

表 8.11 当 $R = -11$ 和 $\alpha = 3/2$ 时, 选取辅助线性算子 (8.41) 和初始猜测解 (8.42), (8.36) 的第一个解在 m 阶近似下的 $u'(0)$, 最优参数 c_0^* 和 σ^*

m	c_0^*	σ^*	E_m	$u'(0)$
10	-1.15185	-1.08120800	1.8×10^{-7}	-1.02375
12	-1.16501	-1.08139600	2.3×10^{-9}	-1.02377
14	-1.25436	-1.08138900	2.5×10^{-10}	-1.02377
16	-1.17126	-1.08137200	3.6×10^{-13}	-1.02377
18	-1.24314	-1.08137198	4.6×10^{-14}	-1.02377
20	-1.17586	-1.08137151	6.9×10^{-17}	-1.02377

表 8.12 当 $R = -11$ 和 $\alpha = 3/2$ 时, 选取辅助线性算子 (8.41) 和初始猜测解 (8.42), (8.36) 的第二个解在 m 阶近似下的 $u'(0)$, 最优参数 c_0^* 和 σ^*

m	c_0^*	σ^*	E_m	$u'(0)$
10	-0.87023	0.293224	4.9×10^{-4}	0.17103
12	-0.96316	0.292173	2.2×10^{-5}	0.16900
14	-0.90210	0.292288	1.1×10^{-6}	0.16926
16	-0.91428	0.292329	5.6×10^{-8}	0.16937
18	-0.94502	0.292321	3.0×10^{-9}	0.16935
20	-0.89768	0.292322	1.8×10^{-10}	0.16935

这样, 我们成功地得到了 (8.36) 的三个解, 分别为 $u'(0) = -1.02377$, $u'(0) = 0.16935$ 和 $u'(0) = 2.76111$, 如图 8.11 所示. 这再次证实, BVPh 1.0 可以通过在初始猜测解中引入多解控制参数找到非线性边值问题的多解.

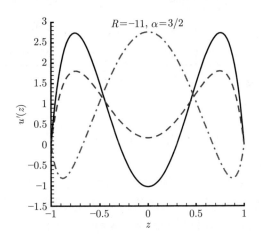

图 8.11 当 $R = -11$ 和 $\alpha = 3/2$ 时, 选取辅助线性算子 (8.41) 和初始猜测解 (8.42), (8.36) 的多解. 实线: 第一个解; 虚线: 第二个解; 点划线: 第三个解

为了获得第三个解, 我们使用 Mathematica 命令 Minimize 并限制 $\sigma > 1$ 来寻找平方残差的最小值. 研究发现, 平方残差 E_m 的最小值确实随着近似阶数的增加而变小, 但速度较慢. 选取最优收敛控制参数 $c_0^* = -0.75716$ 和最优多解控制参数 $\sigma^* = 1.960920$, 即使在 20 阶近似下, 平方残差的最小值仍为 1.6.

为了加快收敛速度, 我们采用 3 阶迭代法. 当 $c_0 = -1/2$ 时, 存在区间 $\Omega_3 \supset [0.5, 3]$, 对于任意 $\sigma \in \Omega_3$, 我们得到相同的收敛同伦近似且 $u'(0) = 2.76111$. 例如, 当 $c_0 = -1/2$ 和 $\sigma = 2$ 时, 控制方程的平方残差迅速变小, 如表 8.13 所示.

表 8.13 当 $R = -11$ 和 $\alpha = 3/2$ 时, 选取 $c_0 = -1/2$ 和 $\sigma = 2$ 的辅助线性算子 (8.41) 和初始猜测解 (8.42), (8.36) 的第三个解在第 m 次迭代近似下的平方残差 E_m 和 $u'(0)$

m	E_m	$u'(0)$
1	6.9×10^3	2.31018
5	1.8×10^2	2.67366
10	4.5	2.75021
15	9.8×10^{-2}	2.75977
20	1.9×10^{-3}	2.76094
25	3.2×10^{-5}	2.76108
30	5.1×10^{-7}	2.76110
35	7.7×10^{-9}	2.76111
40	1.0×10^{-10}	2.76111

8.4 本章小结

本章以三个非线性边值方程为例, 说明了 BVPh 1.0 求解具有多解的非线性边值方程的有效性, 该方程在有限区间 $0 \leqslant z \leqslant a$ 上由 n 阶非线性边值方程 $\mathcal{F}[z, u] = 0$ 控制, 并且满足 n 个线性边界条件 $\mathcal{B}_k[z, u] = \gamma_k\ (1 \leqslant k \leqslant n)$.

为了寻找多解, 我们首次在初始猜测解中引入未知参数, 即多解控制参数. 以 BVPh 1.0 为工具, 可以通过所谓的多解控制参数找到一些 n 阶非线性边值问题的多解, 其最优值由控制方程的平方残差在足够高阶的同伦近似下的最小值确定. 本质上, 未知的多解控制参数在初始猜测解中为我们提供了一个额外的自由度. 注意, 正是同伦分析方法为我们在初始猜测解 $u_0(z)$ 中引入这种未知参数提供了很大的自由度, 其不同的值对应于不同的初始猜测解. 因此, 除了收敛控制参数 c_0 为我们提供了一个保证同伦级数收敛的简便途径之外, 我们在本章中引入了一个新的概念, 即多解控制参数, 它为我们提供了一个寻找多解的简便途径, 尽管如第十章所述, 该参数也会影响同伦级数的收敛. 本章例子表明,选取由控制方程平方残差的最小值确定的最优多解控制参数和最优收敛控制参数 c_0, BVPh 1.0 在有限区间 $[0,a]$ 上可以找到一些 n 阶非线性多点边值问题的多解.

需要注意的是, 与数值包 BVP4c 不同, 我们的方法原则上是一种解析方法, 因此这两个控制参数被用作未知数来获得同伦近似, 直到找到它们的最优值. 这是解析方法相对于数值方法的优势. 从这个角度来看, BVPh 1.0 为我们提供了比数值包 BVP4c 更大的自由度和灵活性, 以保证级数解的收敛性并寻找非线性问题的多解.

基于同伦分析方法, BVPh 1.0 为我们选取不同的基函数来近似有限区间 $0 \leqslant z \leqslant a$ 上 n 阶非线性边值方程 $\mathcal{F}[z,u] = 0$ 的解提供了很大的自由度. 这主要是因为同伦分析方法为我们提供了极大的自由度来选取辅助线性算子 $\mathcal{L}$ 和初始猜测解 $u_0(z)$. 事实上, 正是由于同伦分析方法的这种自由度, 我们才可以在初始猜测解中引入所谓的多解控制参数.

最后, 必须强调的是, n 阶非线性边值方程 (8.1) 和 n 个线性边界条件 (8.2) 是非常普遍的, 以至于很难开发一个对所有条件都有效的软件包. 注意, 如第七章所述, 我们的目标是开发一个对尽可能多 (但不是全部) 的 n 阶非线性多点边值问题都有效的软件包. Mathematica 软件包 BVPh 1.0 为我们提供了一个方便的搜索非线性边值方程多解的解析工具, 尽管未来还需要进一步完善 (参见本章中的问题) 和更多的应用.

需要注意的是, Chebfun 4.0 也为我们提供了 "基于函数而非数的计算" 的能力 [116]. 因此, 通过 Chebfun 为具有奇点的强非线性多点边值问题建立类似的基于同伦分析方法的软件包是非常有趣的, 且 Chebfun 是开源代码.

问题

8.1 具有非线性边界条件的边值问题

开发一种基于同伦分析方法的解析方法, 用于有限区间上的 n 阶非线性边值方程

$$\mathcal{F}[z,u] = 0, \quad 0 \leqslant z \leqslant a,$$

满足 n 个非线性多边界条件

$$\mathcal{B}_k[z,u] = \gamma_k, \quad 1 \leqslant k \leqslant n,$$

其中 $\mathcal{B}_k$ 是非线性算子, γ_k 是常数. 假设上述方程至少存在一个光滑解. 改进 Mathematica 软件包 BVPh 1.0 以解决此类问题.

8.2 耦合的非线性边值问题

开发一种基于同伦分析方法的解析方法, 用于有限区间 $z \in [0,a]$ 上的 n 个耦合非线性边值方程

$$\mathcal{F}_k[z,u] = 0, \quad 1 \leqslant k \leqslant n,$$

满足一些线性 (非线性) 多边界条件, 其中 $n \geqslant 2$. 假设上述方程至少存在一个光滑解. 改进 Mathematica 软件包 BVPh 1.0 以解决此类问题.

8.3 无限区间上的边值问题

开发一种基于同伦分析方法的解析方法, 用于无限区间上的 n 阶非线性边值方程

$$\mathcal{F}[z,u] = 0, \quad 0 \leqslant z < +\infty,$$

满足 n 个非线性多边界条件

$$\mathcal{B}_k[z,u] = \gamma_k, \quad 1 \leqslant k \leqslant n,$$

其中 $\mathcal{B}_k$ 是非线性算子, γ_k 是常数. 假设上述方程至少存在一个光滑解. 改进 Mathematica 软件包 BVPh 1.0 以解决此类问题.

8.4 无限区间上耦合的非线性边值问题

开发一种基于同伦分析方法的解析方法, 用于无限区间 $z \in [0,+\infty)$ 上的 n 个耦合的非线性边值方程

$$\mathcal{F}_k[z,u] = 0, \quad 1 \leqslant k \leqslant n,$$

满足一些线性 (非线性) 边界条件, 其中 $n \geqslant 2$. 假设上述方程至少存在一个光滑解. 改进 Mathematica 软件包 BVPh 1.0 以解决此类问题.

参考文献

[96] Abbasbandy, S., Magyari, E., Shivanian, E.: The homotopy analysis method for multiple solutions of nonlinear boundary value problems. Communications in Nonlinear Science and Numerical Simulation. **14**, 3530–3536 (2009).

[97] Berman, A. S.: Laminar flow in channels with porous walls. J. Appl. Phys. **24**, 1232–1235 (1953).

[98] Graef, J.R., Qian. C. Yang, B.: A three point boundary value problem for nonlinear forth order differential equations. J. Mathematical Analysis and Applications. **287**, 217–233 (2003).

[99] Graef, J.R., Qian. C. Yang, B.. Multiple positive solutions of a boundary value prolem for ordinary differential equations. Electronic J. of Qualitative Theory of Differential Equations. **11**, 1–13 (2004).

[100] Li, Y.J., Nohara, B.T., Liao, S.J.: Series solutions of coupled van der Pol equation by means of homotopy analysis method. J. Mathematical Physics **51**, 063517 (2010). doi:10.1063/1.3445770.

[101] Liao, S.J.: The Proposed Homotopy Analysis Technique for the Solution of Nonlinear Problems. PhD dissertation, Shanghai Jiao Tong University (1992).

[102] Liao, S.J.: A kind of approximate solution technique which does not depend upon small parameters—(II) An application in fluid mechanics. Int. J. Nonlin. Mech. **32**, 815–822 (1997).

[103] Liao, S.J.: An explicit, totally analytic approximation of Blasius viscous flow problems. Int. J. Nonlin. Mech. **34**, 759–778 (1999a).

[104] Liao, S.J.: A uniformly valid analytic solution of 2D viscous flow past a semi-infinite flat plate. J. Fluid Mech. **385**, 101–128 (1999b).

[105] Liao, S.J.: On the analytic solution of magnetohydrodynamic flows of non-Newtonian fluids over a stretching sheet. J. Fluid Mech. **488**, 189–212 (2003a).

[106] Liao, S.J.: Beyond Perturbation –Introduction to the Homotopy Analysis Method. Chapman & Hall/ CRC Press, Boca Raton (2003b).

[107] Liao, S.J.: On the homotopy analysis method for nonlinear problems. Appl. Math. Comput. **147**, 499–513 (2004).

[108] Liao, S.J.: A new branch of solutions of boundary-layer flows over an impermeable stretched plate. Int. J. Heat Mass Tran. **48**, 2529–2539 (2005).

[109] Liao, S.J.: Series solutions of unsteady boundary-layer flows over a stretching flat plate. Stud. Appl. Math. **117**, 2529–2539 (2006).

[110] Liao, S.J.: Notes on the homotopy analysis method - Some definitions and theorems. Commun. Nonlinear Sci. Numer. Simulat. **14**, 983–997 (2009).

[111] Liao, S.J.: On the relationship between the homotopy analysis method and Euler transform. Commun. Nonlinear Sci. Numer. Simulat. **15**, 1421–1431 (2010a). doi:10.1016/j.cnsns.2009.06.008.

[112] Liao, S.J.: An optimal homotopy-analysis approach for strongly nonlinear differential equations. Commun. Nonlinear Sci. Numer. Simulat. **15**, 2003–2016 (2010b).

[113] Liao, S.J., Campo, A.: Analytic solutions of the temperature distribution in Blasius viscous flow problems. J. Fluid Mech. **453**, 411–425 (2002).

[114] Liao, S.J., Tan, Y.: A general approach to obtain series solutions of nonlinear differential equations. Stud. Appl. Math. **119**, 297–355 (2007).

[115] Magyari, E.: Exact abalytic solution of a nonlinear reaction-diffusion model in porous catalysts. Chem. Eng. J. **143**, 167–171 (2008).

[116] Trefethen, L.N.: Computing numerically with functions instead of numbers. Math. in Comp. Sci. **1**, 9–19 (2007).

[117] Xu, H., Lin, Z.L., Liao, S.J., Wu, J.Z., Majdalani, J.: Homotopy-based solutions of the Navier-Stokes equations for a porous channel with orthogonally moving walls. Physics of Fluids. **22**, 053601 (2010). doi:10.1063/1.3392770.

附录 8.1 例 8.3.1 的 BVPh 输入数据

```
(* Input Mathematica package BVPh version 1.0 *)
<<BVPh.txt;

(* Define the physical and control parameters *)
TypeEQ      = 1;
TypeL       = 1;
TypeBase    = 2;
ApproxQ     = 1;
Ntruncated  = 20;

(* Define the governing equation *)
f[z_,u_,lambda_] := u*D[u,{z,2}] - gamma^2;
gamma = 3/5 ;

(* Define Boundary conditions *)
zR       = 1;
OrderEQ  = 2;
BC[1,z_,u_,lambda_] := Limit[D[u,z], z->0 ];
BC[2,z_,u_,lambda_] := u - 1 /. z->zR;

(* Define initial guess *)
U[0] = u[0];
If[TypeL == 1,
   u[0] = sigma + (1-sigma)*z^2,
   u[0] = (sigma+1)/2 + (sigma-1)/2*Cos[kappa*Pi*z];
```

```
    ];
sigma = .;

(* Define output term *)
output[z_,u_,k_]:= Print["output                   = ",
                      u[k] /. z->0//N];

(* Define the auxiliary linear operator *)
omega[1] = Pi/zR;
L[f_] := Module[{temp,numA,numB,i},
If[TypeL == 1,
    temp[1] = D[f,{z,OrderEQ}],
    numA = IntegerPart[OrderEQ/2];
    numB = OrderEQ - 2* numA//Expand;
    temp[0] = D[f,{z,numB}];
    For[i=1, i<=numA, i++,
        temp[1] = D[temp[0],{z,2}]
                    + (kappa*omega[i])^2*temp[0];
        temp[0] = temp[1];
      ];
   ];
temp[1]//Expand
];

(* Print input and control parameters *)
PrintInput[u[z]];

(* Gain 3rd-order approximation *)
BVPh[1,3];

(* Gain squared residual at 3rd-order approx.  *)
GetErr[3];

(* Gain optimal values of c0 and sigma *)
res=Minimize[Err[3],{c0,sigma}];
Print["Minimum square residual = ", res];

(* Set optimal c0 and sigma *)
c0=-1;
```

```
sigma = 3/4;
Print["c0 = ",c0," sigma = ",sigma];

(* Gain 10th-order HAM approximation *)
BVPh[1,10];
```

注意, 对于定义在有限区间上的非线性边值 (特征值) 问题, BVPh 1.0 默认 zL=0.

附录 8.2 例 8.3.2 的 BVPh 输入数据

```
(* Input Mathematica package BVPh version 1.0 *)
<<BVPh.txt;

(* Define the physical and control parameters *)
TypeEQ    = 1;
TypeL     = 1;
TypeBase  = 2;
ApproxQ   = 0;

(* Define the governing equation *)
f[z_,u_,lambda_] := D[u,{z,4}]-beta*z*(1+u^2);
beta = 10 ;

(* Define Boundary conditions*)
zR =1;
OrderEQ = 4;
alpha = 1/5 ;
BC[1,z_,u_,lambda_] := Limit[u, z->0 ];
BC[2,z_,u_,lambda_] := D[u,z] /.  z->zR;
BC[3,z_,u_,lambda_] := D[u,{z,2}] /.  z->zR;
BC[4,z_,u_,lambda_] := Module[{temp},
                       temp[1]=D[u,{z,2}]/.z->0;
                       temp[2]=D[u,{z,2}]/.z-> alpha;
                       temp[1]-temp[2]  //Expand
                       ];

(* Define initial guess *)
```

```
U[0] = u[0];
u[0] = sigma/(2*alpha-3)*((6*alpha-8)*z
       +6*(1-alpha)*z^2+2*alpha*z^3-z^4);
sigma = .;

(* Define output term *)
output[z_,u_,k_]:= Print["output = ",
                   N[ u[k] /. z->1, 24] ];

(* Define the auxiliary linear operator *)
L[f_] := Module[{temp,numA,numB,i},
If[TypeL == 1,
   temp[1] = D[f,{z,OrderEQ}],
   numA = IntegerPart[OrderEQ/2];
   numB = OrderEQ - 2* numA//Expand;
   temp[0] = D[f,{z,numB}];
   For[i=1, i<=numA, i++,
       temp[1] = D[temp[0],{z,2}]
                 + (kappa*omega[i])^2*temp[0];
       temp[0] = temp[1];
      ];
   ];
temp[1]//Expand
];

(* Print input and control parameters *)
PrintInput[u[z]];

(* Gain 3rd-order HAM approximation *)
BVPh[1,3];

(* Gain squared residual at 3rd-order approx. *)
GetErr[3];

(* Gain optimal values of c0 and sigma *)
res=Minimize[Err[3],{c0,sigma}];
Print["Minimum square residual = ", res];

(* Set optimal c0 and sigma *)
```

```
c0=-1;
sigma = 62/100;
Print["c0 = ",c0," sigma = ",sigma];
(*Gain 10th-order HAM approximation *)
BVPh[1,10];
```

注意, 对于定义在有限区间上的非线性边值 (特征值) 问题, BVPh 1.0 默认 zL=0.

附录 8.3 例 8.3.3 的 BVPh 输入数据

```
(* Input Mathematica package BVPh version 1.0 *)
<<BVPh.txt;

(* Define the physical and control parameters *)
TypeEQ = 1;
TypeL = 1;
TypeBase = 2;
ApproxQ = 0;
NgetErr = 1;

(* Define the governing equation *)
f[z_,u_,lambda_]:=D[u,{z,4}]
         + alpha*( z*D[u,{z,3}]+3*D[u,{z,2}] )
         + R( u*D[u,{z,3}] - D[u,z]*D[u,{z,2}] );
alpha = 3/2;
R = -11;

(* Define Boundary conditions *)
zR = 1;
OrderEQ = 4;
BC[1,z_,u_,lambda_] := Limit[u, z->0];
BC[2,z_,u_,lambda_] := Limit[D[u,{z,2}], z->0 ];
BC[3,z_,u_,lambda_] := u - 1 /.  z->zR;
BC[4,z_,u_,lambda_] := D[u,z] /.  z->zR;

(* Define initial guess *)
u[0]=sigma*z+(5-4*sigma)/2*z^3-(3-2*sigma)/2*z^5;
```

```
sigma = . ;

(* Define output term *)
output[z_,u_,k_]:= Print["output = ",
                    D[u[k],z]/.z->0//N];

(* Define the auxiliary linear operator *)
omega[1] = Pi/zR;
omega[2] = Pi/zR;
L[f_] := Module[{temp,numA,numB,i},
If[TypeL == 1,
   temp[1] = D[f,{z,OrderEQ}],
   numA = IntegerPart[OrderEQ/2];
   numB = OrderEQ - 2* numA//Expand;
   temp[0] = D[f,{z,numB}];
   For[i=1, i<=numA, i++,
       temp[1] = D[temp[0],{z,2}]
                 + (kappa*omega[i])^2*temp[0];
       temp[0] = temp[1];
     ];
  ];
temp[1]//Expand
];

(* Print input and control parameters *)
PrintInput[u[z]];

(* Set c0 and sigma *)
c0 =-1/2;
sigma = 2;
Print[" c0 = ",c0, " sigma = ",sigma];

(* Gain approximation by 3rd-order iteration *)
iter[1,40,3];
```

注意, 对于定义在有限区间上的非线性边值 (特征值) 问题, BVPh 1.0 默认 zL=0.

第九章 具有变系数的非线性特征方程

本章运用五种不同类型的例子来说明基于同伦分析方法 (HAM) 的 Mathematica 软件包 BVPh 1.0 对于求解有限区间 $0 \leqslant z \leqslant a$ 上非线性特征方程 $\mathcal{F}[z, u, \lambda] = 0$ 的有效性, 该方程满足 n 个线性边界条件 $\mathcal{B}_k[z, u] = \gamma_k$ ($1 \leqslant k \leqslant n$), 其中 $\mathcal{F}$ 是 n 阶非线性算子, $\mathcal{B}_k$ 是线性算子, γ_k 是常数, $u(z)$ 和 λ 分别表示特征函数和特征值. 这些例子验证了 BVPh 1.0 可以通过不同的初始猜测解和不同类型的基函数找到一些具有奇点和 (或) 多边界条件的强非线性特征方程的多解.

9.1 绪论

在科学和工程领域中, 人们经常需要求解关于特征函数和特征值的非线性微分方程, 并且许多非线性特征方程有多解. 然而, 众所周知, 通过打靶法等数值技术并不容易得到非线性边值问题的多解 [148].

2005 年, 廖世俊应用同伦分析方法 (HAM) [129–143, 150] 成功地求得了边界层方程

$$F''' + \frac{1}{2} F F'' - \beta F'^2 = 0, \quad F(0) = 0, F'(0) = 1, F'(+\infty) = 0$$

的解的两个分支, 其中 $-1 < \beta < +\infty$ 为常数, 并引入额外的未知量 $\delta = F(+\infty)$. 通过求解与该未知量 δ 相关的非线性代数方程, 廖世俊发现了当 $\beta > 1$ 时解的

一个新分支, 该解从来没有被其他解析方法甚至被数值方法得到过, 这主要是因为解的两个分支的 $F'''(0)$ 值之间的差异非常小, 很难区分它们. 此外, 还验证了适当引入未知参数可以找到一些非线性问题的多解.

第八章说明了使用基于同伦分析方法的 Mathematica 软件包 BVPh 1.0 作为解析工具并且在初始猜测解中引入一个未知参数, 即多解控制参数, 可以找到某些强非线性边值问题在有限区间 $z \in [0, a]$ 上的多解. 本章将进一步阐述 BVPh 1.0 对于求解有限区间 $z \in [0, a]$ 上非线性特征值问题的有效性, 该问题满足 n 阶非线性常微分方程

$$\mathcal{F}[z, u, \lambda] = 0, \quad z \in [0, a] \tag{9.1}$$

和 n 个线性边界条件

$$\mathcal{B}_k[z, u] = \gamma_k, \quad 1 \leqslant k \leqslant n, \tag{9.2}$$

其中 $\mathcal{F}$ 为 n 阶非线性算子, $\mathcal{B}_k$ 为线性算子, z 为自变量, $u(z)$ 为有限区间 $z \in [0, a]$ 上的特征函数, λ 为未知特征值, γ_k 和 $a > 0$ 均为有界常数. 注意, 当 $n \geqslant 3$ 时, n 个线性边界条件 (9.2) 可以在区间 $[0, a]$ (包括两个端点) 上的分离点处定义. 线性算子 $\mathcal{B}_k$ 为如下形式

$$\mathcal{B}_k[z, u] = \sum_{i=0}^{n} a_{k,n-i}(z) \frac{d^i u(z)}{dz^i}, \tag{9.3}$$

其中 $a_{k,n-i}(z)$ 为关于 z 的光滑函数. 注意, $a_{k,n-i}(z)$ 中至少有一个是非零的. 假设上述特征值问题至少存在一个光滑的特征函数和特征值.

一般来说, 有限区间 $z \in [0, a]$ 上非线性特征值问题存在许多不同的非零特征函数和特征值, 这依赖于 $u(a), u'(a)$ 等项. 因此, 我们添加如下附加线性边界条件以区分不同的特征函数和特征值,

$$\mathcal{B}_0[z, u] = \gamma_0, \quad z = z^*, \tag{9.4}$$

其中 $z^* \in [0, a]$, γ_0 是常数. 注意, 上述边界条件需与 n 个原始边界条件 (9.2) 线性无关.

一些数值技术被开发以解决非线性特征值问题, 其中之一是 MATLAB 中著名的基于打靶法 [148] 的 BVP4c. 2009 年, 廖世俊 [139] 以三角函数为基函数, 利用同伦分析方法解析求解了受轴向载荷作用的非均匀梁的非线性特征值问题. 2011 年, Abbasbandy 和 Shirzada [118] 以多项式作为基函数, 提出了一

种基于同伦分析方法的线性特征值问题的解析方法, 并且提出了一种由非线性代数方程控制的收敛控制参数 c_0 的不同值来获得多个特征函数的方法.

本章将提出一种基于同伦分析方法的解析方法解决具有多边界条件 (9.2) 的非线性特征方程 (9.1). 我们说明了这种非线性特征值问题可以借助 BVPh 1.0 来解决, 并且由五种不同类型的特征方程证明了 BVPh 1.0 的有效性和普遍性.

9.2 简明数学公式

需要注意的是, 特征函数 $u(z)$ 和特征值 λ 都是未知的. 令 $q \in [0,1]$ 表示同伦参数, $u_0(z)$ 和 λ_0 分别表示特征函数 $u(z)$ 和特征值 λ 的初始猜测值, 其中 $u_0(z)$ 满足 n 个原始边界条件 (9.2) 和附加边界条件 (9.4). 在同伦分析方法的框架中, 我们首先构造两个连续形变 (同伦) $\phi(z;q)$ 和 $\Lambda(q)$, 即当 $q \in [0,1]$ 从 0 增加到 1 时, $\phi(z;q)$ 从初始猜测解 $u_0(z)$ 变化到特征函数 $u(z)$, 以及 $\Lambda(q)$ 从初始猜测值 λ_0 变化到特征值 λ. 构造零阶形变方程

$$(1-q)\mathcal{L}\left[\phi(z;q)-u_0(z)\right] = c_0 q \mathcal{F}[z,\phi(z;q),\Lambda(q)], \quad q \in [0,1] \tag{9.5}$$

以及 n 个线性边界条件

$$\mathcal{B}_k[z,\phi(z;q)] = \gamma_k, \quad 1 \leqslant k \leqslant n \tag{9.6}$$

和附加线性边界条件

$$\mathcal{B}_0[z,\phi(z;q)] = \gamma_0, \tag{9.7}$$

其中 $c_0 \neq 0$ 是所谓的收敛控制参数, $\mathcal{L}$ 是具有性质 $\mathcal{L}(0) = 0$ 的辅助线性算子, 其导数的最高阶数为 n. 初始猜测解 $u_0(z)$ 满足所有的边界条件, 显然,

$$\phi(z;0) = u_0(z), \quad \Lambda(0) = \lambda_0$$

和

$$\phi(z;1) = u(z), \quad \Lambda(1) = \lambda$$

分别为 $q=0$ 和 $q=1$ 时 (9.5) 至 (9.7) 的解. 若选取适当的初始猜测解 $u_0(z)$, 辅助线性算子 $\mathcal{L}$ 和收敛控制参数 c_0, 使得同伦 – Maclaurin 级数

$$\phi(z;q) = u_0(z) + \sum_{m=1}^{+\infty} u_m(z) q^m, \tag{9.8}$$

$$\Lambda(q) = \lambda_0 + \sum_{m=1}^{+\infty} \lambda_m q^m \tag{9.9}$$

在 $q = 1$ 时绝对收敛, 则分别有同伦级数

$$u(z) = u_0(z) + \sum_{m=1}^{+\infty} u_m(z) \tag{9.10}$$

和

$$\lambda = \lambda_0 + \sum_{m=1}^{+\infty} \lambda_m. \tag{9.11}$$

在 M 阶近似下, 有同伦近似

$$u(z) \approx \sum_{m=0}^{M} u_m(z), \quad \lambda \approx \sum_{m=1}^{M} \lambda_{m-1}. \tag{9.12}$$

根据定理 4.15, $u_m(z)$ 和 λ_{m-1} 满足 m 阶形变方程

$$\mathcal{L}\left[u_m(z) - \chi_m u_{m-1}(z)\right] = c_0 \delta_{m-1}(z), \tag{9.13}$$

其中

$$\delta_k(z) = \mathcal{D}_k\{\mathcal{F}[z, \phi(z;q), \Lambda(q)]\}$$

且

$$\chi_k = \begin{cases} 0, & k \leqslant 1, \\ 1, & k > 1. \end{cases}$$

这里, $\mathcal{D}_k$ 是 k 阶同伦导数算子, 定义如下

$$\mathcal{D}_k = \frac{1}{k!} \frac{\partial^k}{\partial q^k}\bigg|_{q=0}.$$

将同伦-Maclaurin 级数 (9.8) 代入 n 个边界条件 (9.6) 和 (9.7) 中, 然后将 q 的相同次幂取等式, 我们有 $n+1$ 个线性边界条件

$$\mathcal{B}_i[z, u_m(z)] = 0, \quad 0 \leqslant i \leqslant n. \tag{9.14}$$

详情请参考第二章和第四章.

需要注意的是, (9.13) 中的 $\delta_k(z)$ 只依赖于非线性算子 $\mathcal{F}$. 借助第四章中同伦导数算子 $\mathcal{D}_k$ 的基本性质, 在大多数情况下我们很容易推导出 $\delta_k(z)$ 的显式表达式. 例如, 当

$$\mathcal{F}[z, u, \lambda] = \mathcal{L}_0(u) + \lambda f(z, u)$$

时, 其中 $\mathcal{L}_0$ 是 n 阶线性算子, $f(z, u)$ 是 z 和 u 的光滑函数, 利用定理 4.1 的同伦导数算子 $\mathcal{D}_n$ 的基本性质、定理 4.2 的线性性质和定理 4.3 的交换性质, 我们可以得到显式表达式

$$\delta_k(z) = \mathcal{L}_0\left[u_k(z)\right] + \sum_{n=0}^{k} \lambda_{k-n} \mathcal{D}_n\{f[z, \phi(z; q)]\} \tag{9.15}$$

以及递推公式

$$\mathcal{D}_0\{f[z, \phi(z; q)]\} = f(z, u_0), \tag{9.16}$$

$$\mathcal{D}_n\{f[z, \phi(z; q)]\} = \sum_{j=0}^{n-1}\left(1 - \frac{j}{n}\right) u_{n-j}(z) \frac{\partial \mathcal{D}_j\{f[z, \phi(z; q)]\}}{\partial u_0}, \tag{9.17}$$

其由定理 4.10 给出. 在这种情况下, 我们总是可以得到任意 n 阶线性算子 $\mathcal{L}_0$ 和任意非线性函数 $f(z, u)$ 的 $\delta_k(z)$ 的显式表达式. 通过 Mathematica 这样的科学计算软件很容易得到非线性算子 $\mathcal{F}[z, u, \lambda]$ 的 $\delta_k(z)$ [119]. 事实上, BVPh 1.0 求解有限区间 $z \in [0, a]$ 上的非线性常微分方程时, 我们无须推导 $\delta_k(z)$ 的表达式, 这是因为它是由软件包自动给出的. 详情请参考第七章附录中的 BVPh 1.0. 注意, m 阶形变方程 (9.13) 和 $n+1$ 个边界条件 (9.14) 都是线性的. 令 $u_m^*(z)$ 表示 (9.13) 的特解. 我们有

$$u_m^*(z) = \chi_m u_{m-1}(z) + c_0 \mathcal{L}^{-1}\left[\delta_{m-1}(z)\right],$$

其中 $\mathcal{L}^{-1}$ 是辅助线性算子 $\mathcal{L}$ 的逆算子. 注意, 特解 $u_m^*(z)$ 包含未知量 λ_{m-1}. (9.13) 的通解 $u_m(z)$ 为

$$u_m(z) = \chi_m u_{m-1}(z) + c_0 \mathcal{L}^{-1}\left[\delta_{m-1}(z)\right] + \sum_{i=1}^{n} A_i \varphi_i(z),$$

其中 A_i 为未知系数, $\varphi_i(z)$ 为已知函数, 且满足方程

$$\mathcal{L}\left[\varphi_i(z)\right] = 0, \quad 1 \leqslant i \leqslant n.$$

未知量 λ_{m-1} 和 n 个未知积分常数 A_i $(1 \leqslant i \leqslant n)$ 完全由 $n+1$ 个线性边界条件 (9.14) 确定. 因此, 给定初始猜测解 $u_0(z)$, 我们可以先求得 $\lambda_0, u_1(z)$, 再求得 $\lambda_1, u_2(z)$, 依此类推. 像 Mathematica 这样的科学计算软件来完成这样的工作是非常有效的.

在同伦分析方法的框架中, 我们有极大的自由度来选取辅助线性算子 $\mathcal{L}$ 和一般特征方程 (9.1) 的基函数. 因此, BVPh 1.0 为我们提供了极大的自由度来选取辅助线性算子 $\mathcal{L}$ 和初始猜测解 $u_0(z)$, 以解决不同类型的非线性特征值问题.

众所周知, 定义在 $z \in [0, a]$ 上的连续函数 $u(z)$ 可以用 Fourier 级数来近似 [122]. Leonhard Euler 解决了一个简单但著名的特征值问题, 即二阶线性微分方程

$$u'' + \lambda u = 0, \quad u(0) = 0, \quad u(a) = 0,$$

其特征函数和特征值为

$$u = A \sin \frac{\kappa \pi z}{a}, \quad \lambda = \left(\frac{\kappa \pi}{a}\right)^2,$$

其中 $\kappa \geqslant 1$ 为正整数, A 为任意常数. 上述线性特征值问题具有无穷多个特征函数和特征值. 需要注意的是, 这是 (9.1) 和 (9.2) 的一个特例. 为了包含这类特征函数, 我们通常为二阶特征值微分方程 (9.1) 选取辅助线性算子

$$\mathcal{L}(u) = u'' + \left(\frac{\kappa \pi}{a}\right)^2 u. \tag{9.18}$$

对于任意常数 C_1, C_2 和正整数 $\kappa \geqslant 1$, 它具有如下性质

$$\mathcal{L}\left(C_1 \cos \frac{\kappa \pi z}{a} + C_2 \sin \frac{\kappa \pi z}{a}\right) = 0.$$

类似地, 对于三阶特征值微分方程 (9.1), 选取辅助线性算子

$$\mathcal{L}(u) = u''' + \left(\frac{\kappa \pi}{a}\right)^2 u'. \tag{9.19}$$

对于任意常数 C_0, C_1, C_2 和正整数 $\kappa \geqslant 1$, 它具有如下性质

$$\mathcal{L}\left(C_0 + C_1 \cos \frac{\kappa \pi z}{a} + C_2 \sin \frac{\kappa \pi z}{a}\right) = 0.$$

一般来说, 对于 n 阶特征值微分方程 (9.1), 选取

$$\mathcal{L}(u) = \left[\prod_{i=1}^{m}\left(\frac{d^2}{dx^2} + \omega_i^2\right)\right] u, \quad n = 2m \tag{9.20}$$

或
$$\mathcal{L}(u) = \left[\prod_{i=1}^{m}\left(\frac{d^2}{dx^2} + \omega_i^2\right)\right]u', \quad n = 2m+1, \tag{9.21}$$

其中 ω_i 是频率, 可能互不相同, 例如

$$\omega_i = i\left(\frac{\kappa\pi}{a}\right), \tag{9.22}$$

也可能相同, 例如

$$\omega_i = \frac{\kappa\pi}{a}. \tag{9.23}$$

这取决于边界条件 (9.2), 这里 $\kappa \geqslant 1$ 为正整数.

需要注意的是, 上述辅助线性算子中不同的 κ 对应一组不同的基函数

$$\left\{\sin\frac{j\kappa\pi z}{a}, \cos\frac{j\kappa\pi z}{a}\,\bigg|\, j = 0, 1, 2, 3, \cdots\right\}.$$

这为我们在有限区间 $z \in [0, a]$ 上寻找二阶非线性特征方程 (9.1) 的多个特征函数和特征值提供了一个简便的途径, 如 §9.3.1, §9.3.2 和 §9.3.3 所述.

众所周知, 有限区间 $z \in [0, a]$ 上的连续函数 $u(z)$ 可以用幂级数

$$u(z) = \sum_{k=0}^{+\infty} a_k z^k$$

或 Chebyshev 级数

$$u(z) = \sum_{k=0}^{+\infty} b_k T_k(z)$$

很好地近似, 其中 $T_k(z)$ 是第一类 Chebyshev 多项式. 所以, 幂级数或 Chebyshev 多项式可用于表示特征函数, 如 §9.3.4 和 §9.3.5 所述. 在这种情况下, 我们只需在有限区间 $z \in [0, a]$ 上为 n 阶特征值问题选取辅助线性算子

$$\mathcal{L}(u) = \frac{d^n u}{dz^n}. \tag{9.24}$$

需要注意的是, $u_m(z)$ 和 λ_{m-1} 包含收敛控制参数 c_0, 它虽然没有物理意义, 但为我们保证同伦级数 (9.10) 和 (9.11) 的收敛性提供了一个简便的途径, 如第二章所述, 并在第四章证明. 令

$$E_m(c_0) = \frac{1}{a}\int_0^a \left\{\mathcal{F}\left[z, \sum_{i=0}^{m} u_i(z), \sum_{i=0}^{m-1}\lambda_i\right]\right\}^2 dz \tag{9.25}$$

表示控制方程 (9.1) 在 m 阶近似下的平均平方残差, 其中积分是通过足够多的离散点以数值方式计算的. 收敛控制参数 c_0 的最优值 c_0^* 由

$$\frac{dE_m(c_0)}{dc_0} = 0$$

确定. 在大多数情况下, 我们选取接近其最优值 c_0^* 的收敛控制参数, 可以快速得到特征函数 $u(z)$ 和特征值 λ 的收敛同伦级数解. 此外, 通过这种方式, 我们可以借助 BVPh 1.0 来求解 $z \in [0,a]$ 上的强非线性特征值问题, 如下所述.

关于辅助线性算子 $\mathcal{L}$ 和初始猜测解 $u_0(z)$ 的选取, 详情请参考第七章.

9.3 例子

下面给出的所有例子均由 BVPh 1.0 求解, BVPh 1.0 在第七章的附录中给出. 此外, 这些例子的 BVPh 1.0 输入数据文件均在本章附录中给出.

在下面的例子中, 我们将截断项的数量设置为 $N_t = 20$, 并用区间 $z \in [0, a]$ 上 50 个等距的离散点来数值计算上述积分. 此外, 若未提及, 则采用 3 阶同伦迭代方法, 即 (7.51) 中 $M = 3$.

9.3.1 轴向载荷作用下的非均匀梁

我们首先考虑在轴向载荷 P 下, 两个支撑上具有任意横截面的非均匀梁 [120, 123, 127, 128], 如图 9.1 所示, 其中 P 为正值时表示压力, 为负值时表示拉力. 设 l 和 θ 分别表示梁的长度和坡度, u 表示梁的挠度. 设 s 表示自然轴的弧坐标, 该轴在直线或未弯曲状态下通过梁每个横截面的质心, $I(s)$ 是横截面上穿过质心的直线的最小惯性矩, E 是材料的杨氏模量. 假设所有惯性主轴均平行, 这样梁就不会发生扭转变形. 从数学上讲, 该问题满足方程

$$EI(s)\frac{d\theta}{ds} + Pu = 0, \quad u(0) = u(l) = 0,$$

其中 $I(s)$ 为惯性矩. 假设 $I(s)$ 和 $I'(s)$ 是有界连续函数, 其中 ' 表示对 s 求导. 通过对原始控制方程关于 s 求导, 并由 $\sin\theta = du/ds$, 我们得到了非线性屈曲方程

$$(EI\theta')' + P\sin\theta = 0.$$

将边界条件 $u(0) = u(l) = 0$ 代入原始控制方程, 我们得到了等价边界条件

$$\theta'(0) = \theta'(l) = 0,$$

即梁两端弯矩为零.

图 9.1 受轴向载荷 P 作用, 具有可变惯性矩 $I(s)$ 的梁

定义自然轴的无量纲弧坐标为

$$z = \frac{\pi s}{l},$$

并写作

$$I = I_0 \mu(z),$$

其中 $\mu(z)$ 是 I 的分布函数, $I_0 > 0$ 是参考惯性矩. 例如, I_0 可定义为

$$I_0 = \frac{1}{l}\int_0^l I(s)ds = \frac{1}{\pi}\int_0^\pi I(z)dz,$$

尽管这不是必要的. 此外, 所考虑的问题满足方程

$$\mu(z)\theta''(z) + \mu'(z)\theta'(z) + \lambda \sin\theta(z) = 0, \quad \theta'(0) = \theta'(\pi) = 0, \qquad (9.26)$$

其中 ′ 表示对 z 求导, 且

$$\lambda = \frac{P}{EI_0}\left(\frac{l}{\pi}\right)^2 \qquad (9.27)$$

称为轴向载荷参数. 对于给定的 λ, 可以直接获得轴向载荷

$$P = \lambda(EI_0)\left(\frac{\pi}{l}\right)^2.$$

因此, $\lambda > 0$ 对应压力 P, $\lambda < 0$ 对应拉力 P.

具有变系数 $\mu(z)$ 和 $\mu'(z)$ 的方程 (9.26) 只是 (9.1) 的一个特例, 即

$$\mathcal{F}[z, \theta, \lambda] = \mu(z)\theta''(z) + \mu'(z)\theta'(z) + \lambda \sin\theta.$$

显然, 不同非零特征函数 $\theta(z)$ 的 $\theta(0)$ 值是不同的. 所以, 这为我们区分非零特征函数提供了附加边界条件

$$\theta(0) = \gamma, \qquad (9.28)$$

$u(z)$ 和 λ 的 M 阶近似由 (9.12) 给出,其中 $\theta_m(z)$ 和 λ_{m-1} 满足 m 阶形变方程

$$\mathcal{L}[\theta_m(z) - \chi_m \theta_{m-1}(z)] = c_0 \delta_{m-1}(z) \tag{9.29}$$

和边界条件

$$\theta_m'(0) = 0, \quad \theta_m'(\pi) = 0, \quad \theta_m(0) = 0, \tag{9.30}$$

其中

$$\delta_k(z) = \mu(z)\theta_k''(z) + \mu'(z)\theta_k'(z) + \sum_{i=0}^{k} \lambda_{k-i} \mathcal{D}_i\{\sin[\phi(z;q)]\} \tag{9.31}$$

由定理 4.1 给出,$\mathcal{D}_i\{\sin[\phi(z;q)]\}$ 由定理 4.8 和定理 4.10 描述的递推公式得出.

9.3.1.1 均匀梁

我们先考虑均匀梁在 $\mu(z) = 1$ 的情况,即

$$\mathcal{F}[z,\theta,\lambda] = \theta'' + \lambda \sin\theta. \tag{9.32}$$

由于上式为二阶微分方程,我们选取辅助线性算子 (9.18),其包含正整数 κ. 为了满足 (9.26) 中的两个原始边界条件和附加边界条件 (9.28),我们选取初始猜测解 $\theta_0(z)$,即

$$\theta_0'(0) = \theta_0'(\pi) = 0, \quad \theta_0(0) = \gamma,$$

其中满足上述条件的函数有很多,如

$$\theta_0(z) = \gamma \cos(\kappa z) \tag{9.33}$$

和

$$\theta_0(z) = \sigma - (\sigma - \gamma)\cos(\kappa z), \tag{9.34}$$

这里,σ 是实数. 这为我们选取初始猜测解提供了一个额外的自由度. 注意,当 $\sigma = 0$ 时,初始猜测解 (9.33) 是 (9.34) 的一个特例. 我们称 σ 为多解控制参数,因为它为我们寻找多个特征函数提供了一个简便的途径,如下所述.

非线性特征值问题借助 BVPh 1.0 来求解. 不失一般性,我们首先考虑 $\gamma = 1$ 和 $\kappa = 1$ 的情况. 选取 $\kappa = 1$ 的初始猜测解 (9.33),控制方程关于收敛控制参数 c_0 的平方残差在高达 3 阶近似下的曲线如图 9.2 所示,这表明最优收敛控制

参数 c_0^* 的值接近 -1. 选取 $c_0 = -1$, 即使是无迭代的 3 阶同伦近似结果也相当精确: 平方残差仅为 4.1×10^{-14}. 在第二次迭代时, 利用 3 阶迭代公式, 平方残差变小至 1.3×10^{-26}, 该数字非常小, 没有必要再进行更多的迭代. 在这种情况下, 特征值 λ 相当快地收敛至 1.137069, 如表 9.1 所示.

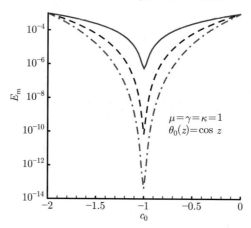

图 9.2 当 $\mu = \gamma = \kappa = 1$ 和 $\theta_0(z) = \cos z$ 时, 控制方程关于 c_0 的平方残差. 实线: 1 阶近似; 虚线: 2 阶近似; 点划线: 3 阶近似

表 9.1 当 $\gamma = \kappa = 1$, $\theta_0(z) = \cos z$, $c_0 = -1$ 时, 3 阶迭代法的第一类特征值和平方残差

迭代次数 m	λ	平方残差 E_m
1	1.137069	4.1×10^{-14}
2	1.137069	1.3×10^{-26}

需要注意的是, 初始猜测解 (9.33) 是 $\sigma = 0$ 时 (9.34) 的特例, 我们尝试使用不同 σ 值的初始猜测解 (9.34), 例如 $\sigma = 2, 3, 4$ 等. 令人惊讶的是, 当 $\sigma = 2$ 和 $\sigma = 3$ 时, 选取适当的收敛控制参数 c_0, 可以得到一种新类型的解. 例如, 在 $\sigma = 3$ 的情况下, 高达 3 阶的同伦近似的平方残差与 c_0 的关系曲线如图 9.3 所示, 这表明最优收敛控制参数 c_0 约为 -1.25. 事实上, 由 $c_0 = -1$ 的 3 阶同伦迭代方法得到了特征值 $\lambda = -1.951368$ 的收敛近似值, 其平方残差迅速变小, 如表 9.2 所示. 注意, 第八章在初始猜测解中引入一个类似的参数, 以找出有限区间 $z \in [0, a]$ 上非线性常微分方程的多解. 这揭示了我们称其为多解控制参数的原因.

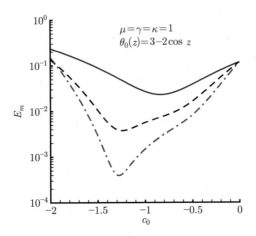

图 9.3 当 $\mu = \gamma = \kappa = 1$ 和 $\theta_0(z) = 3 - 2\cos z$ 时，控制方程关于 c_0 的平方残差。实线：1 阶近似；虚线：2 阶近似；点划线：3 阶近似

表 9.2 当 $\gamma = \kappa = 1$, $\theta_0(z) = 3 - 2\cos z$, $c_0 = -1$ 时，3 阶迭代法的第二类特征值和平方残差

迭代次数 m	λ	平方残差 E_m
1	-1.922288	1.5×10^{-3}
2	-1.954534	4.6×10^{-6}
3	-1.951247	3.6×10^{-8}
4	-1.951353	1.2×10^{-9}
5	-1.951371	9.7×10^{-12}
6	-1.951368	2.2×10^{-14}
7	-1.951368	5.0×10^{-16}
8	-1.951368	1.2×10^{-17}
9	-1.951368	2.0×10^{-18}
10	1.951368	1.9×10^{-18}

因此，在 $\mu = \gamma = \kappa = 1$ 的情况下，存在两种不同类型的特征值：一个是正特征值 $\lambda = 1.137069$，另一个是负特征值 $\lambda = -1.951368$，分别对应两种不同类

型的特征函数 $\theta(z)$, 如图 9.4 所示. 它们还给出了完全不同的位移, 如图 9.5 所示, 其中 x 和 u 分别表示梁的水平位移和垂直位移, 定义为

$$x(z) = \int_0^z \cos\theta(s)ds, \quad u(z) = \int_0^z \sin\theta(s)ds,$$

对于 $\gamma = \theta(0)$ 的其他值, BVPh 1.0 可用类似的方式获得两种不同类型的特征函数和特征值.

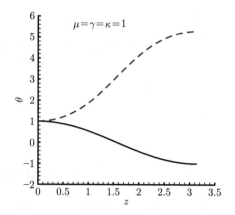

图 9.4 当 $\mu = \gamma = \kappa = 1$ 时, 两个不同的特征函数. 实线: 当 $\theta_0(z) = \cos z$ 时, $\lambda = 1.137069$ 的第一个特征函数; 虚线: 当 $\theta_0(z) = 3 - 2\cos z$ 时, $\lambda = -1.951368$ 的第二个特征函数

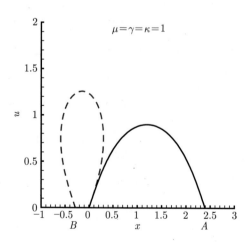

图 9.5 当 $\mu = \gamma = \kappa = 1$ 时, 两个不同特征函数的位移. 实线: 当 $\theta_0(z) = \cos z$ 时, $\lambda = 1.137069$ 的位移; 虚线: 当 $\theta_0(z) = 3 - 2\cos z$ 时, $\lambda = -1.951368$ 的位移

在上述关于 $x(z)$ 和 $u(z)$ 的表达式中,我们将梁的一端 $z = 0$ 处视为一个固定点,即 $x(0) = 0, u(0) = 0$. 令 $(x_A, 0)$ 和 $(x_B, 0)$ 表示梁另一端的坐标,其分别对应于第一类和第二类特征函数,如图 9.5 所示. 显然,x_A 和 x_B 依赖于 γ 的取值. 我们发现,对于第一类特征函数,当 γ 从 0 增加到 π 时,x_A 从 $x_A = \pi$ 单调减小. 对于第二类特征函数,当 γ 从 π 减小到 0 时,x_B 从 $x_B = -\pi$ 单调增加. 特别地,对于第一类特征函数,当 $\gamma = 2.281319 = \gamma^*$ 时,$x_A = -1.4 \times 10^{-6}$,特征值 $\lambda = 2.183380 = \lambda^*$. 对于第二类特征函数,当 $\gamma = 0.860274 = \tilde{\gamma}$ 时,$x_B = 7.2 \times 10^{-7}$,特征值 $\lambda = -2.183380 = \tilde{\lambda}$. 有趣的是,

$$\lambda^* = -\tilde{\lambda}, \quad \gamma^* + \tilde{\gamma} \approx \pi, \tag{9.35}$$

此外,这两种不同类型的特征值给出的位移几乎相同,如图 9.6 所示.

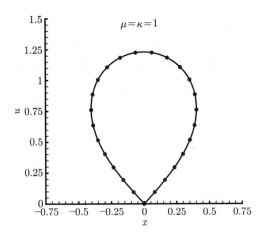

图 9.6 当 $\mu = \kappa = 1$ 时,两种不同类型特征函数的位移. 实线:$\gamma^* = 2.281319$,正特征值 $\lambda^* = 2.183380$ 的第一类特征函数的位移;实心圆:$\tilde{\gamma} = 0.860274$,负特征值 $\tilde{\lambda} = -2.183380$ 的第二类特征函数的位移

在一般情况下,令 λ^* 表示当 $\theta(0) = \gamma^*$ 时的第一类特征函数的特征值,$\tilde{\lambda}$ 表示当 $\theta(0) = \tilde{\gamma}$ 时的第二类特征函数的特征值. 我们发现,只要

$$\gamma^* + \tilde{\gamma} = \pi,$$

就有

$$\lambda^* = -\tilde{\lambda},$$

并且两个不同的特征函数对应的位移是对称的, 如图 9.7 和图 9.8 所示. 因此, 在一般情况下关系式 (9.35) 成立.

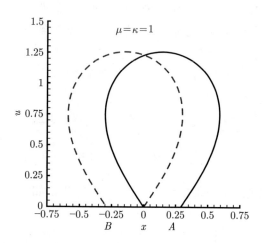

图 9.7 当 $\mu = \kappa = 1$ 时, 两种不同类型的特征函数的位移. 实线: $\gamma^* = \pi - 1$, 特征值 $\lambda^* = 1.951368$ 的第一类特征函数的位移; 虚线: $\tilde{\gamma} = 1$, 特征值 $\tilde{\lambda} = -1.951368$ 的第二类特征函数的位移

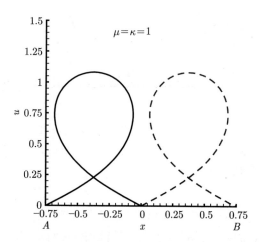

图 9.8 当 $\mu = \kappa = 1$ 时, 两种不同类型的特征函数的位移. 实线: $\gamma^* = \pi - 1/2$, 特征值 $\lambda^* = 3.202901$ 的第一类特征函数的位移; 虚线: $\tilde{\gamma} = 1/2$, 特征值 $\tilde{\lambda} = -3.202901$ 的第二类特征函数的位移

根据这种对称性, 我们选取初始猜测解

$$\theta_0(z) = \gamma \cos(\kappa z)$$

为第一类特征函数, 并选取初始猜测解

$$\theta_0(z) = \pi - (\pi - \gamma)\cos(\kappa z)$$

为第二类特征函数, 分别对应于 (9.34) 中 $\sigma = 0$ 和 $\sigma = \pi$ 的情况. 通过上述初始猜测解和 3 阶同伦迭代方法, 我们借助 BVPh 1.0 获得了特征值与 γ 的关系曲线 ($\mu = 1$ 和 $\kappa = 1$), 如图 9.9 所示.

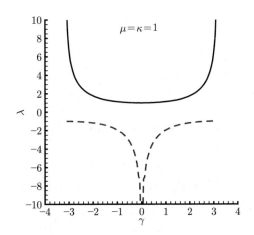

图 9.9 当 $\mu = \kappa = 1$ 时, 特征值与 $\gamma = \theta(0)$ 的关系. 实线: 第一类特征函数的正特征值; 虚线: 第二类特征函数的负特征值

此外, 上述特征函数和特征值的对称性可以用数学方法证明. 假设特征函数 $\theta(z)$ 和特征值 λ 满足

$$\theta'' + \lambda \sin\theta = 0, \theta'(0) = \theta'(\pi) = 0.$$

令 $\tilde{\theta}(z) = \pi - \theta(z)$ 和 $\tilde{\lambda} = -\lambda$. 那么

$$\tilde{\theta}'' + \tilde{\lambda}\sin\tilde{\theta} = -\theta'' - \lambda\sin(\pi - \theta) = -(\theta'' + \lambda\sin\theta) = 0$$

和

$$\tilde{\theta}'(0) = -\theta'(0) = 0, \quad \tilde{\theta}'(\pi) = -\theta'(\pi) = 0.$$

此外
$$\tilde{x}(z) = \int_0^z \cos\tilde{\theta}(z)dz = \int_0^z \cos(\pi-\theta)dz = -\int_0^z \cos\theta = -x(z)$$
和
$$\tilde{u}(z) = \int_0^z \sin\tilde{\theta}(z)dz = \int_0^z \sin(\pi-\theta)dz = \int_0^z \sin\theta = u(z)$$
也成立. 即两种不同类型的特征函数对应的位移是对称的. 因此, 我们有以下定理:

定理 9.1 若特征函数 $\theta(z)$ 和特征值 λ 满足
$$\theta'' + \lambda\sin\theta = 0, \theta'(0) = \theta'(\pi) = 0,$$
则特征函数 $\pi - \theta(z)$ 和特征值 $-\lambda$ 也满足该方程.

这种对称性甚至对具有任意 $\mu(z)$ 的非均匀梁也成立, 如下所述:

定理 9.2 若特征函数 $\theta(z)$ 和特征值 λ 满足
$$\mu(z)\theta''(z) + \mu'(z)\theta'(z) + \lambda\sin\theta = 0, \quad \theta'(0) = \theta'(\pi) = 0,$$
则特征函数 $\pi - \theta(z)$ 和特征值 $-\lambda$ 也满足该方程.

从数学上讲, 这种对称性确实成立. 然而, 对于具有负特征值的第二类特征函数在物理学上使我们感到困惑, 因为负特征值对应于拉力! 这是一个众所周知的力学知识, 如果压力大于临界值, 那么受压力 ($P > 0$) 作用的梁可能有很大的挠度. 但是, 人们传统上认为, 这样的挠度永远不会发生在受拉力 ($P < 0$) 作用的梁上. 然而, 从数学角度来看, 如前所述, 确实存在具有负特征值的第二类特征函数. 那么, 具有负特征值的第二类特征函数的物理意义是什么呢?

让我们再次考虑 $\mu = \gamma = \kappa = 1$ 的情况. 相应的两个特征函数如图 9.4 所示, 其位移如图 9.5 所示. 假设在 $t = 0$ 时突然施加轴向力 P 前, $\theta(z) = 0$. 那么, 对于大于临界值的压力, 梁以一种 "自然" 的方式偏转, 使得梁在 $z = \pi$ 处的一端从 $x = \pi$ 处开始向左移动, 直到 $x = x_A$. 这一现象已经在教科书的物理实验中观察到. 然而, 对于大于临界值的拉力, 梁在 $z = \pi$ 处的端点须突然移到梁在 $z = 0$ 处端点的左侧, 再移动到 $x = x_B$ 的位置. 在物理学中, 这种突然形变是 "非自然的", 并且比 "自然的" 形变需要更多的能量, 因此在实际中几乎不会发生. 即便如此, 负特征值的第二类特征函数仍具有物理意义. 因此, 当梁受到足够大的拉力时, 理论上梁也可能具有很大的挠度, 尽管这种现象, 如果没有突然的、巨大的外部干扰, 很难观察到.

需要注意的是, 对于小的 θ, $\sin\theta \approx \theta$. 然而, 线性化方程

$$\theta'' + \lambda\theta = 0, \quad \theta'(0) = \theta'(\pi) = 0$$

没有这种对称性, 因为其特征值总是正的. 这说明我们可能会因为将非线性方程线性化而丢失很多解. 换句话说, 非线性方程可能具有比线性化方程更有趣的性质.

对于其他的 κ 值, 我们通过 BVPh 1.0 以类似的方式得到收敛的特征函数和特征值. 例如, 在 $\gamma = 1$ 的情况下, 当 $\kappa = 2$ 和 $\kappa = 3$ 时, 我们得到了收敛的正特征值和负特征值, 如表 9.3 所示. 特征函数分别如图 9.10 和图 9.11 所示. 理论上, 给定 $\gamma = \theta(0)$, 每个 $\kappa \geqslant 1$ 都存在一个正特征值和一个负特征值. 因此, 对于给定的 $\gamma = \theta(0)$, 存在着无穷多个特征函数和特征值. 这些特征函数和特征值都可以借助 BVPh 1.0 以类似的方式找到.

表 9.3 当 $\gamma = 1$, $\kappa > 1$ 时, 均匀梁的多个特征值

κ	正特征值	负特征值
2	4.548275	-7.805465
3	10.233617	-17.562313

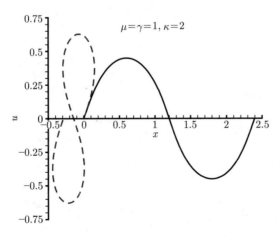

图 9.10 当 $\mu = \gamma = 1$ 和 $\kappa = 2$ 时, 不同的特征函数的位移. 实线: $\theta_0(z) = \cos(2z)$, 正特征值 $\lambda = 4.548275$ 的位移; 虚线: $\theta_0(z) = \pi - (\pi - 1)\cos(2z)$, 负特征值 $\lambda = -7.805465$ 的位移

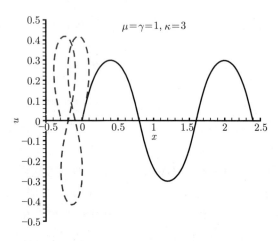

图 9.11 当 $\mu = \gamma = 1$ 和 $\kappa = 3$ 时,不同的特征函数的位移. 实线: $\theta_0(z) = \cos(3z)$, 正特征值 $\lambda = 10.233617$ 的位移; 虚线: $\theta_0(z) = \pi - (\pi - 1)\cos(3z)$, 负特征值 $\lambda = -17.562313$ 的位移

因此,选取不同的初始猜测解和(或)不同的辅助线性算子,BVPh 1.0 可以找出有限区间 $z \in [0, a]$ 上非线性特征方程的多解,如上所述. 注意,多解控制参数 σ 为我们提供了一个找出多个特征函数和特征值的简便途径. 类似地,我们也可以把辅助线性算子 (9.18) 中的正整数 κ 视为一种多解控制参数.

需要强调的是,据作者所知,与负特征值对应的 (9.32) 的 "非自然" 特征函数从未被报道过,尽管在实际中几乎没有发生过. 这表明了 BVPh 1.0 对有限区间 $z \in [0, a]$ 上具有多解的非线性特征方程的有效性和潜力.

9.3.1.2 非均匀梁

为了证明 BVPh 1.0 对于非线性特征值问题的普遍有效性,我们进一步考虑惯性矩非均匀分布的梁

$$\mu(z) = 1 + \frac{\cos(4z)}{2\sqrt{1 + \exp(z^2) + \sin z^2}}. \tag{9.36}$$

需要注意的是,其平均惯性矩为

$$\frac{1}{\pi}\int_0^\pi \left[1 + \frac{\cos(4z)}{2\sqrt{1 + \exp(z^2) + \sin z^2}}\right] dz \approx 1.002,$$

这与均匀梁 $\mu(z) = 1$ 的情况相当接近.

在这种情况下,控制方程 (9.26) 包含变系数 $\mu(z)$ 和 $\mu'(z)$,因此比均匀梁的方程 (9.32) 复杂得多. 即便如此,借助 BVPh 1.0,我们仍然选取与均匀梁相

同的辅助线性算子 (9.18). 此外, 我们选取相同的初始猜测解 $\theta_0(z) = \gamma \cos(\kappa z)$ 和 $\theta_0(z) = \pi - (\pi - \gamma)\cos(\kappa z)$ 来分别获得正负特征值的两种不同类型的特征函数.

不失一般性, 我们考虑 $\gamma = 1$ 和 $\kappa = 1$ 的情况. 选取 $c_0 = -1$, 可以得到两类收敛的特征函数. 其中, 第一类特征函数的正特征值 $\lambda = 1.1061$, 第二类的负特征值 $\lambda = -1.8158$. 如上所述, 对于任意的 $\mu(z)$ 和给定的 $\gamma = \theta(0)$, 存在具有正特征值和负特征值两种类型的特征函数. 以上均可由 BVPh 1.0 以类似的方式获得. 这说明了 BVPh 1.0 对具有强非线性的复杂特征方程的有效性和普遍性.

9.3.2 Gelfand 方程

我们进一步考虑 Gelfand 方程 [121, 126, 146]

$$u'' + (K-1)\frac{u'}{z} + \lambda e^u = 0, u(0) = 0, u(1) = 0, \qquad (9.37)$$

其中 ′ 表示对 z 求导, $K \geqslant 1$ 是常数, $u(z)$ 和 λ 分别表示特征函数和特征值.

这是 (9.1) 的一个特例, 即

$$\mathcal{F}[z, u, \lambda] = u'' + (K-1)\frac{u'}{z} + \lambda e^u.$$

需要注意的是, (9.37) 是强非线性的, 因为它包含指数项 $\exp(u)$. 此外, 项 $u'(z)/z$ 使得 (9.37) 在 $z = 0$ 处包含一个奇点. 尽管 $u'(z)/z$ 在 $z \to 0$ 时的极限为常数, 但这种奇点给数值技术, 如 BVP4c 中的打靶法, 带来了困难. 然而, BVPh 1.0 可以轻松解决这种奇点, 因为科学计算软件 Mathematica 为我们提供了 "基于函数而非数的计算" 的能力 [149].

由于不同非零特征函数有不同的 $u'(0)$ 值, 我们添加附加边界条件

$$u'(0) = A, \qquad (9.38)$$

其中 A 为给定常数, 以区分不同的特征函数.

$u(z)$ 和 λ 的 M 阶近似由 (9.12) 给出, 其中 $u_m(z)$ 和 λ_{m-1} 满足 m 阶形变方程

$$\mathcal{L}[u_m(z) - \chi_m u_{m-1}(z)] = c_0 \delta_{m-1}(z) \qquad (9.39)$$

和边界条件

$$u_m(0) = 0, u_m(1) = 0, u'_m(0) = 0, \qquad (9.40)$$

其中

$$\delta_k(z) = u_k''(z) + (K-1)\frac{u_k'(z)}{z} + \sum_{i=0}^{k} \lambda_{k-i} \mathcal{D}_i \{\exp[\phi(x;q)]\} \quad (9.41)$$

由定理 4.1 给出, 项 $\mathcal{D}_i\{\exp[\phi(x;q)]\}$ 由定理 4.7 和定理 4.10 的递推公式求得.

特征函数 $u(z)$ 由三角函数和多项式组成的混合基表示, 如 §7.2.3 所述. 尽管 (9.37) 与 (9.26) 有很大不同, 但是我们仍选取相同的辅助线性算子 (9.18), 其中 $\kappa = 1$. 此外, 为了满足 (9.37) 中的两个原始边界条件和附加边界条件 (9.38), 我们选取初始猜测解

$$u_0(z) = \frac{A}{2}[1 + \cos(\pi z)]. \quad (9.42)$$

BVPh 1.0 可求解这类非线性特征方程. 对于给定的 A, 收敛控制参数 c_0 的最优值是由 (9.25) 定义的控制方程 (9.37) 的平方残差的最小值确定.

不失一般性, 我们首先考虑 $K = A = 1$ 的情况. 图 9.12 展示了控制方程在高达 3 阶同伦近似下的平方残差与 c_0 的关系, 这表明最优收敛控制参数的值在 -0.55 附近. 事实上, 我们通过 $c_0 = -3/5$ 和 $N_t = 20$ 的 3 阶迭代法获得了快速收敛的特征函数和特征值 $\lambda = 0.866215$, 如表 9.4 所示, 这与 $K = 1$ 时的封闭解 [126]

$$u(z) = \frac{1}{2}e^{-A}\left\{\ln\left[2e^A + 2\sqrt{e^A(e^A-1)} - 1\right]\right\}^2 \quad (9.43)$$

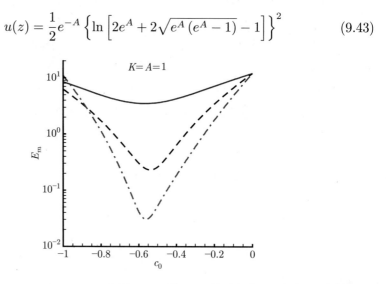

图 9.12　当 $K = A = 1$ 时, Gelfand 方程 (9.37) 关于 c_0 的平方残差. 实线: 1 阶近似; 虚线: 2 阶近似; 点划线: 3 阶近似

表 9.4 当 $K=1$ 和 $A=1$ 时，选取 $c_0=-3/5$ 和 $N_t=20$，Gelfand 方程 (9.37) 的特征值和平方残差

迭代次数 m	特征值 λ	平方残差 E_m
1	0.779831	4.1×10^{-2}
2	0.866491	3.5×10^{-5}
3	0.866221	1.1×10^{-7}
4	0.866215	4.3×10^{-9}
5	0.866215	4.1×10^{-9}
6	0.866215	4.1×10^{-9}

给出的精确特征值 $\lambda=0.866215$ 吻合良好。给定 A 的其他值，BVPh 1.0 以类似的方式求得了收敛的特征值和特征函数，这与上述精确公式吻合良好，如图 9.13 所示。

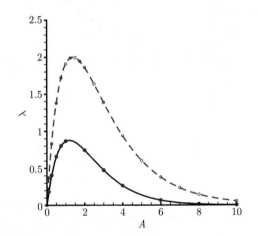

图 9.13 当 $K=1$ 和 $K=2$ 时，Gelfand 方程 (9.37) 的特征值与 A 的关系。实线：$K=1$；虚线：$K=2$；实心圆：精确解 (9.43)

在 $K=2$ 的情况下，通过 3 阶同伦迭代方法，BVPh 1.0 以类似的方式得到了收敛的特征函数和特征值，如表 9.5 和图 9.13 所示。类似地，在 $A=1$ 和 $A=2$ 的情况下，我们还得到了不同 K 值的收敛特征函数和特征值，如图 9.14 和表 9.6 所示。

表 9.5 当 $K=2$ 时，Gelfand 方程 (9.37) 的特征值

A	特征值 λ	A	特征值 λ
0.05	0.192644	2	1.860353
0.10	0.371137	2.5	1.635358
0.25	0.829569	3	1.386745
0.50	1.378161	4	0.936157
0.75	1.719382	5	0.602776
1.00	1.909210	6	0.378467
1.25	1.990053	7	0.234285
1.50	1.993891	8	0.143900
1.75	1.944705	10	0.053500

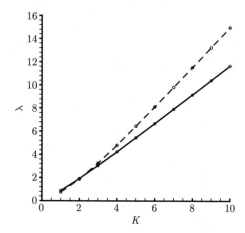

图 9.14 当 $A=1$ 和 $A=2$ 时，Gelfand 方程 (9.37) 的特征值与 K 的关系。实线：$A=1$；虚线：$A=2$

表 9.6 当 $A=1$ 和 $A=2$ 时，Gelfand 方程 (9.37) 的特征值

K	特征值 $\lambda(A=1)$	特征值 $\lambda(A=2)$
1	0.8662	0.7436
2	1.9092	1.8604
3	3.0460	3.2522
4	4.2328	4.8059
5	5.4471	6.4428
6	6.6775	8.1209
7	7.9177	9.8198
8	9.1642	11.5299
9	10.4149	13.2462
10	11.6684	14.9663

像 Mathematica 这样的科学计算软件很容易解决当 $z \to 0$ 时的奇异项 $u'(z)/z$，因为它将 z 看作函数而非数，此外，当 $z \to 0$ 时 $u'(z)/z$ 的极限为常数.

由于 (9.37) 包含指数项 $\exp(u)$ 和奇异项 $u'(z)/z$，这个例子证实了 BVPh 1.0 对于具有奇点的强非线性特征方程的有效性和普遍性.

9.3.3 具有奇点和变系数的方程

需要注意的是，BVPh 1.0 对 n 阶非线性特征方程 (9.1) 有效，该方程满足 n 个线性边界条件 (9.2)，这些条件在形式上相当广泛. 为了证明 BVPh 1.0 的普遍有效性，我们求解以下有限区间 $0 < z < \pi$ 上变系数的非线性特征方程

$$\sqrt{1+z^2}u'' + \frac{\cos(\pi z)u'}{z} + \lambda\left[\frac{e^u}{1+z^2} + (1+z)\sin u\right] = \sin\left(z^2 + e^{-z}\right), \quad (9.44)$$

满足两个边界条件

$$u'(0) = 0, u(\pi) - u'(\pi) = \frac{3}{5}, \quad (9.45)$$

其中 ′ 表示对 z 求导，$u(z)$ 和 λ 分别是未知的特征函数和特征值.

上述问题是 (9.1) 的一个特例，即

$$\mathcal{F}[z,u,\lambda] = \sqrt{1+z^2}u'' + \frac{\cos(\pi z)u'}{z} + \lambda\left[\frac{e^u}{1+z^2} + (1+z)\sin u\right] - \sin\left(z^2 + e^{-z}\right).$$

特别地，它包含变系数

$$\sqrt{1+z^2}, \frac{\cos(\pi z)}{z}, \frac{1}{1+z^2}, (1+z), -\sin\left(z^2 + e^{-z}\right)$$

和强非线性项 $\exp(u)$ 及 $\sin u$. 此外，$u'(z)/z$ 项在 $z = 0$ 处有一个奇点. 该奇点导致了数值技术（如 BVP4c 的打靶法）的困难. 因此，上述方程是相当复杂的. 幸运的是，$u'(z)/z$ 当 $z \to 0$ 时的极限为常数，因此在同伦分析方法的框架中，我们可以通过像 Mathematica 这样的科学计算软件轻松解决，其将 z 视为函数而非数.

所以对于非零特征函数 $u(z)$，$u(0) \neq 0$. 我们使用

$$u(0) = A \quad (9.46)$$

作为附加边界条件来区分不同的非零特征函数.

$u(z)$ 和 λ 的 M 阶近似由 (9.12) 给出，其中 $u_m(z)$ 和 λ_{m-1} 满足 m 阶形变方程

$$\mathcal{L}[u_m(z) - \chi_m u_{m-1}(z)] = c_0 \delta_{m-1}(z) \quad (9.47)$$

和边界条件
$$u'_m(0) = 0, u_m(\pi) - u'_m(\pi) = 0, \tag{9.48}$$

其中
$$\delta_k(z) = \sqrt{1+z^2}u''_k + \frac{\cos(\pi z)u'_k}{z} - (1-\chi_{k+1})\sin\left(z^2+e^{-z}\right)$$
$$+ \sum_{i=0}^{k}\lambda_{k-i}\left[\frac{\mathcal{D}_i\{\exp[\phi(z;q)]\}}{1+z^2} + (1+z)\mathcal{D}_i\{\sin[\phi(z;q)]\}\right] \tag{9.49}$$

由定理 4.1 给出，$\mathcal{D}_i\{\exp[\phi(z;q)]\}$ 和 $\mathcal{D}_i\{\sin[\phi(z;q)]\}$ 分别由定理 4.7 和定理 4.8，或定理 4.10 的递推公式给出.

BVPh 1.0 成功地解决了上述具有奇点的强非线性特征值问题. 由于该方程定义在有限区间 $z \in [0,\pi]$ 上，我们可以用 §7.2.3 中提到的混合基来表示特征函数. 因此，尽管 (9.44) 比 (9.26) 和 (9.37) 更复杂，但我们仍选取相同的辅助线性算子 (9.18)，其中 $\kappa = 1$. 此外，选取初始猜测解

$$u_0(z) = \frac{5A+3}{10} + \frac{5A-3}{10}\cos z, \tag{9.50}$$

其满足两个原始边界条件 (9.45) 和附加边界条件 (9.46).

不失一般性，我们先考虑 $A = 1/2$ 的情况. 图 9.15 展现了控制方程 (9.44) 在高达 3 阶近似下的平方残差，这表明最优收敛控制参数约为 -0.4. 事实上，选取 $c_0 = -2/5$，通过 $N_t = 30$ 的 3 阶同伦迭代方法，我们获得了快速收敛的特

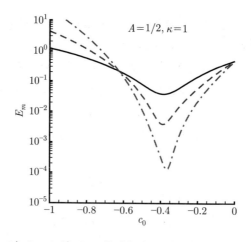

图 9.15 当 $A = 1/2$ 时，(9.44) 关于 c_0 的平方残差. 实线: 1 阶近似; 虚线: 2 阶近似; 点划线: 3 阶近似

征值和特征函数, 如表 9.7 和图 9.16 所示. 注意, 当 $m \geqslant 4$ 时, 控制方程 (9.44) 的平方残差在 $E_m = 5.3 \times 10^{-8}$ 处停止变小. 然而, 如果使用更多的截断项 (即更大的 N_t 值), 则会得到更小的平方残差.

表 9.7 当 $c_0 = -2/5$, $N_t = 30$ 以及 $A = 1/2$ 和 $\kappa = 1$ 时, 初始猜测解 (9.50) 的 3 阶同伦迭代方法的特征值和平方残差

迭代次数 m	特征值 λ	平方残差 E_m
1	0.373901	3.3×10^{-4}
2	0.380270	5.4×10^{-7}
3	0.379978	5.6×10^{-8}
4	0.379957	5.3×10^{-8}
5	0.379956	5.3×10^{-8}
6	0.379956	5.3×10^{-8}
7	0.379956	5.3×10^{-8}
8	0.379956	5.3×10^{-8}

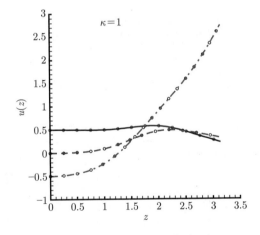

图 9.16 当 $\kappa = 1$ 时, (9.44) 和 (9.45) 的特征函数. 实线: 当 $c_0 = -2/5$, $A = 1/2$ 时的第 10 次迭代近似; 虚线: 当 $c_0 = -2/5$, $A = 0$ 时的第 10 次迭代近似; 点划线: 当 $c_0 = -2/5$, $A = -1/2$ 时的第 15 次迭代近似; 空心圆: 第 3 次迭代近似 ($A = 1/2$ 和 $A = 0$) 或第 8 次迭代近似 ($A = -1/2$)

给定 A 的其他值, BVPh 1.0 以类似的方式求得了收敛的特征值和特征函数. 例如, 当 $A = 1/2$, $A = 0$ 和 $A = -1/2$ 时的特征函数, 如图 9.16 所示. 我们

发现, $A = 1/2$ 和 $A = -1/2$ 的特征函数之间不存在对称性. 此外, 选取 $\kappa = 1$, 特征值与 A 的关系曲线如图 9.17 所示. 我们发现, 当 $\kappa = 1$ 时, 有两个特征值分支. 特征值的第一个分支为正, 且当 $A \to +\infty$ 时, 其整体趋于零, 如表 9.8 所示; 特征值的第二个分支的取值范围为区间 $c_0^L < c_0 < c_0^R$, 其中 c_0^L 接近 -3.2, $-0.49 < c_0^R < -0.33$. 与特征值的第一个分支不同的是, 特征值的第二个分支在 $u(0) = A$ 的某些数值下为负, 如表 9.9 所示. 根据我们的计算, 选取 $\kappa = 1$, 当 $c_0 < c_0^R$ 时, 似乎不存在收敛的特征函数和特征值. 图 9.17 表明, $c_0 = c_0^L$ 和 $c_0 = c_0^R$ 处可能存在某些类型的奇点.

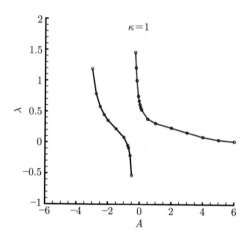

图 9.17 当 $\kappa = 1$ 时, (9.44) 和 (9.45) 关于 A 的特征值

表 9.8 当 $\kappa = 1$ 时, (9.44) 的特征值的第一个分支

A	特征值 λ	A	特征值 λ
-0.33	1.607740	0.05	0.565095
-0.32	1.559068	0.1	0.528119
-0.31	1.508751	0.5	0.379956
-0.30	1.457289	1	0.311333
-0.25	1.205111	2	0.238667
-0.20	1.001034	3	0.162146
-0.10	0.748010	4	0.089831
-0.05	0.670271	6	0.020306
0	0.611237	10	0.000797

9.3 例 子

表 9.9 当 $\kappa = 1$ 时, (9.44) 的特征值的第二个分支

A	特征值 λ	A	特征值 λ
-3.2	2.356321	-2.0	0.355912
-3.1	1.517056	-1.5	0.221140
-3.0	1.187195	-1.0	0.081523
-2.9	0.984646	-0.7	-0.089925
-2.75	0.784556	-0.6	-0.212886
-2.5	0.578778	-0.5	-0.532902
-2.25	0.448467	-0.49	-0.634153

所有上述结果都是利用 $\kappa = 1$ 的辅助线性算子 (9.18) 和初始猜测解 (9.50) 获得的. 如前所述, 我们可以选取不同的辅助线性算子和不同的初始猜测解来找到多解. 注意, 初始猜测解 (9.50) 可以概括为

$$u_0(z) = \frac{5A+3}{10} + \frac{5A-3}{10}\cos(\kappa z), \tag{9.51}$$

其中 $\kappa \geqslant 1$ 是奇数. 选取上述初始猜测解和 $\kappa = 3$ 及 $\kappa = 5$ 时的辅助线性算子 (9.18), 且 $c_0 = -1/5$, 我们在 $A = 1$ 时获得了多个特征函数和特征值 $\lambda = 7.3500\ (\kappa = 3)$ 以及 $\lambda = 19.9043\ (\kappa = 5)$, 如图 9.18 和表 9.10 所示. 这表明具有边界条件 (9.45) 的特征方程 (9.44) 可能有无穷多个特征函数和特征值.

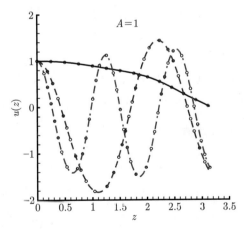

图 9.18 当 $A = 1$ 时, 选取不同的 κ 值, (9.44) 和 (9.45) 的多个特征函数. 实线: 当 $\kappa = 1$ 和 $c_0 = -1/5$ 时的第 8 次迭代近似; 虚线: 当 $\kappa = 3$ 和 $c_0 = -1/5$ 时的第 30 次迭代近似; 点划线: 当 $\kappa = 5$ 和 $c_0 = -1/5$ 时的第 30 次迭代近似; 空心圆: 第 10 次迭代近似 ($\kappa = 3$ 和 $\kappa = 5$) 或第 3 次迭代近似 ($\kappa = 1$)

表 9.10 当 $A=1$ 时，选取 $\kappa=3$, $N_t=40$, $c_0=-1/5$ 和 $\kappa=5$, $N_t=50$, $c_0=-1/5$, (9.44) 的特征值和平方残差

迭代次数 m	$\lambda\,(\kappa=3)$	$E_m\,(\kappa=3)$	$\lambda\,(\kappa=5)$	$E_m\,(\kappa=5)$
1	6.2210	18.96	13.4273	126.40
3	6.5234	1.57	18.7510	5.73
5	7.1834	0.25	19.6540	0.37
7	7.3451	2.10×10^{-2}	19.8553	2.10×10^{-2}
10	7.3555	1.60×10^{-4}	19.9021	1.70×10^{-4}
15	7.3500	7.10×10^{-7}	19.9044	9.10×10^{-7}
20	7.3500	6.30×10^{-7}	19.9043	8.40×10^{-7}
25	7.3500	6.30×10^{-7}	19.9043	8.40×10^{-7}

值得注意的是, 在初始猜测解 (9.51) 和辅助线性算子 (9.18) 中, 只需选取不同的 κ 值即可获得多个特征函数和特征值. 因此, 以三角函数为基函数, 选取不同的初始猜测解和不同的辅助线性算子, 我们可以在同伦分析方法的框架中找出一些非线性特征方程的多解.

需要注意的是, BVPh 1.0 可以很容易解决当 $z\to 0$ 时的奇异项 $u'(z)/z$, 因为科学计算软件 Mathematica 将 z 视为函数而非数字, 且当 $z\to 0$ 时 $u'(z)/z$ 的极限为常数. 除此之外, 选取不同 κ 值的辅助线性算子 (9.18) 和初始猜测解 (9.51), 我们可以简单地找到多解. 因此, 我们也可以将 κ 视为一种多解控制参数. 最后, 收敛控制参数 c_0 为保证强非线性控制方程 (9.44) 同伦级数解的收敛性提供了一个简便的途径.

因此, 尽管 (9.44) 没有物理意义, 但是这个例子证实了 BVPh 1.0 对具有强非线性和奇点的复杂特征值问题的普遍有效性.

9.3.4 具有多解的多点边值问题

某些具有多解的非线性边值问题可以转化为特征方程来求解. 为了说明 BVPh 1.0 的普遍有效性, 我们考虑具有多边界条件的四阶非线性微分方程

$$u'''' = \beta z\left(1+u^2\right),\ u(0)=u'(1)=u''(1)=0,\ u''(0)-u''(\alpha)=0, \quad (9.52)$$

其中 $\alpha\in(0,1)$ 和 β 为给定常数. Graef 等人 [124,125] 证明, 当 $\alpha=1/5$ 和 $\beta=10$ 时, 上述方程至少存在两个正解. 注意, 与前三个例子不同的是, 该边值方程是四阶的, 并且边界条件定义在三个不同的点上, 即它是多点边值问题.

令 $u(z) = \lambda\theta(z)$,且 $\lambda = u(1)$,原方程 (9.52) 变为

$$\lambda\theta'''' = \beta z\left(1 + \lambda^2\theta^2\right), \quad \theta(0) = \theta'(1) = \theta''(1) = 0, \quad \theta''(0) - \theta''(\alpha) = 0, \tag{9.53}$$

附加边界条件为

$$\theta(1) = 1. \tag{9.54}$$

我们将 λ 视为未知特征值,且 (9.53) 是 (9.1) 的一个特例,即

$$\mathcal{F}[z,\theta,\lambda] = \lambda\theta'''' - \beta z\left(1 + \lambda^2\theta^2\right).$$

$\theta(z)$ 和 λ 的 M 阶近似由 (9.12) 给出,其中 $\theta_m(z)$ 和 λ_{m-1} 满足 m 阶形变方程

$$\mathcal{L}\left[\theta_m(z) - \chi_m\theta_{m-1}(z)\right] = c_0\delta_{m-1}(z) \tag{9.55}$$

和多边界条件

$$\theta_m(0) = \theta'_m(1) = \theta''_m(1) = 0, \; \theta''_m(0) - \theta''_m(\alpha) = 0, \tag{9.56}$$

以及附加边界条件

$$\theta_m(1) = 0, \tag{9.57}$$

其中

$$\delta_k(z) = \sum_{i=0}^{k}\lambda_{k-i}\theta_i'''' - (1-\chi_{k+1})\beta z$$
$$- \beta z\sum_{i=0}^{k}\left(\sum_{j=0}^{i}\lambda_j\lambda_{i-j}\right)\left(\sum_{r=0}^{k-i}\theta_r\theta_{k-i-r}\right) \tag{9.58}$$

由第四章中的定理 4.1 给出.

上述具有多边界条件的非线性特征值问题是通过 BVPh 1.0 来解决的. 因为 (9.53) 包含 βz 项,我们可用 z 的多项式来表示特征函数 $\theta(z)$. 因此,我们选取辅助线性算子

$$\mathcal{L}(\theta) = \theta'''' \tag{9.59}$$

和初始猜测解

$$\theta_0(z) = \frac{1}{2\alpha-3}\left[2(3\alpha-4)z + 6(1-\alpha)z^2 + 2\alpha z^3 - z^4\right], \tag{9.60}$$

其满足 (9.53) 中的四个原始边界条件和附加边界条件 (9.54).

不失一般性, 我们先考虑 $\alpha = 1/5$ 和 $\beta = 10$ 的情况. 我们发现, 在 1 阶近似下, 附加边界条件 (9.57), 即 $\theta_1(1) = 0$, 给出了非线性代数方程

$$\lambda_0^2 - 2.77904\lambda_0 + 1.35864 = 0, \tag{9.61}$$

该方程有两个不同的解, 分别为 $\lambda_0 = 0.63313$ 和 $\lambda_0 = 2.14591$. 注意, 该代数方程与收敛控制参数 c_0 无关. 选取上述代数方程的一个解作为特征值的初始猜测值, 我们得到了特征函数 $\theta(z)$ 的 m 阶同伦近似和特征值 λ 的 $m - 1$ 阶同伦近似: 它们都包含收敛控制参数 c_0. 我们发现, 当 $\lambda_0 = 0.63313$ 时, 控制方程 (9.53) 的平方残差 E_m 在区间 $c_0 \in (-3, 0)$ 上随着 m 的增加而变小. 此外, 最优收敛控制参数 c_0 约为 $-3/2$, 如图 9.19 所示. 事实上, 选取 $\lambda_0 = 0.63313$ 和 $c_0 = -3/2$, (9.53) 的平方残差在 10 阶同伦近似下单调减小至相当小的值 5.3×10^{-16}. 因此, 我们得到了第一个特征函数 $\theta(z)$ 和第一个特征值 $\lambda = 0.627315$, 如表 9.11 所示. 类似地, 选取 $\lambda_0 = 2.14591$, 我们可以发现, 平方残差 E_m 在区间 $c_0 \in (-0.7, 0)$ 上随着 m 的增加而减小, 且最优收敛控制参数 c_0 约为 $-1/2$, 如图 9.20 所示. 事实上, 选取 $c_0 = -1/2$ 和 $\lambda_0 = 2.14591$, 在 20 阶近似下, 平方残差减小至 1.7×10^{-16}, 我们得到第二个特征函数和第二个特征值 $\lambda = 2.24118$, 如表 9.12 所示. 我们发现, 虽然两个特征值明显不同, 但相应的两个特征函数却相当接近, 如图 9.21 所示. 一般来说, 用数值方法来区分这种非常接近的特征函数是相当困难的.

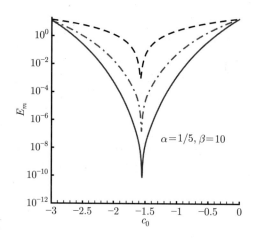

图 9.19 当 $\alpha = 1/5$ 和 $\beta = 10$ 时, 选取特征值的第一个初始猜测值 $\lambda_0 = 0.63313$, 控制方程 (9.53) 关于 c_0 的平方残差 E_m. 虚线: 1 阶; 点划线: 3 阶; 实线: 5 阶

表 9.11 当 $\lambda_0 = 0.63313$, $c_0 = -3/2$, $\alpha = 1/5$ 和 $\beta = 10$ 时, (9.53) 的特征值和平方残差 E_m

近似阶数 m	特征值 λ	平方残差 E_m
2	0.627446	2.3×10^{-3}
4	0.627318	1.0×10^{-6}
6	0.627315	6.9×10^{-10}
8	0.627315	5.7×10^{-13}
10	0.627315	5.3×10^{-16}

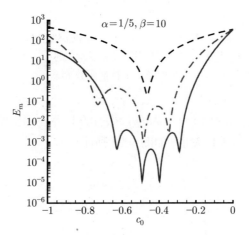

图 9.20 当 $\alpha = 1/5$ 和 $\beta = 10$ 时, 选取特征值的第二个初始猜测值 $\lambda_0 = 2.14591$, 控制方程 (9.53) 关于 c_0 的平方残差 E_m. 虚线: 1 阶; 点划线: 3 阶; 实线: 5 阶

表 9.12 当 $\lambda_0 = 2.14591$, $c_0 = -1/2$, $\alpha = 1/5$ 和 $\beta = 10$ 时, (9.53) 的特征值和平方残差 E_m

近似阶数 m	特征值 λ	平方残差 E_m
4	2.24105	6.9×10^{-3}
8	2.24118	1.6×10^{-6}
12	2.24118	6.3×10^{-10}
16	2.24118	3.2×10^{-13}
20	2.24118	1.7×10^{-16}

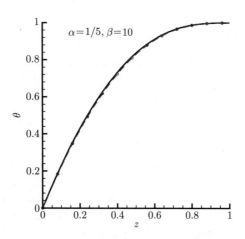

图 9.21 当 $\alpha = 1/5$ 和 $\beta = 10$ 时, 选取特征值的不同初始猜测值 λ_0, (9.53) 和 (9.54) 的两个特征函数的比较. 实线: $\lambda_0 = 0.63313$ 和 $c_0 = -3/2$ 的第一个特征函数 $\theta(z)$; 带空心圆的虚线: $\lambda_0 = 2.14591$ 和 $c_0 = -1/2$ 的第二个特征函数 $\theta(z)$

需要注意的是, 由于 $u(z) = \lambda\theta(z)$, 微分方程存在两个明显不同的解 $u(z)$, 其分别对应于两个不同的 λ 值, 如图 9.22 所示.

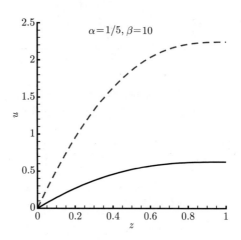

图 9.22 当 $\alpha = 1/5$ 和 $\beta = 10$ 时, 原方程 (9.52) 的两个解. 实线: $\lambda_0 = 0.63313$ 和 $c_0 = -3/2$ 的第一个解 $u(z)$; 虚线: $\lambda_0 = 2.14591$ 和 $c_0 = -1/2$ 的第二个解 $u(z)$

类似地, 对于不同的 α 和 β 值, 我们可以获得非线性多点边值方程 (9.52) 的两个解. 这说明具有多解的非线性边值问题可以转化为非线性特征值问题. 注意, 如 §8.3.2 所述, 也可以将 (9.52) 视为非线性边值方程来直接求解. 还要注意的是, (9.52) 是一个四阶非线性边值方程, 其边界条件在两个端点和区间 $z \in (0,1)$ 上的一个分离点处成立. 因此, 无论我们把它看作是一个正常的非线性边值问题还是非线性特征值问题, (9.52) 都可以通过 BVPh 1.0 来求解.

类似地, 如 §8.3.2 所述, BVPh 1.0 可以轻松解决多边界条件, 因为这些边界条件可以通过像 Mathematica 这样的科学计算软件以相同的方式处理. 这主要是因为科学计算软件为我们提供了 "基于函数而非数的计算" 的能力 [149].

因此, 这个例子验证了 BVPh 1.0 对于具有多边界条件的高阶非线性特征值问题的有效性和普遍性.

9.3.5 具有复系数的流动稳定性 Orr-Sommerfeld 方程

作为最后一个例子, 我们考虑著名的 Orr-Sommerfeld 方程 [144,147], 其用于研究平面泊肃叶流动的稳定性, 该方程为

$$\left(\mathcal{D}^2 - \alpha^2\right)^2 u - i(\alpha R)\left[(U_0 - \lambda)\left(\mathcal{D}^2 - \alpha^2\right) - \mathcal{D}^2 U_0\right] u = 0, \quad (9.62)$$

满足边界条件

$$u'(0) = u'''(0) = 0, \ u(1) = 0, \ u'(1) = 0, \quad (9.63)$$

其中 ′ 表示对 z 求导, $i = \sqrt{-1}$ 是虚数单位, 算子 $\mathcal{D}$ 由 $\mathcal{D}u = u''$ 定义, R 表示雷诺数, λ 为复特征值, $U_0 = 1 - z^2$ 是平面泊肃叶流动的精确解. 速度的二维扰动与

$$u(z) \exp[i\alpha(x - \lambda t)]$$

成正比, 其中 α 为实数, λ 为复数. 因此, 当 λ 的虚部为正, 即 $\text{Im}(\lambda) > 0$ 时, 流动变得不稳定. 详情请参考 [144] 和 [147].

上述具有复系数的特征值问题也是 (9.1) 的一个特例, 即

$$\mathcal{F}[z, u, \lambda] = \left(\mathcal{D}^2 - \alpha^2\right)^2 u - i(\alpha R)\left[(U_0 - \lambda)\left(\mathcal{D}^2 - \alpha^2\right) - \mathcal{D}^2 U_0\right] u.$$

注意, 如果 $\bar{u}(z)$ 是一个特征函数, 那么 $u(z) = \bar{u}(z)/\bar{u}(0)$ 也是同一个特征值 λ 的特征函数, 这是因为 (9.62) 是线性的. 因此, 我们有附加边界条件

$$u(0) = 1.$$

$u(z)$ 和 λ 的 M 阶近似由 (9.12) 给出, 其中 $u_m(z)$ 和 λ_{m-1} 满足 m 阶形变方程

$$\mathcal{L}\left[u_m(z) - \chi_m u_{m-1}(z)\right] = c_0 \delta_{m-1}(z) \quad (9.64)$$

和边界条件

$$u_m'(0) = u_m'''(0) = 0, \ u_m(1) = 0, \ u_m'(1) = 0, \qquad (9.65)$$

以及附加边界条件

$$u_m(0) = 0, \qquad (9.66)$$

其中 c_0 是收敛控制参数，$\mathcal{L}$ 是辅助线性算子，$u_0(z)$ 是 $u(z)$ 的初始猜测解，并且

$$\begin{aligned}\delta_n(z) = \left(\mathcal{D}^2 - \alpha^2\right)^2 u_n(z) - i(\alpha R)\left[U_0\left(\mathcal{D}^2 - \alpha^2\right) - \mathcal{D}^2 U_0\right] u_n(z) \\ + i(\alpha R) \sum_{k=0}^n \lambda_k \left(\mathcal{D}^2 - \alpha^2\right) u_{n-k}(z)\end{aligned} \qquad (9.67)$$

由第四章中的定理 4.1 给出.

这里，我们只考虑具有对称性 $u(z) = u(-z)$ 的特征函数 $u(z)$. 由于 $u(z)$ 定义在有限区间 $z \in [-1, 1]$ 上，所以它可以用 z 的多项式表示. 因此，我们选取辅助线性算子

$$\mathcal{L}(u) = u''''. \qquad (9.68)$$

注意，临界特征值通常与特征函数的最简形式有关，如 §9.3.1 的第一个例子所述. 因此，我们选取初始猜测解

$$u_0(z) = \left(1 - z^2\right)^2, \qquad (9.69)$$

它是满足所有边界条件 (9.63) 和附加边界条件 $u(0) = 1$ 的 z 的最简多项式.

BVPh 1.0 成功地解决了这个问题. 不失一般性，我们首先考虑 $\alpha = 1$ 与不同雷诺数 R 的情况. 注意，(9.62) 包含虚数 $i = \sqrt{-1}$. 幸运的是，在同伦分析方法的框架中，我们有极大的自由度来选取收敛控制参数 c_0，使得 c_0 可以为复数，如定理 5.3 和定理 5.4 所述. 因此，我们在这里使用具有复收敛控制参数 c_0 的 3 阶同伦迭代公式. 例如，在 $R = 100$ 和 $\alpha = 1$ 的情况下，选取复收敛控制参数 $c_0 = (-1+i)/2$，我们得到了收敛特征值

$$\lambda = 0.478494 - 0.162944i,$$

如表 9.13 所示. 这验证了，在同伦分析方法的框架中，收敛控制参数 c_0 确实可以是复数. 此外，我们发现，对于给定的 R，总是可以找到一个适当的 c_0，其

形式为

$$c_0 = \frac{-1+i}{\rho}, \quad \rho \geqslant 1,$$

表 9.13 当 $R = 100$ 和 $\alpha = 1$ 时，选取 $c_0 = (-1+i)/2$ 的 3 阶迭代公式，(9.62) 的特征值和平方残差

迭代次数 m	特征值 λ	平方残差 E_m
1	$0.473522 - 0.158606i$	2330
3	$0.478861 - 0.161875i$	88.8
5	$0.478652 - 0.163091i$	3.77
10	$0.478490 - 0.162945i$	7.0×10^{-4}
15	$0.478494 - 0.162944i$	9.3×10^{-8}
20	$0.478494 - 0.162944i$	1.8×10^{-11}
25	$0.478494 - 0.162944i$	6.2×10^{-15}
30	$0.478494 - 0.162944i$	2.3×10^{-18}

它的 3 阶同伦迭代方法收敛，如表 9.14 所示. 对于从 $R = 100$ 至临界值 $R = 5814.83$ 的不同雷诺数，收敛特征函数的实部和虚部分别如图 9.23 和图 9.24 所示. 我们发现，对于小雷诺数 R，特征值的虚部为负，即 $\text{Im}(\lambda) < 0$，对应于稳定的黏性流动. 当 $\alpha = 1, R = 5814.83$ 时，特征值

$$0.261233 + 2.2495 \times 10^{-9} i$$

的虚部相当接近于零. 此外，当 $\alpha = 1, R > 5814.83$ 时，有 $\text{Im}(\lambda) > 0$，流动变得不稳定. 因此，上述特征值和特征函数对应于最不稳定的黏性流动. 注意，与 [147] 不同的是，我们不需要计算其他更高阶数的特征函数和特征值，这些对流动的稳定性并不重要.

表 9.14 当 $\alpha = 1$ 时，选取复收敛控制参数 c_0 的 3 阶同伦迭代公式，不同雷诺数 R 的收敛特征值. 特征函数为 z 至 $o(z^{N_t})$ 的多项式

R	特征值 λ	c_0 ($i = \sqrt{-1}$)	N_t
100	$0.478494 - 0.162944i$	$(-1+i)/2$	90
200	$0.430714 - 0.116810i$	$(-1+i)/4$	90
500	$0.380566 - 0.0704922i$	$(-1+i)/10$	90
1000	$0.346285 - 0.0421283i$	$(-1+i)/20$	150
2000	$0.312100 - 0.0197987i$	$(-1+i)/40$	200
3000	$0.292289 - 0.0101846i$	$(-1+i)/60$	300
5000	$0.268131 - 0.0017503i$	$(-1+i)/100$	300
5500	$0.263762 - 0.00060763i$	$(-1+i)/100$	300
5800	$0.261348 - 0.00002691i$	$(-1+i)/100$	300
5814	$0.261239 - 1.5004 \times 10^{-6}i$	$(-1+i)/100$	300
5814.83	$0.261233 + 2.2495 \times 10^{-9}i$	$(-1+i)/100$	300
5815	$0.261231 + 3.1000 \times 10^{-7}i$	$(-1+i)/100$	300
5825	$0.261154 + 0.00001837i$	$(-1+i)/100$	300
6000	$0.259816 + 0.00032309i$	$(-1+i)/120$	300

需要注意的是，雷诺数 R 从 0 增加至临界值 $R_c \approx 5814.83$ 时，特征函数实部的变化很小，如图 9.23 所示，但虚部变化很大，如图 9.24 所示. 特别地，当 $\alpha = 1$ 和 $R = 5814.83$ 时，特征函数的虚部在大区间 $-0.7 \leqslant z \leqslant 0.7$ 上几乎接近于零. 因此，揭示这一有趣结果的物理意义将是有价值的.

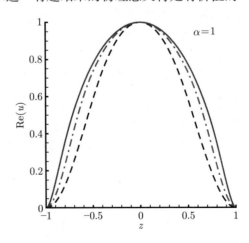

图 9.23 当 $\alpha = 1$ 时，Orr-Sommerfeld 方程 (9.62) 的特征函数的实部. 虚线: $R = 100$; 点划线: $R = 1000$; 实线: $R = 5814.83$

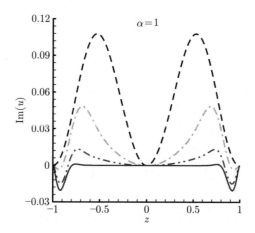

图 9.24 当 $\alpha = 1$ 时,Orr-Sommerfeld 方程 (9.62) 的特征函数的虚部. 虚线: $R = 100$; 点划线: $R = 1000$; 双点划线: $R = 3000$; 实线: $R = 5814.83$

临界雷诺数 R_c 由 Orszag [147] 定义,是存在不稳定特征模的最小 R 值. 对于平面泊肃叶流动,Orszag [147] 报告了临界雷诺数 $R_c = 5772.22$ 且 $\alpha_c = 1.02056$. 通过使用 3 阶同伦迭代方法的 BVPh 1.0 和复收敛控制参数 $c_0 = (-1+i)/100$,在 $R_c = 5772.22$ 和 $\alpha_c = 1.02056$ 的情况下,我们得到了特征值

$$\lambda = 0.26943 - 3.085 \times 10^{-9} i,$$

这与 [147] 给出的数值结果吻合良好.

因此,该例子验证了 BVPh 1.0 对具有复系数的 Orr-Sommerfeld 方程的有效性和普遍性.

9.4 本章小结

本章论述了 BVPh 1.0 对于求解有限区间 $0 \leqslant z \leqslant a$ 上非线性特征方程 $\mathcal{F}[z, u, \lambda] = 0$ 的有效性,该方程满足 n 个线性边界条件 $\mathcal{B}_k[z, u] = \gamma_k$ ($1 \leqslant k \leqslant n$),其中 $\mathcal{F}$ 表示 n 阶非线性算子,$\mathcal{B}_k$ 表示线性算子,γ_k 为常数,$u(z)$ 和 λ 分别表示特征函数和特征值. 我们讨论了五种不同类型的例子,如轴向载荷作用下的非均匀梁、Gelfand 方程、具有变系数的特征方程、具有多解的多点边值问题以及著名的具有复系数的流动稳定性 Orr-Sommerfeld 方程. 这些例子表明,BVPh 1.0 通过不同的初始猜测解和不同类型的基函数,可以找到某些具有奇点和多边界条件的强非线性特征方程的多解.

第一个例子说明了非线性特征方程的多解可以借助不同的初始猜测解来找

到. 特别地, 我们可能是首次成功地得到梁方程 (9.26) 具有负特征值的特征函数, 甚至还证明了一般的非均匀梁方程 (9.26) 确实存在这种具有负特征值的 "非自然" 特征函数. 此外, 这种具有负特征值的特征函数表明, 受足够大拉力作用的梁也可能具有较大的挠度, 如同受足够大压力作用的梁. 然而, 这种 "非自然" 偏转在 $t = 0$ 时需要施加突然并且巨大的扰动, 这需要比 "自然" 偏转大得多的能量, 因此在实际应用中几乎不会发生. 据作者所知, 这种非均匀梁方程的具有负特征值的 "非自然" 特征函数从未被报道过. 事实上, 任何一种全新的方法总会带来一些新的和 (或) 不同的东西. 这表明了 BVPh 1.0 对于具有多解和奇点的强非线性特征方程的巨大潜力和有效性.

此外, 我们验证了 BVPh 1.0 对于具有强非线性和奇点 (例 9.3.2), 和 (或) 不同的多个边界条件 (例 9.3.3), 和 (或) 变系数和高阶导数 (例 9.3.4), 和 (或) 复系数 (例 9.3.5) 特征值问题的有效性.

在同伦分析方法的框架中, 我们有极大的自由度来选取不同的辅助线性算子和初始猜测解, 以便在不同的基函数中得到收敛的特征函数, 正如这些例子所述. 在前三个例子中, 三角函数和多项式的组合被用作混合基函数来表示特征函数, 以及辅助线性算子 (9.18) 和初始猜测解 (9.34) 和 (9.51), 它们包含正整数 κ. 参数 κ 也可以被看作是一个多解控制参数, 这是因为选取不同的 κ 值可以得到多个特征函数和特征值.

使用 z 的多项式作为基函数来表示最后两个四阶特征方程的特征函数, 并选取简单的辅助线性算子

$$\mathcal{L}(u) = u''''.$$

例 9.3.4 只有有限个特征函数, 因为它最初是具有多个边界条件的非线性边值问题. 注意, 例 9.3.4 的多解是通过多个初始猜测值 λ_0 获得的, 且由非线性代数方程 (9.61) 控制. 这展示了求得非线性边值问题多解的另一种方法. 例 9.3.5 也使用了多项式, 这主要是因为我们只对 $\mathrm{Im}(\lambda)$ 最大的特征值对应的特征函数感兴趣. 以多项式为基函数可以轻松得到这种独特的特征函数, 如例 9.3.5 所述.

这些例子验证了 BVPh 1.0 对于具有奇点和 (或) 多边界条件和 (或) 复系数的复杂、强非线性边值问题的有效性和普遍性.

最后, 虽然 BVPh 1.0 是针对有限区间 $z \in [0, a]$ 上的一般特征方程 $\mathcal{F}[z, u, \lambda] = 0$ 开发的, 但这并不意味着它可以解决这种形式的所有特征方程. 如第七章中提到的, 我们的目标是为尽可能多的非线性边值问题开发一个有效的 Mathematica 软件包. 当然, 今后还需要进一步的改进 (见本章的问题) 和更多的应用. 即便如此, Mathematica 软件包 BVPh 1.0 为我们提供了一个有用的替代工

具来研究科学和工程中的许多非线性特征值问题.

需要注意的是, Chebfun 4.0 也为我们提供了"基于函数而非数的计算"的能力 [149]. 因此, 借助 Chebfun 为具有奇点的强非线性多点边值问题建立类似的基于同伦分析方法的软件包应该是有价值的, 且 Chebfun 是开源代码.

问题

9.1 具有非线性边界条件的特征值问题

开发一种基于同伦分析方法的解析方法, 用于 n 阶非线性特征方程

$$\mathcal{F}[z,u,\lambda] = 0, \quad 0 \leqslant z \leqslant a,$$

满足 n 个非线性多边界条件

$$\mathcal{B}_k[z,u,\lambda] = \gamma_k, \quad 1 \leqslant k \leqslant n,$$

其中 $\mathcal{B}_k$ 是非线性算子, γ_k 是常数. 假设上述方程至少存在一个光滑解. 改进第七章中给出的 Mathematica 软件包 BVPh 1.0 以解决此类问题.

9.2 耦合的非线性特征值问题

开发一种基于同伦分析方法的解析方法, 用于有限区间 $z \in [0,a]$ 上 n 个耦合的非线性特征方程

$$\mathcal{F}_k[z,u,\lambda_1,\lambda_2,\cdots,\lambda_n] = 0, \quad 1 \leqslant k \leqslant n,$$

满足某些线性 (非线性) 多边界条件, 其中 $n \geqslant 2$. 假设上述方程至少存在一个光滑解. 通常, 此类问题会提供 Mathematica 软件包.

9.3 无限区间上的特征值问题

开发一种基于同伦分析方法的解析方法, 用于无限区间上的 n 阶非线性特征方程

$$\mathcal{F}[z,u,\lambda] = 0, \quad 0 \leqslant z < +\infty,$$

满足 n 个多点非线性边界条件

$$\mathcal{B}_k[z,u,\lambda] = \gamma_k, \quad 1 \leqslant k \leqslant n,$$

其中 $\mathcal{B}_k$ 是非线性算子, γ_k 是常数. 假设上述方程至少存在一个光滑解. 通常, 此类问题会提供 Mathematica 软件包.

9.4 无限区间上耦合的非线性特征值问题

开发一种基于同伦分析方法的解析方法, 用于无限区间 $z \in [0, +\infty)$ 上 n 个耦合的非线性特征方程

$$\mathcal{F}_k[z, u, \lambda_1, \lambda_2, \cdots, \lambda_n] = 0, \quad 1 \leqslant k \leqslant n,$$

满足一些线性 (非线性) 多边界条件, 其中 $n \geqslant 2$. 假设上述方程至少存在一个光滑解. 通常, 此类问题会提供 Mathematica 软件包.

参考文献

[118] Abbasbandy, S., Shirzadi, A.: A new application of the homotopy analysis method: Solving the Sturm-Liouville problems. Commun. Nonlinear Sci. Numer. Simulat. **16**, 112–126 (2011).

[119] Abell, M.L., Braselton, J.P.: Mathematica by Example (3rd Edition). Elsevier Academic Press. Amsterdam (2004).

[120] Boley, A.B.: On the accuracy of the Bernoulli-Euler theory for beams of variable section. J. Appl. Mech. ASME **30**, 373–378 (1963).

[121] Boyd, J.P.: An analytical and numerical Study of the two-dimensional Bratu equation. Journal of Scientific Computing. **1**, 183–206 (1986).

[122] Boyd, J.P.: Chebyshev and Fourier Spectral Methods. DOVER Publications, Inc. New York (2000).

[123] Chang, D., Popplewell, N.: A non-uniform, axially loaded Euler-Bernoulli beam having complex ends. Q.J. Mech. Appl. Math. **49**, 353–371 (1996).

[124] Graef, J.R., Qian. C. Yang, B.: A three point boundary value problem for nonlinear forth order differential equations. J. Mathematical Analysis and Applications. **287**, 217–233 (2003).

[125] Graef, J.R., Qian. C. Yang, B.: Multiple positive solutions of a boundary value prolem for ordinary differential equations. Electronic J. of Qualitative Theory of Differential Equations. **11**, 1–13 (2004).

[126] Jacobsen, J., Schmitt, K.: The Liouville-Bratu-Gelfand problem for radial operators. Journal of Differential Equations. **184**, 283–298 (2002).

[127] Katsikadelis, J.T., Tsiatas, G.C.: Non-linear dynamic analysis of beams with variable stiffness. J. Sound and Vibration **270**, 847–863 (2004).

[128] Lee, B.K., Wilson, J.F., Oh, S.J.: Elastica of cantilevered beams with variable cross sections. Int. J. Non-linear Mech. **28**, 579–589 (1993).

[129] Liao, S.J.: The Proposed Homotopy Analysis Technique for the Solution of Nonlinear Problems. PhD dissertation, Shanghai Jiao Tong University (1992).

[130] Liao, S.J.: A kind of approximate solution technique which does not depend upon small parameters (II) –An application in fluid mechanics. Int. J. Nonlin. Mech. **32**, 815–822 (1997).

[131] Liao, S.J.: An explicit, totally analytic approximation of Blasius viscous flow problems. Int. J. Nonlin. Mech. **34**, 759–778 (1999a).

[132] Liao, S.J.: A uniformly valid analytic solution of 2D viscous flow past a semi-infinite flat plate. J. Fluid Mech. **385**, 101–128 (1999b).

[133] Liao, S.J.: On the analytic solution of magnetohydrodynamic flows of non-Newtonian fluids over a stretching sheet. J. Fluid Mech. **488**, 189–212 (2003a).

[134] Liao, S.J.: Beyond Perturbation—Introduction to the Homotopy Analysis Method. Chapman & Hall/ CRC Press, Boca Raton (2003b).

[135] Liao, S.J.: On the homotopy analysis method for nonlinear problems. Appl. Math. Comput. **147**, 499–513 (2004).

[136] Liao, S.J.: A new branch of solutions of boundary-layer flows over an impermeable stretched plate. Int. J. Heat Mass Tran. **48**, 2529–2539 (2005).

[137] Liao, S.J.: Series solutions of unsteady boundary-layer flows over a stretching flat plate. Stud. Appl. Math. **117**, 2529–2539 (2006).

[138] Liao, S.J.: Notes on the homotopy analysis method—Some definitions and theorems. Commun. Nonlinear Sci. Numer. Simulat. **14**, 983–997 (2009a).

[139] Liao, S.J.: Series solution of deformation of a beam with arbitrary cross section under an axial load. ANZIAM J. **51**, 10–33 (2009b).

[140] Liao, S.J.: On the relationship between the homotopy analysis method and Euler transform. Commun. Nonlinear Sci. Numer. Simulat. **15**, 1421–1431 (2010a). doi:10.1016/j.cnsns.2009.06.008.

[141] Liao, S.J.: An optimal homotopy-analysis approach for strongly nonlinear differential equations. Commun. Nonlinear Sci. Numer. Simulat. **15**, 2003–2016 (2010b).

[142] Liao, S.J., Campo, A.: Analytic solutions of the temperature distribution in Blasius viscous flow problems. J. Fluid Mech. **453**, 411–425 (2002).

[143] Liao, S.J., Tan, Y.: A general approach to obtain series solutions of nonlinear differential equations. Stud. Appl. Math. **119**, 297–355 (2007).

[144] Lin, C.C.: The Theory of Hydrodynamics Stability. Cambridge University Press. Cambridge (1955).

[145] Mason, J.C., Handscomb, D.C.: Chebyshev Polynomials. Chapman & Hall/CRC Press, Boca Raton (2003).

[146] McGough, J.S.: Numerical continuation and the Gelfand problem. Applied Mathematics and Computation. **89**, 225–239 (1998).

[147] Orszag, S.A.: Accurate solution of the Orr-Sommerfeld stability equation. J. Fluid Mech. **50**, 689–703 (1971).

[148] Shampine, L.F., Gladwell, I., Thompson, S.: Solving ODEs with MATLAB. Cambridge University Press. Cambridge (2003).

[149] Trefethen, L.N.: Computing numerically with functions instead of numbers. Math. in Comp. Sci. **1**, 9–19 (2007).

[150] Xu, H., Lin, Z.L., Liao, S.J., Wu, J.Z., Majdalani, J.: Homotopy-based solutions of the Navier-Stokes equations for a porous channel with orthogonally moving walls. Physics of Fluids. **22**, 053601 (2010). doi:10.1063/1.3392770.

附录 9.1 例 9.3.1 的 BVPh 输入数据

```
(*Input Mathematica package BVPh version 1.0 *)
<<BVPh.txt;

(* Define the physical and control parameters *)

TypeEQ       = 2;
TypeL        = 2;
TypeBase     = 2;
ApproxQ      = 1;
Ntruncated   = 20;

(* Define the governing equation *)
mu[z_] := 1;
f[z_,u_,lambda_] := mu[z]*D[u,{z,2}]
         + D[mu[z],z]*D[u,z]+lambda*Sin[u];

(* Define Boundary conditions *)
zR      = Pi;
OrderEQ = 2;
BC[0,z_,u_,lambda_] := Limit[u - gamma, z -> 0];
BC[1,z_,u_,lambda_] := Limit[D[u,z], z -> 0];
BC[2,z_,u_,lambda_] := D[u,z] /. z -> Pi;

(* Define initial guess *)
u[0]  = sigma - (sigma - gamma)*Cos[kappa*z];
kappa = 1;
sigma = Pi;
gamma = 1;
```

```
(* Define output term *)
output[z_,u_,k_]:= Print["output              = ",
                D[u[k],z] /. z->0//N];

(* Define the auxiliary linear operator *)
omega[1] = Pi/zR;
L[f_] := Module[{temp,numA,numB,i},
If[TypeL == 1,
   temp[1] = D[f,{z,OrderEQ}],
   numA = IntegerPart[OrderEQ/2];
   numB = OrderEQ - 2* numA//Expand;
   temp[0] = D[f,{z,numB}];
   For[i=1, i<=numA, i++,
      temp[1] = D[temp[0],{z,2}]
                + (kappa*omega[i])^2*temp[0];
      temp[0] = temp[1];
      ];
   ];
temp[1]//Expand
];

(* Print input and control parameters *)
PrintInput[u[z]];

(* Use 3rd-order iteration approach *)
iter[1,6,3];
```

注意, 对于在有限区间上定义的非线性边值 (特征值) 问题, BVPh 1.0 默认 zL=0.

附录 9.2 例 9.3.2 的 BVPh 输入数据

```
(* Input Mathematica package BVPh version 1.0 *)
<<BVPh.txt;

(* Define the physical and control parameters *)
TypeEQ       =   2;
TypeL        =   2;
TypeBase     =   2;
```

```
ApproxQ      =   1;
Ntruncated   =   20;

(* Define the governing equation *)
f[z_,u_,lambda_] := D[u,{z,2}]
         + (K-1)*D[u,z]/z + lambda * Exp[u]  ;
K = 1;

(* Define Boundary conditions *)
zR       = 1;
OrderEQ  = 2;
BC[0,z_,u_,lambda_] := Limit[D[u,z] - A, z -> 0];
BC[1,z_,u_,lambda_] := Limit[u, z -> 0];
BC[2,z_,u_,lambda_] := u /.  z -> 1;
A = 1;

(* Define initial guess *)
u[0]  =  A/2*(1 + Cos[Pi*z]);

(* Define output term *)
output[z_,u_,k_]:= Print["output = ",
               D[u[k],z] /. z->0//N];

(* Define the auxiliary linear operator *)
omega[1] = Pi/zR;
L[f_] := Module[{temp,numA,numB,i},
If[TypeL == 1,
   temp[1] = D[f,{z,OrderEQ}],
   numA = IntegerPart[OrderEQ/2];
   numB = OrderEQ - 2* numA//Expand;
   temp[0] = D[f,{z,numB}];
   For[i=1, i<=numA, i++,
      temp[1] = D[temp[0],{z,2}]
              + (kappa*omega[i])^2*temp[0];
      temp[0] = temp[1];
      ];
   ];
temp[1]//Expand
];
```

```
(* Print input and control parameters *)
PrintInput[u[z]];

(* Use 3rd-order iteration approach *)
iter[1,6,3];
```

注意，对于在有限区间上定义的非线性边值(特征值)问题，BVPh 1.0 默认 zL=0.

附录 9.3 例 9.3.3 的 BVPh 输入数据

```
(* Input Mathematica package BVPh version 1.0 *)
<<BVPh.txt;

(* Define the physical and control parameters *)
TypeEQ       = 2;
TypeL        = 2;
TypeBase     = 2;
ApproxQ      = 1;
Ntruncated   = 30;

(* Define the governing equation *)
L0[z_, u_]:= Sqrt[1+z^2]*D[u,{z,2}]
             + Cos[Pi*z]*D[u,z]/z;
F[z_,u_]  := Exp[u]/(1+z^2) + (1+z)*Sin[u];
g[z_]     := Sin[z^2 + Exp[-z]];
f[z_,u_,lambda_]:= L0[z,u] + lambda*F[z,u] - g[z];

(* Define Boundary conditions *)
zR       = Pi;
OrderEQ  = 2;
BC[0,z_,u_,lambda_] := Limit[u - A, z->0];
BC[1,z_,u_,lambda_] := Limit[D[u,z], z->0];
BC[2,z_,u_,lambda_] := u-D[u,z] - 3/5 /. z->zR;
A =1/2;

(* Define initial guess *)
u[0] = (5*A+3)/10 + (5*A-3)/10*Cos[kappa*z];
```

```
kappa = 1;

(* Define output term *)
output[z_,u_,k_] := Print["output             = ",
                    u[k] /. z -> zR //N ];

(* Define the auxiliary linear operator *)
omega[1] = Pi/zR;
L[f_] := Module[{temp,numA,numB,i},
If[TypeL == 1,
   temp[1] = D[f,{z,OrderEQ}],
   numA = IntegerPart[OrderEQ/2];
   numB = OrderEQ - 2* numA//Expand;
   temp[0] = D[f,{z,numB}];
   For[i=1, i<=numA, i++,
      temp[1] = D[temp[0],{z,2}]
                 + (kappa*omega[i])^2*temp[0];
      temp[0] = temp[1];
      ];
   ];
temp[1]//Expand
];

(* Print input and control parameters *)
PrintInput[u[z]];

(* Gain approximations by 3rd-order iteration *)
iter[1,6,3];
```

注意,对于在有限区间上定义的非线性边值(特征值)问题,BVPh 1.0 默认 zL=0。

附录 9.4 例 9.3.4 的 BVPh 输入数据

```
(* Input Mathematica package BVPh version 1.0 *)
<<BVPh.txt;

(* Define the physical and control parameters *)
TypeEQ      = 2;
```

```
TypeL        = 1;
TypeBase     = 2;
ApproxQ      = 0;
Ntruncated   = 20;

(* Define the governing equation *)
f[z_,u_,lambda_] := lambda*D[u,{z,4}]-beta*z
                    beta*z*u^2*lambda^2;
beta = 10;

(* Define Boundary conditions *)
zR       = 1;
OrderEQ  = 4;
BC[0,z_,u_,lambda_] := u - 1 /. z -> 1;
BC[1,z_,u_,lambda_] := Limit[u, z -> 0 ];
BC[2,z_,u_,lambda_] := D[u,z] /. z -> 1;
BC[3,z_,u_,lambda_] := D[u,{z,2}] /. z -> 1;
BC[4,z_,u_,lambda_] := Module[{temp},
                 temp[1] = D[u,{z,2}] /. z -> 0;
                 temp[2] = D[u,{z,2}] /. z -> alpha;
                 temp[1]-temp[2]//Expand
                 ];
alpha = 1/5;

(* Define initial guess *)
u[0] = sigma/(2*alpha-3)*((6*alpha-8)*z
    + 6*(1-alpha)*z^2+2*alpha*z^3-z^4);
sigma = 1;

(* Define output term *)
output[z_,u_,k_]:= Print["output            = ",
               D[u[k],z] /. z->0//N]

(* Define the auxiliary linear operator *)
omega[1] = Pi/zR;
omega[2] = Pi/zR;
L[f_] := Module[{temp,numA,numB,i},
If[TypeL == 1,
    temp[1] = D[f,{z,OrderEQ}],
```

```
        numA = IntegerPart[OrderEQ/2];
        numB = OrderEQ - 2* numA//Expand;
        temp[0] = D[f,{z,numB}];
        For[i=1, i<=numA, i++,
            temp[1] = D[temp[0],{z,2}]
                     + (kappa*omega[i])^2*temp[0];
            temp[0] = temp[1];
            ];
        ];
temp[1]//Expand
];


(* Print input and control parameters *)
PrintInput[u[z]];


(* Gain HAM approx. by 3rd-order iteration *)
iter[1,6,3];
```

注意, 对于在有限区间上定义的非线性边值 (特征值) 问题, BVPh 1.0 默认 zL=0.

附录 9.5 例 9.3.5 的 BVPh 输入数据

```
(* Input Mathematica package BVPh version 1.0 *)
<<BVPh.txt;

(* Define the physical and control parameters *)
TypeEQ      = 2;
TypeL       = 1;
TypeBase    = 2;
ApproxQ     = 0;
ErrReq      = 10^(-20);
Nupdate     = 10000;
NtermMax    = 90;
ComplexQ    = 1;

(* Define the governing equation *)
L0[u_,z_]  := D[u,{z,2}]-alpha^2*u;
L02[u_,z_] := L0[L0[u,z],z];
```

```
U0   =  1 - z^2;
U02  =  D[U0,{z,2}];
f[z_,u_,lambda_] := L02[u,z]
              - I*alpha*R*(U0*L0[u,z]-U02*u)
              + I*lambda*R*L0[u,z];
alpha = 1 ;
R     = 100 ;

(* Define Boundary conditions *)
zR        = 1;
OrderEQ   = 4;
BC[0,z_,u_,lambda_] := Limit[u - 1 , z->0 ];
BC[1,z_,u_,lambda_] := Limit[D[u,z], z->0 ];
BC[2,z_,u_,lambda_] := Limit[D[u,{z,3}], z->0 ];
BC[3,z_,u_,lambda_] := u /. z->zR;
BC[4,z_,u_,lambda_] := D[u,z] /. z->zR;

(* Define initial guess *)
U[0] = u[0];
u[0] = ( 1 - z^2 )^2;

(* Define the auxiliary linear operator *)
L[f_] := D[f,{z,4}];

(* Define output term *)
output[z_,u_,k_]:= Print["output              = ",
                D[u[k],{z,2}] /. z->0//N];

(* Print input and control parameters *)
PrintInput[u[z]];

(* Set convergence-control parameter *)
c0 = (-1+I)/2;

(* Gain HAM approx. by 3rd-order iteration *)
iter[1,31,3];
```

注意，对于在有限区间上定义的非线性边值(特征值)问题，BVPh 1.0 默认 zL=0。

第十章　具有无穷多解的边界层流动

本章运用基于同伦分析方法的 Mathematica 软件包 BVPh 1.0 得到了一个非线性边值方程在无限区间内呈指数和代数衰减的解. 特别是本章应用同伦分析方法首次发现了无穷多个呈代数衰减的解, 这说明了同伦分析方法在求解非线性边值问题上的原创性和有效性.

10.1　绪论

本章阐述了 Mathematica 软件包 BVPh 1.0 对于求解无限区间内非线性边值问题的有效性. 该问题由一个 n 阶非线性常微分方程

$$\mathcal{F}[z, u] = 0, \quad 0 \leqslant z < +\infty \tag{10.1}$$

控制, 满足一些线性边界条件, 其中 $\mathcal{F}$ 表示非线性算子, $u(z)$ 表示一个光滑解. 假设 $u(z)$ 随着 $z \to +\infty$ 呈指数或代数形式衰减.

第八章和第九章说明了 BVPh 1.0 为求解有限区间 $z \in [0, a]$ 上具有奇点和多点边界条件的 n 阶强非线性边值 (特征值) 问题的多解提供了一个有效的工具. 本章将进一步说明 BVPh 1.0 对于求解无限区间内的这类非线性常微分方程的有效性.

2005 年, 廖世俊 [160] 成功地应用同伦分析方法 [152–165, 168, 171] 求解了无限区间内的非线性边值方程

$$F''' + \frac{1}{2} F\, F'' - \beta F'^2 = 0, \quad F(0) = 0,\ F'(0) = 1,\ F'(+\infty) = 0,$$

其中 $-1 < \beta < +\infty$ 为一个常数. 引入未知量 $\delta = F(+\infty)$, 当 $\beta > 1$ 时, 廖世俊 [160] 发现了呈指数衰减的解的一个新分支, 该分支从来没有被其他解析方法甚至数值方法得到过, 这主要是因为呈指数衰减的解的两个分支的 $F''(0)$ 值之间的差异非常小, 很难进行区分. 此外, 基于同伦分析方法, 廖世俊和 Magyari [167] 于 2006 年发现了一种边界层流的呈代数衰减的解的新分支. 确实, 一种全新的方法总会带来一些新的和 (或) 不同的东西, 所有这些都说明了同伦分析方法的原创性和有效性. 此外, 廖世俊 [160], 廖世俊和 Magyari [167] 还说明了同伦分析方法可用于求解无限区间内呈指数或代数衰减的解的非线性边界值问题.

不失一般性, 我们在这里考虑在 $x > 0$ 和 $y > 0$ 区域内的二维边界层黏性流动, 其中 (x, y) 表示笛卡儿坐标系, 且该流动完全是由于不渗透平板在其平面内 $y = 0$ 处的运动而产生的. 令 $U_w(x) = a(x + b)^\kappa$ 表示平板的运动速度, 其中 $a > 0, b > 0$ 为给定常数. 假设边界层方程是适用的, 则该流动可以用以下偏微分方程

$$u\frac{\partial u}{\partial x} + v\frac{\partial u}{\partial y} = \nu \frac{\partial^2 u}{\partial y^2},$$

$$\frac{\partial u}{\partial x} + \frac{\partial v}{\partial y} = 0$$

来描述, 满足边界条件

$$u = a(x+b)^\kappa, \quad v = 0, \quad y = 0$$

以及

$$u = 0, \quad y \to +\infty,$$

其中 ν 为运动黏度, u, v 分别表示沿 x, y 正方向的速度分量.

令 ψ 表示流函数, 利用相似变换

$$\psi = \sqrt{a\nu}(x+b)^{(\kappa+1)/2} f(\eta), \quad \eta = \sqrt{\frac{a}{\nu}}\, y\, (x+b)^{(\kappa-1)/2}, \qquad (10.2)$$

原始的偏微分方程变成以下非线性常微分方程

$$f''' + \frac{1+\kappa}{2} f f'' - \kappa f'^2 = 0, \quad f(0) = 0,\ f'(0) = 1,\ f'(+\infty) = 0. \qquad (10.3)$$

详情参见 Banks [151]. 上述方程分别具有封闭解

$$f(\eta) = 1 - \exp(-\eta), \quad \kappa = 1 \qquad (10.4)$$

和
$$f(\eta) = \sqrt{6}\,\tanh\left(\frac{\eta}{\sqrt{6}}\right), \quad \kappa = -1/3. \tag{10.5}$$

除此之外,当 $-1/2 < \kappa < +\infty$ 时也存在解.

在无限区间 $z \in [0,+\infty)$ 上非线性边值方程 (10.3) 有两种类型的解:一种在无穷远处呈指数衰减 [166], 另一种则呈代数衰减 [167]. 方程 (10.3) 的这两种类型的解可以通过 BVPh 1.0 得到, 如下所述.

10.2 呈指数衰减的解

在物理上, 大多数边界层流动都在无穷远处呈指数趋向于均匀流动. 在数学上, 该结论被封闭解 (10.4) 和 (10.5) 所证实. 这种呈指数衰减的解可以通过同伦分析方法得到, 如廖世俊和 Pop [166] 的论文所述. 这里, 我们基于 BVPh 1.0 求解, 求解过程见附录 7.1, 相应的输入数据文件见附录 10.1.

为了简单起见, 定义非线性算子

$$\mathcal{N}[\phi(\eta;q)] = \phi''' + \frac{1+\kappa}{2}\,\phi\,\phi'' - \kappa\,(\phi')^2. \tag{10.6}$$

在同伦分析方法的框架下, 构造如下零阶形变方程

$$(1-q)\mathcal{L}\left[\phi(\eta;q) - f_0(\eta)\right] = q\,c_0\,\mathcal{N}\left[\phi(\eta;q)\right], \tag{10.7}$$

满足边界条件

$$\phi(0;q) = 0, \quad \phi'(0;q) = 0, \quad \phi'(+\infty;q) = 0, \tag{10.8}$$

其中 ′ 表示对 η 求导, $f_0(\eta)$ 为初始猜测解, $\mathcal{L}$ 为辅助线性算子, c_0 为收敛控制参数. 以上方程定义了一种从初始猜测解 $f_0(\eta)$ ($q = 0$) 变化到原始方程 (10.3) ($q = 1$) 之解 $f(\eta)$ 的连续变化 $\phi(\eta;q)$. $\phi(\eta;q)$ 的同伦–Maclaurin 级数可表示为

$$\phi(\eta;q) = f_0(\eta) + \sum_{n=1}^{+\infty} f_n(\eta)\,q^n,$$

其中

$$f_n(\eta) = \mathcal{D}_n\left[\phi(\eta;q)\right] = \frac{1}{n!}\left.\frac{\partial^n \phi(\eta;q)}{\partial q^n}\right|_{q=0} \tag{10.9}$$

以及 $\mathcal{D}_n$ 被称为 n 阶同伦导数算子.

假设初始猜测解 $f_0(\eta)$、辅助线性算子 $\mathcal{L}$, 特别是收敛控制参数 c_0 选取合适, 使得上述同伦 – Maclaurin 级数在 $q = 1$ 时绝对收敛. 那么, 同伦级数解为

$$f(\eta) = f_0(\eta) + \sum_{n=1}^{+\infty} f_n(\eta), \tag{10.10}$$

其中 $f_n(\eta)$ 由 n 阶形变方程

$$\mathcal{L}[f_n(\eta) - \chi_n f_{n-1}(\eta)] = c_0\, \delta_{n-1}(\eta) \tag{10.11}$$

控制, 且满足下列边界条件

$$f_n(0) = 0, \quad f_n'(0) = 0, \quad f_n'(+\infty) = 0, \tag{10.12}$$

其中

$$\chi_m = \begin{cases} 0, & m \leqslant 1, \\ 1, & m > 1, \end{cases}$$

以及

$$\delta_m(\eta) = \mathcal{D}_m\{\mathcal{N}[\phi(\eta;q)]\} = f_m''' + \frac{1+\kappa}{2} \sum_{i=0}^{m} f_{m-i} f_i'' - \kappa \sum_{i=0}^{m} f_{m-i}' f_i' \tag{10.13}$$

由定理 4.1 得到. 方程 (10.11) 由定理 4.15 给出. 详情请参考第四章.

方程 (10.3) 中呈指数衰减的解, 可表示为

$$f(\eta) = \sum_{n=0}^{+\infty} A_n(\eta) \exp(-n\eta), \tag{10.14}$$

其中 $A_n(\eta)$ 为待定多项式. 这为我们提供了 $f(\eta)$ 的解表达. 为了通过 BVPh 1.0 获得这种呈指数衰减的解, 我们分别选取初始猜测解

$$f_0(\eta) = \sigma + (1 - 2\sigma)\, e^{-\eta} - (1 - \sigma) e^{-2\eta} \tag{10.15}$$

和辅助线性算子

$$\mathcal{L}(f) = f''' - f', \tag{10.16}$$

其中 $\sigma = f(+\infty)$ 为未知参数. 值得注意的是, 初始猜测解 (10.15) 满足所有的边界条件 (10.3) 且 $f_0(+\infty) = \sigma$. 显然, 一个更好的 σ 值对应于一个更好的初

始猜测解. 因此, 未知参数 σ 为我们选取初始猜测解 $f_0(\eta)$ 提供了另一个自由度. 注意, 辅助线性算子 (10.16) 具有如下性质

$$\mathcal{L}[B_0 + B_1 \exp(-\eta) + B_2 \exp(\eta)] = 0. \tag{10.17}$$

此外, 对于任意一个在无穷远处呈指数衰减的给定函数 $f^*(\eta)$, 函数

$$f^*(\eta) + B_0 + B_1 \exp(-\eta) + B_2 \exp(\eta)$$

中的未知积分常数 B_0, B_1, B_2 可由方程 (10.3) 中的三个边界条件唯一确定. 换句话说, 辅助线性算子 $\mathcal{L}$ 通过以下方式进行选取: 对于呈指数衰减的解, 解 $f(\eta)$ 以 (10.14) 的形式表示, 并且所有积分常数都是唯一确定的. 在一般情况下, 对于初始猜测解和辅助线性算子 $\mathcal{L}$ 的选取, 请参考 §7.1.3 和 §7.1.4.

令

$$f_n^*(\eta) = \chi_n f_{n-1}(\eta) + c_0 \mathcal{L}^{-1}[\delta_{n-1}(\eta)]$$

表示 (10.11) 的一个特解. 通解可表示为

$$f_n(\eta) = \chi_n f_{n-1}(\eta) + c_0 \mathcal{L}^{-1}[\delta_{n-1}(\eta)] \\ + B_0 + B_1 \exp(-\eta) + B_2 \exp(\eta), \tag{10.18}$$

其中积分常数 B_0, B_1 和 B_2 可由方程 (10.12) 中的三个边界条件唯一确定. 详情请参考廖世俊和 Pop [166].

我们先考虑当 $\kappa = -1/3$ 的情况. 相应的同伦近似解由 BVPh 1.0 获得, 其中包含两个未知参数: 初始猜测解 (10.15) 中的 σ 和收敛控制参数 c_0. 基于 Mathematica 中的命令 Minimize, 我们发现对于 5 阶同伦近似, (10.3) 在区间 $\eta \in [0, 10]$ 上的平均平方残差的最小值为 6.8×10^{-6}, 对应于最优收敛控制参数 $c_0^* = -1.2411$ 和最优参数 $\sigma^* = 3.0435$. 实际上, 选取 $c_0 = -5/4$ 和 $\sigma = 3$, (10.3) 在区间 $\eta \in [0, 10]$ 上的平均平方残差也迅速变小, 如表 10.1 所示. 值得注意的是, $f''(0)$ 迅速收敛到精确值 $f''(0) = 0$, 并且 $f(\eta)$ 的 10 阶同伦近似解与精确解吻合良好, 如图 10.1 所示.

需要注意的是, 初始猜测解 (10.15) 中的未知参数 σ 在此用于搜索最优初始猜测解, 可以将其看作一种收敛控制参数. 因此, 这里有两个收敛控制参数: 零阶形变方程 (10.7) 中的 c_0 和初始猜测解 (10.15) 中的 σ. 值得注意的是, 第八章和第九章中, 初始猜测解中的未知参数 σ 被当作所谓的多解控制参数用于寻找有限区间 $z \in [0, a]$ 上非线性常微分方程的多解. 因此, 初始猜测解中的未知参数既可以当作多解控制参数来找出多解, 也可以当作收敛控制参数来控制

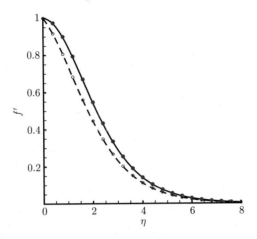

图 10.1 方程 (10.3) 呈指数衰减的解 $f'(\eta)$. 实心圆: 当 $\kappa = -1/3$ 时, 选取 $c_0 = -5/4$ 和 $\sigma = 3$ 的 10 阶同伦近似; 实线: 当 $\kappa = -1/3$ 时的精确解; 空心圆: 当 $\kappa = -1/4$ 时, 选取 $c_0 = -5/4$ 和 $\sigma = 11/4$ 的 10 阶同伦近似; 虚线: 当 $\kappa = -1/4$ 时, 选取 $c_0 = -5/4$ 和 $\sigma = 11/4$ 的 30 阶同伦近似

表 10.1 当 $\kappa = -1/3$ 时, 选取 $c_0 = -5/4$ 和 $\sigma = 3$, m 阶 (呈指数衰减) 同伦近似对应的 (10.3) 在区间 $\eta \in [0, 10]$ 上的平均平方残差

近似阶数 m	$f''(0)$	平均平方残差 E_m
2	7.0×10^{-2}	5.2×10^{-3}
4	-2.1×10^{-2}	3.7×10^{-5}
6	-1.4×10^{-2}	3.8×10^{-6}
8	-8.9×10^{-3}	1.8×10^{-6}
10	-4.8×10^{-3}	6.8×10^{-7}
15	-6.6×10^{-4}	3.9×10^{-8}
20	8.2×10^{-5}	1.4×10^{-9}

同伦级数解的收敛. 这主要是因为, 基于同伦分析方法, BVPh 1.0 为我们选择初始猜测解提供了极大的自由.

类似地, 基于 BVPh 1.0, 我们可以得到 $-1/2 < \kappa < +\infty$ 时 (10.3) 的精确同伦近似解 $f(\eta)$. 例如, 当 $\kappa = -1/4$ 时, 对于 5 阶同伦近似, (10.3) 在区间 $\eta \in [0, 10]$ 上的平均平方残差的最小值为 6.5×10^{-6}, 对应于两个最优的收敛控

制参数 $c_0^* = -1.2133$ 和 $\sigma^* = 2.7369$. 通过选取 $c_0 = -5/4$ 和 $\sigma = 11/4$, 我们确实得到了精确的同伦级数解, 其 $f''(0)$ 收敛至 -0.1620, 如表 10.2 所示. 此外, $f'(\eta)$ 的 10 阶近似更加精确, 如图 10.1 所示. 因此, 基于 BVPh 1.0, 我们成功地得到了 (10.3) 呈指数衰减的解. 为了简单起见, 在下一节中, 我们用 $f_{\exp}(\eta)$ 表示这种呈指数衰减的解.

表 10.2 当 $\kappa = -1/4$ 时, 选取 $c_0 = -5/4$ 和 $\sigma = 11/4$, m 阶 (呈指数衰减) 同伦近似对应的 (10.3) 在区间 $\eta \in [0, 10]$ 上的平均平方残差

近似阶数 m	$f''(0)$	平均平方残差 E_m
5	-0.1692	2.9×10^{-6}
10	-0.1639	1.4×10^{-7}
15	-0.1623	4.5×10^{-9}
20	-0.1620	8.7×10^{-11}
25	-0.1620	7.8×10^{-12}
30	-0.1620	1.1×10^{-12}

还要注意的是, 运用变换

$$f(\eta) = \sqrt{\frac{2}{\kappa+1}}\, F(z), \quad z = \sqrt{\frac{\kappa+1}{2}}\, \eta,$$

方程 (10.3) 变为

$$F''' + F F'' - \varepsilon F'^2 = 0,\ F(0) = 0,\ F'(0) = 1,\ F'(+\infty) = 0,$$

其中 $\varepsilon = 2\kappa/(1+\kappa)$, ′ 表示对 z 求导. 基于同伦分析方法, 廖世俊和 Pop [166] 得到了 3 阶近似

$$F''(0) = -\frac{145293 + 231153\varepsilon + 94999\varepsilon^2 + 12395\varepsilon^3}{15120(3+\varepsilon)^{5/2}}, \tag{10.19}$$

其与整个区间 $0 \leqslant \varepsilon < +\infty$ 上的数值结果吻合良好. 详情请参考廖世俊和 Pop [166]. 因此, 对于 (10.3) 呈指数衰减的解 $f_{\exp}(\eta)$, 我们有如下 3 阶同伦近似

$$f''_{\exp}(0) = \sqrt{\frac{\kappa+1}{2}} F''(0)$$

$$= -\frac{145293 + 898185\kappa + 1740487\kappa^2 + 1086755\kappa^3}{15120\sqrt{2}(3+5\kappa)^{5/2}}, \tag{10.20}$$

其在无限区间 $\kappa \in (0, +\infty)$ 上是有效的.

这个例子验证了 BVPh 1.0 对于无限区间内解在无穷远处呈指数衰减的非线性边值常微分方程的有效性.

10.3 呈代数衰减的解

正如 Magyari, Pop 和 Keller [169] 所指出的那样, 当 $\kappa = -1/3$ 时, (10.3) 具有无穷多个封闭解

$$f(\eta) = (36\mu)^{1/3}\left[\frac{Bi'(t_0)Ai'(t) - Ai'(t_0)Bi'(t)}{Bi'(t_0)Ai(t) - Ai'(t_0)Bi(t)}\right], \tag{10.21}$$

其中 $Ai(t)$ 和 $Bi(t)$ 为两种 Airy 函数,

$$t_0 = \left(\sqrt{6}\,\mu\right)^{-2/3}, \quad t = t_0\,(1 + \mu\,\eta),$$

以及

$$\mu = f''(0) \geqslant f''_{\exp}(0), \tag{10.22}$$

这里, $f_{\exp}(\eta)$ 表示 (10.3) 在无穷远处呈指数衰减的解, 这在上一节中提到过. 因为当 $\kappa = -1/3$ 时 $f''_{\exp}(0) = 0$, 则 $\mu = f''(0) \geqslant 0$ 取任意值, 解 (10.21) 都满足 (10.3) ($\kappa = -1/3$). 文献 [169] 中证明了当 $\mu = f''(0)$ 趋近于 0 时, 解 (10.21) 等同于封闭解 (10.5), 但对于 $\mu > 0$ 的其他任何值, 解 (10.21) 在无穷远处呈代数衰减. 换句话说, 当 $\kappa = -1/3$ 时, 呈指数衰减的解

$$f_{\exp}(\eta) = \sqrt{6}\tanh(\eta/\sqrt{6})$$

为呈代数衰减的解 (10.21) 的极限情况, 此时 $f''(0) \to f''_{\exp}(0) = 0$.

BVPh 1.0 也适用于此类呈代数衰减的边界层流动. 基于 BVPh 1.0, 我们发现不仅在 $\kappa = -1/3$ 处而且在整个区间 $-1/2 < \kappa < 0$ (10.3) 上确实存在无穷多个呈代数衰减的解, 如下所述.

我们先考虑在无穷远处呈代数衰减的解 $f(\eta)$ 的渐近性质. 将渐近表达式写为

$$f' \sim \eta^b, \ \text{即}\ f \sim \frac{1}{b+1}\eta^{b+1}, \quad \eta \to +\infty,$$

其中 b 为待定常数. 将这些渐近表达式代入 (10.3) 并平衡主要项, 我们有

$$b = \frac{2\kappa}{1-\kappa}, \quad \text{即} \quad 1+b = \frac{1+\kappa}{1-\kappa} = \beta.$$

显然, 为了满足边界条件 $f'(+\infty) = 0$, b 必须为负数, 对应于 $-1/2 < \kappa < 0$, 由此得到

$$\frac{1}{3} < \beta < 1.$$

因此, 该呈代数衰减的解具有以下渐近性质

$$f \sim \eta^\beta, \quad \frac{1}{3} < \beta < 1, \quad \eta \to +\infty.$$

为了避免上述渐近表达式在 $\eta = 0$ 时的奇异性, 我们使用以下变换

$$\xi = 1 + \alpha\,\eta, \quad f(\eta) = \alpha^{-1}\,g(\xi),$$

其中 $\alpha > 0$ 为常数, 则原始方程 (10.3) 变为

$$\alpha^2\,g'''(\xi) + \frac{1+\kappa}{2} g\,g'' - \kappa(g')^2 = 0, \quad g(1) = 0,\ g'(1) = 1,\ g'(+\infty) = 0, \tag{10.23}$$

其中 ′ 表示对 ξ 求导. 由于 (10.3) 具有无穷多解, 取决于 $\mu = f''(0) > f''_{\exp}(0)$, 因此我们应添加一个附加的边界条件 $f''(0) = \mu$, 即

$$g''(1) = \frac{\mu}{\alpha}, \tag{10.24}$$

以便区分这些不同的呈代数衰减的解. 考虑到上面提到的渐近性质, 我们的目标是获得如下形式的呈代数衰减的解

$$g(\xi) = a_{0,0}\,\xi^\beta + \sum_{m=0}^{+\infty}\sum_{n=0}^{+\infty} a_{m,n}\,\xi^{-(1-\beta)m-n}, \tag{10.25}$$

这为我们提供了呈代数衰减的解 $g(\xi)$ 的解析表达.

这种非线性边值常微分方程 (10.23) 的呈代数衰减的解, 在附加的边界条件 $g''(1) = \mu/\alpha$ 下, 可通过 BVPh 1.0 得到, 如下所示. 相应的输入数据文件见附录 10.2.

定义一个非线性算子

$$\check{\mathcal{N}}[u] = \alpha^2\,u''' + \frac{1+\kappa}{2} u\,u'' - \kappa(u')^2,$$

其中 ′ 表示对 ξ 求导. 令 $q \in [0,1]$ 表示嵌入变量. 在同伦分析方法的框架下, 我们先构造一种连续的变化 (或形变) $\check{\phi}(\xi;q)$. 当 $q=0$ 时, $\check{\phi}(\xi;q) = g_0(\xi)$, 当 $q=1$ 时, $\check{\phi}(\xi;q) = g(\xi)$. 这种连续变化由零阶形变方程

$$(1-q)\check{\mathcal{L}}\left[\check{\phi}(\xi;q) - g_0(\xi)\right] = c_0\, q\, \check{H}(\xi)\, \check{\mathcal{N}}\left[\check{\phi}(\xi;q)\right] \tag{10.26}$$

定义, 满足边界条件

$$\check{\phi}(\xi;q) = 0, \quad \check{\phi}'(\xi;q) = 1, \quad \check{\phi}''(\xi;q) = \mu/\alpha, \quad \xi = 1 \tag{10.27}$$

和

$$\check{\phi}'(\xi;q) = 0, \quad \xi \to +\infty, \tag{10.28}$$

其中 $\check{\mathcal{L}}$ 为辅助线性算子, $g_0(\xi)$ 为 $g(\xi)$ 的初始猜测解, c_0 为收敛控制参数, $\check{H}(\xi)$ 为辅助函数, ′ 表示对 ξ 求导. 假设初始猜测解 $g_0(\xi)$, 辅助函数 $\check{H}(\xi)$, 尤其是收敛控制参数 c_0 选取合适, 同伦-Maclaurin 级数

$$\check{\phi}(\xi;q) = g_0(\xi) + \sum_{m=1}^{+\infty} g_m(\xi)\, q^m$$

在 $q=1$ 时绝对收敛, 我们有同伦级数解

$$g(\xi) = g_0(\xi) + \sum_{m=1}^{+\infty} g_m(\xi), \tag{10.29}$$

其中 $g_m(\xi)$ 由 m 阶形变方程

$$\check{\mathcal{L}}\left[g_m(\xi) - \chi_m\, g_{m-1}(\xi)\right] = c_0\, \check{H}(\xi)\, \check{\delta}_{m-1}(\xi) \tag{10.30}$$

控制, 满足边界条件

$$g_m(1) = 0, \quad g_m'(1) = 0, \quad g_m''(1) = 0, \quad g_m'(+\infty) = 0, \tag{10.31}$$

其中

$$\check{\delta}_n(\xi) = \alpha^2\, g_m''' + \frac{1+\kappa}{2}\sum_{i=0}^{m} g_{m-i}\, g_i'' - \kappa \sum_{i=0}^{m} g_{m-i}'\, g_i' \tag{10.32}$$

由定理 4.1 得到. 方程 (10.30) 由定理 4.15 给出. 详情请参考第四章.

解表达式 (10.25) 在选择初始猜测解 $g_0(\xi)$ 和辅助线性算子 (10.34) 中起着重要作用. 为了满足解表达 (10.25), 我们选取如下形式的初始猜测解

$$g_0(\xi) = a_0 \xi^\beta + a_1 \xi^{\beta-1} + a_2 \, \xi^{2(\beta-1)},$$

其中未知系数 a_0, a_1 和 a_2 由两个边界条件 $g(1)=0, g'(1)=1$ 和一个附加边界条件 $g''(1)=\mu/\alpha$ 确定. 因此, 我们有

$$g_0(\xi) = \frac{4-3\beta+\alpha^{-1}\mu}{2-\beta}\xi^\beta - \frac{3-3\beta+\alpha^{-1}\mu}{1-\beta}\xi^{\beta-1}$$
$$+ \frac{2-2\beta+\alpha^{-1}\mu}{(1-\beta)(2-\beta)}\xi^{2\beta-2}, \qquad (10.33)$$

其自动满足边界条件 $g'(+\infty)=0$, 且在无穷远处呈代数衰减.

为了满足解表达 (10.25), 我们选取如下辅助线性算子 $\check{\mathcal{L}}$

$$\check{\mathcal{L}}(u) = u''' + B_1(\xi)u'' + B_2(\xi)u' + B_3(\xi)u,$$

该三阶线性微分方程

$$\check{\mathcal{L}}(u) = 0$$

的通解在无穷远处呈代数衰减, 等同于

$$\check{\mathcal{L}}\left[C_1\,\xi^\beta + C_2\,\xi^{\beta-1} + C_3\,\xi^{2(\beta-1)}\right] = 0.$$

将 $u=\xi^\beta$, $u=\xi^{\beta-1}$ 和 $u=\xi^{2(\beta-1)}$ 代入 $\check{\mathcal{L}}(u)=0$ 中, 可得到关于 $B_1(\xi), B_2(\xi)$ 和 $B_3(\xi)$ 的三个线性代数方程, 由此可以确定 $B_1(\xi), B_2(\xi)$ 和 $B_3(\xi)$ 的值. 通过这种方式, 我们可以得到辅助线性算子

$$\check{\mathcal{L}}(u) = \xi^3\, u''' - 2(2\beta-3)\xi^2 u'' + (\beta-1)(5\beta-6)\xi u' - 2\beta(\beta-1)^2 u. \quad (10.34)$$

令

$$g_m^*(\xi) = \chi_m\, g_{m-1}(\xi) + c_0 \check{\mathcal{L}}^{-1}\left[\check{H}(\xi)\check{\delta}_{m-1}(\xi)\right]$$

表示 (10.30) 的特解, 其中 $\check{\mathcal{L}}^{-1}$ 为 $\check{\mathcal{L}}$ 的逆算子, 则其通解为

$$g_m(\xi) = g_m^*(\xi) + C_{m,1}\,\xi^\beta + C_{m,2}\,\xi^{\beta-1} + C_{m,3}\,\xi^{2(\beta-1)},$$

其中积分常数 $C_{m,1}, C_{m,2}$ 和 $C_{m,3}$ 由 $\xi=1$ 处的三个边界条件 (10.31) 确定. 值得注意的是, 无穷远处的边界条件, 即 $g_m'(+\infty)=0$, 是自动满足的. 因此, 选取合适的基函数, 能够轻松地满足无穷远处的边界条件. 这主要是因为像

Mathematica 这样的科学计算软件为我们提供了"基于函数而非数的计算"的能力 [170].

此外, 为了满足所谓的解表达 (10.25), 方程 (10.30) 的右端项 $\check{H}(\xi)\check{\delta}_{m-1}(\xi)$ 不能包含

$$\xi^\beta, \xi^{\beta-1}, \xi^{2(\beta-1)}.$$

基于此, 我们必须选取辅助函数 $\check{H}(\xi) = \xi$. 详情请参考廖世俊和 Magyari [167].

为了验证 BVPh 1.0 对于无限区间内非线性常微分方程的呈代数衰减的解的有效性, 我们先考虑当 $\kappa = -1/3$ 时的情况, 其封闭解 (10.21) 是已知的. 根据 (10.23), 我们有精确的关系式 $\alpha^2 g'''(1) = \kappa$, 可用于检验同伦近似的精确性. 注意, c_0 和 α 都是未知的, 其最优值由控制方程 (10.23) 的平方残差的最小值确定. 例如, 当 $\mu = f''(0) = 1$ 时, (10.23) 在区间 $\xi \in [1,20]$ 上的 5 阶同伦近似处的平均平方残差通过选取最优收敛控制参数 $c_0^* = -8.1789$ 和最优参数 $\alpha^* = 0.4185$ 具有最小值 4.2×10^{-4}. 确实, 通过选取 $c_0 = -8$ 和 $\alpha = 2/5$, 相应的同伦近似快速收敛到精确解 (10.21), 如表 10.3 和图 10.2 所示. 这说明了 BVPh 1.0 对于无限区间 $\eta \in [0, +\infty)$ 上非线性边值常微分方程的呈代数衰减的解的有效性.

表 10.3 当 $\kappa = -1/3$ 和 $\mu = 1$ 时, 选取 $c_0 = -8, \alpha = 2/5$ 和 $\check{H}(\xi) = \xi$, m 阶 (呈代数衰减) 同伦近似对应的 (10.23) 在区间 $\xi \in [1, 20]$ 上的平均平方残差

近似阶数 m	$\alpha^2 g'''(1)$	平均平方残差 E_m
2	−0.4546	9.8×10^{-3}
4	−0.3428	1.6×10^{-3}
6	−0.3341	1.7×10^{-4}
8	−0.3334	2.8×10^{-5}
10	−0.3333	1.1×10^{-5}
20	−0.3333	6.1×10^{-7}
30	−0.3333	6.2×10^{-8}
50	−0.3333	1.4×10^{-9}

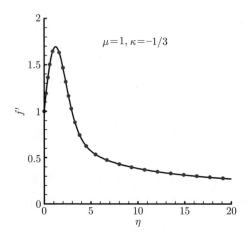

图 10.2 当 $\kappa = -1/3$ 和 $\mu = 1$ 时, (10.3) 呈代数衰减的解 $f'(\eta)$. 实线: 精确解 (10.21); 实心圆: 选取 $c_0 = -8, \alpha = 2/5$ 和 $\check{H}(\xi) = \xi$ 时的 20 阶同伦近似

值得注意的是, (10.33) 中的未知参数 α 为我们选取初始猜测解提供了一个额外的自由度, 这提供了一种通过控制方程平方残差的最小值确定最优初始猜测解的便捷方法. 因此, 我们可以将 α 也视为收敛控制参数. 显然, 如上所述, 在同伦分析方法框架中, 使用两个收敛控制参数 c_0 和 α 提高了保证同伦级数解收敛的能力.

类似地, BVPh 1.0 可用于在整个区间 $\kappa \in (-1/2, 0)$ 上得到其他 κ 值的呈代数衰减的解. 例如, 当 $\kappa = -1/4, \mu = 1$ 时, (10.23) 在区间 $\xi \in [1, 20]$ 上的 5 阶同伦近似的平均平方残差通过选取最优收敛控制参数 $c_0^* = -6.8702$ 和最优参数 $\alpha^* = 0.4133$ 具有最小值 1.7×10^{-3}. 选取 $c_0 = -7$ 和 $\alpha = 2/5$, 相应的同伦近似快速收敛, 如表 10.4 和图 10.3 所示. 类似地, 当 $\kappa = -1/4$ 时, 我们得到 $\mu = f''(0) > f''_{\exp}(0) = -0.1620$ 取其他值时的呈代数衰减的解, 如图 10.3 所示. 可以发现, 当 $\kappa = -1/4$ 时, 边界层方程 (10.3) 存在无穷多个呈代数衰减的解 $f(\eta)$, 具有 $f''(0) \geqslant f''_{\exp}(0) = -0.1620$ 的性质, 其中 $f_{\exp}(\eta)$ 是相应的呈指数衰减的解.

上述结论具有普遍意义: 据廖世俊和 Magyari [167] 指出, 当 $-1/2 < \kappa < 0$ 时, 边界层方程 (10.3) 存在无穷多个具有 $f''(0) \geqslant f''_{\exp}(0) = -0.1620$ 的性质的呈代数衰减的解 $f(\eta)$, 其中 $f_{\exp}(\eta)$ 是相应的呈指数衰减的解. 换句话说, 不仅在 $\kappa = -1/3$ 处, 而且在整个区间 $-1/2 < \kappa < 0$ 上, 边界层方程 (10.3) 呈指数衰减的解 $f_{\exp}(\eta)$ 是无穷多个呈代数衰减的解 $f(\eta)$ 的极限情况, 如图 10.4 所示.

表 10.4 当 $\kappa = -1/4$ 和 $\mu = 1$ 时, 选取 $c_0 = -7, \alpha = 2/5$ 和 $\check{H}(\xi) = \xi$, m 阶 (呈代数衰减) 同伦近似对应的 (10.23) 在区间 $\xi \in [1, 20]$ 上的平均平方残差

近似阶数 m	$\alpha^2 \, g'''(1)$	平均平方残差 E_m
2	−0.2699	2.6×10^{-2}
4	−0.2503	5.1×10^{-3}
6	−0.2500	9.0×10^{-4}
8	−0.2500	2.4×10^{-4}
10	−0.2500	1.0×10^{-4}
20	−0.2500	9.9×10^{-6}
30	−0.2500	1.1×10^{-6}
40	−0.2500	1.6×10^{-7}
50	−0.2500	3.7×10^{-8}

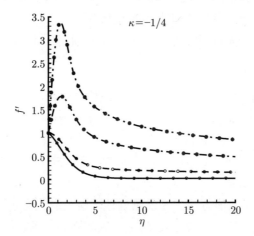

图 10.3 当 $\kappa = -1/4$ 时, (10.3) 呈代数衰减的解 $f'(\eta)$. 实线: 50 阶同伦近似; 空心圆: 20 阶同伦近似; 实心圆: 30 阶同伦近似; 实线: $\mu = -0.15$; 虚线: $\mu = 0$; 点划线: $\mu = 1$; 双点划线: $\mu = 3$

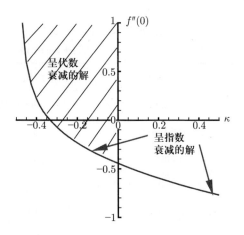

图 10.4 对于 (10.3) 的呈指数和代数衰减的解, $f''(0)$ 随着 κ 的变化. 实线: 呈指数衰减的解; 阴影区域: 呈代数衰减的解

值得注意的是, 这些呈代数衰减的解只有在 $\kappa = -1/3$ 时才具有封闭的表达式. 需要强调的是, 边界层方程 (10.3) 在整个区间 $-1/2 < \kappa < 0$ (除了 $\kappa = -1/3$) 上新的呈代数衰减的解是首次通过同伦分析方法找到的 [167]. 这显示了同伦分析方法的巨大潜力和普遍有效性.

以上所有结果都是通过 BVPh 1.0 得到的, 附录 10.2 中给出了相应的输入数据.

10.4 本章小结

本章使用 BVPh 1.0 获得了无限区间内非线性常微分方程控制的边界层流动呈指数和代数衰减的解. 因此, 许多边界层流动问题都可以通过 BVPh 1.0 以类似的方式进行求解.

值得注意的是, 数值方法不可能精确地求解具有无穷远处边界条件的无限区间的问题. 许多数值包如 BVP4c 将无限区间视为一种奇点, 并在实际计算中将其替换为有限区间. 然而, 基于科学计算软件, BVPh 1.0 可以精确地求解无限区间内的非线性常微分方程. 这主要是因为科学计算软件为我们提供了 "基于函数而非数的计算" 的能力 [170]. 因此, 如本章所述, 我们可以通过选择合适的基函数轻松地解决无穷远处的边界条件.

还需要注意的是, BVPh 1.0 可用于在无限区间内获得非线性边值常微分方程呈指数和代数衰减的解. 这主要是因为, 基于同伦分析方法, BVPh 1.0 为我们提供了极大的自由度和灵活性来选取不同类型的辅助线性算子和初始猜测解. 此外, 收敛控制参数 c_0 为我们提供了一种便捷的方式来保证同伦级数解的收敛性.

此外, (10.15) 中的未知参数 σ 和 (10.33) 中的 α 也为我们选取初始猜测解提供了一个额外的自由度, 这为找出最优初始猜测解提供了一种便捷的方法. 与 c_0 一样, σ 和 α 都可以看作收敛控制参数. 如本章所述, 在同伦分析方法的框架下, σ 和 α 与 c_0 结合, 它们中的每一个都进一步提高了保证同伦级数解收敛的能力.

需要强调的是, (10.3) 的无穷多个呈代数衰减的解只有在 $\kappa = -1/3$ 时才具有封闭解. 基于同伦分析方法, 廖世俊和 Magyari [167] 在整个区间 $-1/2 < \kappa < 0$ (除了 $\kappa = -1/3$) 上首次发现了 (10.3) 新的呈代数衰减的解. 事实上, 通过同伦分析方法, 廖世俊 [161] 已经找到了一些其他边界层流动问题的新解, 而这些解从未被数值方法报道过, 甚至被数值方法忽略. 确实, 一种全新的方法总会带来一些新的和 (或) 不同的东西. 所有这些都显示了同伦分析方法的原创性.

BVPh 1.0 可以作为解析工具来求解无限区间内的许多非线性常微分方程, 尤其是那些与边界层流动有关的, 它甚至可以用于求解无限区间内的一些非线性偏微分方程, 例如第十一章中所述的非相似边界层流动问题, 以及第十二章所述的非定常边界层流动问题. 然而, 这些并不意味着 BVPh 1.0 对无限区间内的所有边值问题都有效. 如第七章中提到的, 我们的目标是为尽可能多的非线性边值问题开发一个软件包. 因此, 今后还需要进一步的改进和应用.

需要注意的是, Chebfun 4.0 也为我们提供了 "基于函数而非数的计算" 的能力 [170]. 因此, 通过 Chebfun 为具有奇点的强非线性多点边值问题建立类似的基于同伦分析方法的软件包是有价值的.

参考文献

[151] Banks, W.H.H.: Similarity solutions of the boundary-layer equations for a stretching wall. Journal de Mecanique theorique et appliquee. **2**, 375–392 (1983).

[152] Li, Y.J., Nohara, B.T., Liao, S.J.: Series solutions of coupled van der Pol equation by means of homotopy analysis method. J. Mathematical Physics **51**, 063517 (2010). doi:10.1063/1.3445770.

[153] Liao, S.J.: The Proposed Homotopy Analysis Technique for the Solution of Nonlinear Problems. PhD dissertation, Shanghai Jiao Tong University (1992).

[154] Liao, S.J.: A kind of approximate solution technique which does not depend upon small parameters—(II) An application in fluid mechanics. Int. J. Nonlin. Mech. **32**,

815–822 (1997).

[155] Liao, S.J.: An explicit, totally analytic approximation of Blasius viscous flow problems. Int. J. Nonlin. Mech. **34**, 759–778 (1999a).

[156] Liao, S.J.: A uniformly valid analytic solution of 2D viscous flow past a semi-infinite flat plate. J. Fluid Mech. **385**, 101–128 (1999b).

[157] Liao, S.J.: On the analytic solution of magnetohydrodynamic flows of non-Newtonian fluids over a stretching sheet. J. Fluid Mech. **488**, 189–212 (2003a).

[158] Liao, S.J.: Beyond Perturbation—Introduction to the Homotopy Analysis Method. Chapman & Hall/ CRC Press, Boca Raton (2003b).

[159] Liao, S.J.: On the homotopy analysis method for nonlinear problems. Appl. Math. Comput. **147**, 499–513 (2004).

[160] Liao, S.J.: A new branch of solutions of boundary-layer flows over an impermeable stretched plate. Int. J. Heat Mass Tran. **48**, 2529–2539 (2005).

[161] Liao, S.J.: Series solutions of unsteady boundary-layer flows over a stretching flat plate. Stud. Appl. Math. **117**, 2529–2539 (2006).

[162] Liao, S.J.: Notes on the homotopy analysis method—Some definitions and theorems. Commun. Nonlinear Sci. Numer. Simulat. **14**, 983–997 (2009).

[163] Liao, S.J.: On the relationship between the homotopy analysis method and Euler transform. Commun. Nonlinear Sci. Numer. Simulat. **15**, 1421–1431 (2010a). doi:10.1016/j.cnsns.2009.06.008.

[164] Liao, S.J.: An optimal homotopy-analysis approach for strongly nonlinear differential equations. Commun. Nonlinear Sci. Numer. Simulat. **15**, 2003–2016 (2010b).

[165] Liao, S.J., Campo, A.: Analytic solutions of the temperature distribution in Blasius viscous flow problems. J. Fluid Mech. **453**, 411–425 (2002).

[166] Liao, S.J., Pop, I.: Explicit analytic solution for similarity boundary layer equations. Int. J. Heat and Mass Transfer. **47**, 75–85 (2004).

[167] Liao, S.J., Magyari, E.: Exponentially decaying boundary layers as limiting cases of families of algebraically decaying ones. Z. angew. Math. Phys. **57**, 777–792 (2006).

[168] Liao, S.J., Tan, Y.: A general approach to obtain series solutions of nonlinear differential equations. Stud. Appl. Math. **119**, 297–355 (2007).

[169] Magyari, E., Pop, I., Keller, B.: New analytical solutions of a well known boundary value problem in fluid mechanics. Fluid Dyn. Res. **33**, 313–317 (2003).

[170] Trefethen, L.N.: Computing numerically with functions instead of numbers. Math. in Comp. Sci. **1**, 9–19 (2007).

[171] Xu, H., Lin, Z.L., Liao, S.J., Wu, J.Z., Majdalani, J.: Homotopy-based solutions of the Navier-Stokes equations for a porous channel with orthogonally moving walls. Physics of Fluids. **22**, 053601 (2010). doi:10.1063/1.3392770.

附录 10.1　呈指数衰减的解的 BVPh 输入数据

```
(* Input Mathematica package BVPh version 1.0 *)
<<BVPh.txt;

(* Define the physical and control parameters *)
TypeEQ     = 1;
ApproxQ    = 0;
ErrReq     = 10^(-10);
zRintegral = 10;

(* Define the governing equation *)
kappa = -1/4;
f[z_,u_,lambda_] := D[u,{z,3}]
        +(1+kappa)/2*u*D[u,{z,2}]-kappa*D[u,z]^2 ;

(* Define Boundary conditions *)
zR  = infinity;
OrderEQ = 3;
BC[1,z_,u_,lambda_] := Limit[u, z -> 0 ];
BC[2,z_,u_,lambda_] := Limit[D[u,z] - 1,  z -> 0 ];
BC[3,z_,u_,lambda_] := Limit[D[u,z], z -> zR ];

(* Define initial guess *)
temp[1] = (1-2*sigma);
temp[2] = (1-sigma);
u[0]   = sigma+temp[1]*Exp[-z]-temp[2]*Exp[-2*z];

(* Define output term *)
output[z_,u_,k_]:= Print["output         = ",
             D[u[k],{z,2}] /. z->0//N];

(* Defines the auxiliary linear operator *)
L[u_] := D[u,{z,3}] -  D[u,z];

(* Print input and control parameters *)
PrintInput[u[z]];

(* Set optimal c0 and sigma *)
```

```
c0       =-5/4 ;
sigma = 3;
Print[" c0  =  ",c0, "     sigma  =  ",sigma];

(*  Gain up to 10th-order approxiamtion *)
BVPh[1,10];
```

附录 10.2 呈代数衰减的解的 BVPh 输入数据

```
(* Input Mathematica package BVPh version 1.0 *)
<<BVPh.txt;

(* Define the physical and control parameters *)
TypeEQ       = 1;
ApproxQ      = 0;
ErrReq       = 10^(-10);
zRintegral   = 10;
H[z_]        := z;

(* Define the governing equation *)
kappa = -1/3;
alpha =  2/5;
f[z_,u_,lambda_]:=alpha^2*D[u,{z,3}]
        +(1+kappa)/2*u*D[u,{z,2}]-kappa*D[u,z]^2;

(* Define Boundary conditions *)
zL  = 1;
zR  = infinity;
OrderEQ  = 3;
BC[1,z_,u_,lambda_] := Limit[u, z -> 1 ];
BC[2,z_,u_,lambda_] := Limit[D[u,z] - 1,   z -> 1 ];
BC[3,z_,u_,lambda_] := Limit[D[u,{z,2}] - mu/alpha,
                             z -> 1 ];

(* Define initial guess *)
mu = 1;
beta = (1+kappa)/(1-kappa);
u[0] = Module[{temp},
    temp[1] = (4-3*beta+mu/alpha)/(2-beta);
```

```
temp[2] = (3-3*beta+mu/alpha)/(1-beta);
temp[3] = (2-2*beta+mu/alpha)/(1-beta)/(2-beta);
temp[1]*z^beta - temp[2]*z^(beta-1)
         + temp[3]*z^(2*beta-2)
];

(* Define output term *)
output[z_,u_,k_]:= Print["output           = ",
           alpha^2*D[u[k],{z,3}] /. z->1//N];

(* Define the auxiliary linear operator *)
(L[u_] := Module[{temp},
temp[1] = 2*(2*beta-3);
temp[2] = (beta-1)*(5*beta-6);
temp[3] = 2*beta*(beta-1)^2;
z^3*D[u,{z,3}] - temp[1]*z^2*D[u,{z,2}]
          + temp[2]*z*D[u,z] - temp[3]*u
];

(* Print input and control parameters *)
PrintInput[u[z]];

(*  Exact solution when kappa = -1/3 *)
Uexact  = Module[{temp,t,t0,Ai,Bi,Ait,Bit},
Ai = AiryAi[t];
Bi = AiryBi[t];
Ait = D[Ai,t];
Bit = D[Bi,t];
Ait0 = Ait /. t -> t0;
Bit0 = Bit /. t -> t0;
temp[1] = Bit0*Ait - Ait0*Bit;
temp[2] = Bit0*Ai  - Ait0*Bi ;
t0 = (Sqrt[6]*mu)^(-2/3);
t  = t0*(1 + mu*x);
(36*mu)^(1/3)*temp[1]/temp[2]//Expand
];
Uzexact = D[Uexact,x];

(* Coordinate transform *)
```

```
Wz[k_] := Uz[k] /. z -> 1 + alpha*x;

(*  Set optimal c0   *)
c0 = -8;
Print["   c0  =  ",c0];

(* Gain up to 10th-order approximation *)
BVPh[1,30];
```

注意, 对于在无限区间上定义的非线性边值问题, 我们必须在 BVPh 1.0 的输入数据中设置 zR = infinity.

第十一章 非相似边界层流动

本章阐述了基于同伦分析方法的 Mathematica 软件包 BVPh 1.0 求解与非相似边界层流动相关的非线性偏微分方程的有效性. 结果表明, BVPh 1.0 可以用于求解非相似边界层流动问题, 其求解方法与求解由非线性常微分方程控制的相似边界层流动所采用的方法类似. 换句话说, 在同伦分析方法的框架下, 求解非相似边界层流动与求解相似边界层流动一样简单. 这表明了 BVPh 1.0 对于求解某些非线性偏微分方程的有效性, 特别是对于那些与边界层流动相关的偏微分方程.

11.1 绪论

第二部分的前几章阐述了基于同伦分析方法的 Mathematica 软件包 BVPh 1.0 为我们提供了一个解析工具来求解无论是有限区间 $z \in [0,a]$ 还是无限区间 $z \in [b,+\infty)$ 上由非线性常微分方程控制的边值问题, 其中 $a > 0$ 和 $b \geqslant 0$ 为有界常数. 本章和下一章将说明 BVPh 1.0 甚至可以求解某些非线性偏微分方程, 尤其是与边界层流动相关的偏微分方程.

不失一般性, 我们在这里考虑源于一种非相似边界层流动的非线性偏微分方程. 自从 Prandtl [210] 在 1904 年提出了黏性流体边界层流动的革命性概念, 边界层理论 [175, 183, 208, 214, 215, 220] 得到了很大的发展并几乎被应用于流体力学的所有领域 [219]. 当存在相似解时, 边界层流动由非线性常微分方程控制. 然而, 当不存在这种相似性时, 就必须直接求解非线性偏微分方程, 这比求解常微分方程要难得多. 这主要是因为该领域的大多数研究人员都专注于相似边界层流动问题: 目前已经发表了数千篇与相似边界层流动 [175, 183, 208, 214, 215, 220] 相关的文章, 但是与相似边界层流动的大量出版物相比, 关于非相似流动 [176, 177, 179, 180, 209, 218] 的文章就要少得多.

例如, 我们在这里考虑牛顿流体在拉伸平板上的非相似边界层流动 [174, 178], 沿着平板施加两个方向相反, 大小相同的力. 因此, 平板上有一个静止点, 将其定义为坐标系的原点. x 轴和 y 轴分别沿着和垂直于平板. 流体在远离平板的地方静止 (即 $y \to +\infty$). 由于流动的对称性, 我们只考虑四分之一平面 $x \geqslant 0$ 和 $y \geqslant 0$ 上的流动. 令 $U_w(x)$ 表示平板的拉伸速度, (u, v) 表示速度分量, ν 表示流体的运动黏度. 正如 Prandtl [210] 所提到的那样, 垂直流动方向上的速度变化远大于流动方向上的速度变化, 因此在平板附近存在一个相当薄的边界层. 在边界层理论的框架下, 这种黏性流动由如下方程

$$\frac{\partial u}{\partial x} + \frac{\partial v}{\partial y} = 0, \tag{11.1}$$

$$u \frac{\partial u}{\partial x} + v \frac{\partial u}{\partial y} = \nu \frac{\partial^2 u}{\partial y^2} \tag{11.2}$$

控制, 满足边界条件

$$u = U_w(x), \quad v = 0 \quad y = 0 \tag{11.3}$$

和

$$u = 0, \quad \frac{\partial v}{\partial x} = 0 \quad x = 0, \tag{11.4}$$

$$u \to 0, \quad y \to +\infty. \tag{11.5}$$

相似解只存在于 $U_w(x)$ 的某些特殊情况下. 继 Görtler [181], 我们定义所谓的主函数

$$\Delta(x) = \frac{U'_w(x)}{U_w^2(x)} \int_0^x U_w(\xi) d\xi. \tag{11.6}$$

当 $\Delta(x)$ 等于常数 β 时, 即

$$\frac{U'_w(x)}{U_w^2(x)} \int_0^x U_w(\xi) d\xi = \beta, \tag{11.7}$$

存在相似解

$$\psi = \sqrt{\nu \int_0^x U_w(\xi) d\xi}\, g(\eta), \quad \eta = \frac{U_w(x)}{\sqrt{\nu \int_0^x U_w(\xi) d\xi}}\, y, \tag{11.8}$$

其中 ψ 为流函数, $g(\eta)$ 由如下非线性常微分方程

$$g''' + \frac{1}{2}gg'' - \beta g'^2 = 0, \quad g(0) = 0, \quad g'(0) = 1, \quad g'(+\infty) = 0 \qquad (11.9)$$

控制, 相应的相似边界层流动的局部表面摩擦系数为

$$C_f = \frac{\tau_w}{\frac{1}{2}\rho U_w^2(x)} = 2q''(0)\sqrt{\frac{\nu}{\int_0^x U_w(\xi)d\xi}}. \qquad (11.10)$$

例如, 当 $U_w(x) = a\, x^\lambda$ 时, 其中 $a(1+\lambda) > 0$, 有

$$\beta = \lambda/(1+\lambda),$$

使得相似性标准 (11.7) 满足, 因此存在相似性解

$$\psi = \sqrt{\frac{a\,\nu}{1+\lambda}}\, x^{\frac{\lambda+1}{2}}g(\eta), \quad \eta = \sqrt{\frac{a(1+\lambda)}{\nu}}\, x^{\frac{\lambda-1}{2}}y. \qquad (11.11)$$

这种情况下, 两个耦合偏微分方程 (11.1) 和 (11.2) 转化为了一个常微分方程 (11.9), 这比求解原始方程容易得多. 因此, 从数学的角度来看, 当存在相似解时, 问题将被大大简化.

对于相似边界层流动, 不同 x 处的所有速度剖面都是相似的. 然而, 对于非相似流动, 这种相似性会消失 [176, 177, 179, 180, 209, 216–218, 221]. 从物理上讲, 非相似边界层流动在实际中更加普遍, 因此比相似边界层流动更重要.

当相似解不存在时, 必须求解非线性偏微分方程. 传统上, 有两种不同的求解方法: 解析方法和数值方法. 一方面, 数值方法被广泛应用于研究非相似边界层流动. 如文献 [211, 213] 中所示, 可以使用数值方法在大量离散点处获得近似结果. 然而, 其必须用有限区间代替无限区间, 这导致了数值结果中存在一些额外的不精确性和不确定性. 另一方面, 通过解析方法可以求解无限区间的非线性偏微分方程. 然而, 遗憾的是, 使用传统的解析方法 (如摄动方法), 很难得到对整个区间内所有物理变量都有效且精确的解析近似值. 这主要是因为摄动方法往往依赖于小的物理变量或参数, 因此摄动结果往往无法对所有物理参数 (变量) 都有效. 目前, Cimpean 等人 [177] 结合数值技术, 应用摄动方法求解了多孔介质中垂直平板上的自由对流非相似边界层问题. 如大多数摄动解一样, 他们的结果仅对小或大的摄动量 x 有效.

此外, 针对非相似边界层问题的所谓 "局部相似性方法" [209, 217] 是基于这样一个假设, 即控制方程中的非相似项非常小, 以至于可以将它们视为零, 因此原始偏微分方程变为了常微分方程. 然而, Sparrow 等人 [217] 指出, "局部

相似性方法"给出的结果具有"不确定的准确性", Massoudi [209] 指出, "局部相似性方法" 一般只对小变量有效. 这很容易理解, 因为非相似项肯定不是零, 必须加以考虑. Sparrow 等人 [216, 217, 221] 引入了所谓的 "局部非相似性方法", Massoudi [209] 应用其求解了非牛顿流体在楔形体上的非相似流动. 通过沿自由流速度的无量纲变量 ξ 对原始控制方程进行求导, Masoudi [209] 对于动量和能量方程给出了两个附加的辅助非线性偏微分方程, 然后将这两个偏微分方程中的变量 ξ 视为一个常数, 从而将它们简化为一个常微分方程组, 最后使用数值技术求解了更复杂的四个方程组. 尽管有学者 [218, 221] 指出, 在某些情况下, "局部非相似性方法" 的结果与数值解或级数解非常吻合, 遗憾的是, Masoudi [209] 只给出了 ξ 很小时的数值结果.

上述尝试揭示了非相似边界层流动在数学上求解的困难. 这可能就是非相似边界层流动的研究远少于相似边界层流动的原因, 尽管前者不管在理论上还是在应用上都比后者更重要.

借助流函数 ψ, (11.1) 自动满足, 然后 (11.2) 变为如下的非线性偏微分方程

$$\nu \frac{\partial^3 \psi}{\partial y^3} + \frac{\partial \psi}{\partial x} \frac{\partial^2 \psi}{\partial y^2} - \frac{\partial \psi}{\partial y} \frac{\partial^2 \psi}{\partial x \partial y} = 0, \qquad (11.12)$$

满足边界条件

$$\psi = 0, \quad \frac{\partial \psi}{\partial y} = U_w(x), \quad y = 0,$$
$$\frac{\partial \psi}{\partial y} \to 0, \quad y \to +\infty. \qquad (11.13)$$

基于变换

$$\eta = \frac{y}{\nu^{1/2} \sigma(x)}, \quad \psi = \nu^{1/2} \sigma(x) f(\eta, x), \qquad (11.14)$$

其中 $\sigma(x) > 0$ 是稍后选取的实函数, 我们有速度

$$u = \frac{\partial f}{\partial \eta}, \quad v = \nu^{1/2} \left[\sigma'(x) \left(\eta \frac{\partial f}{\partial \eta} - f \right) - \sigma(x) \frac{\partial f}{\partial x} \right].$$

然后, 控制方程 (11.12) 变为

$$\frac{\partial^3 f}{\partial \eta^3} + \frac{1}{2} [\sigma^2(x)]' f \frac{\partial^2 f}{\partial \eta^2} + \sigma^2(x) \left(\frac{\partial f}{\partial x} \frac{\partial^2 f}{\partial \eta^2} - \frac{\partial f}{\partial \eta} \frac{\partial^2 f}{\partial x \partial \eta} \right) = 0, \qquad (11.15)$$

满足边界条件

$$f(0, x) = 0, \quad f_\eta(0, x) = U_w(x), \quad f_\eta(+\infty, x) = 0, \qquad (11.16)$$

其中 f_η 表示 $f(\eta, x)$ 关于 η 的偏导数。值得注意的是，(11.15) 是具有变系数 $[\sigma^2(x)]'/2$ 和 $\sigma^2(x)$ 的非线性偏微分方程。

这里存在无穷多个不满足相似性标准 (11.7) 的平板拉伸速度 $U_w(x)$。不失一般性，这里我们考虑 $U_w(x) = \check{U}_w(\xi)$ 的情况，其中 $\xi = \Gamma(x)$ 定义了一种变换。然后，(11.15) 变成

$$\frac{\partial^3 f}{\partial \eta^3} + \sigma_1(\xi)\, f\, \frac{\partial^2 f}{\partial \eta^2} + \sigma_2(\xi) \left(\frac{\partial f}{\partial \xi} \frac{\partial^2 f}{\partial \eta^2} - \frac{\partial f}{\partial \eta} \frac{\partial^2 f}{\partial \xi \partial \eta} \right) = 0, \qquad (11.17)$$

满足边界条件

$$f(0, \xi) = 0, \quad f_\eta(0, \xi) = \check{U}_w(\xi), \quad f_\eta(+\infty, \xi) = 0, \qquad (11.18)$$

其中

$$\sigma_1(\xi) = \frac{1}{2} [\sigma^2(x)]', \quad \sigma_2(\xi) = \Gamma'(x)\, \sigma^2(x), \qquad (11.19)$$

其中 x 用 ξ 表示，即 $x = \Gamma^{-1}(\xi)$。例如，当 $\sigma(x) = \sqrt{1+x}$ 和 $\xi = \Gamma(x) = x/(1+x)$ 时，我们有 $\sigma_1(\xi) = 1/2$ 和 $\sigma_2(\xi) = 1 - \xi$，详情请参考廖世俊 [200]。

非相似边界层流动的局部表面摩擦系数由下式给出

$$C_f(x) = \frac{\tau(x)}{\frac{1}{2}\rho U_w^2(x)} = \frac{2\nu^{1/2}}{\sigma(x)\, U_w^2(x)} \left. \frac{\partial^2 f}{\partial \eta^2} \right|_{\eta \to 0}. \qquad (11.20)$$

因此，获得 $f_{\eta\eta}(0, \xi)$ 的精确结果很重要。替换边界层厚度 $\bar{\delta}(x)$ 由下式给出

$$\bar{\delta}(x) = \frac{1}{U_w(x)} \int_0^{+\infty} u(x, y)\, dy. \qquad (11.21)$$

11.2 简明数学公式

在同伦分析方法 [188–206] 的框架下，由偏微分方程 (11.17) 控制的非相似流动可以在没有任何额外假设的情况下进行求解，如廖世俊 [199] 所示。这与上面提到的所有解析方法完全不同。如下所述，这种非相似边界层流动甚至可以通过 BVPh 1.0 以与第十章中提到的求解相似流动类似的方式进行求解。

根据 (11.17)，我们定义如下非线性算子

$$\mathcal{N}(f) = \frac{\partial^3 f}{\partial \eta^3} + \sigma_1(\xi)\, f\, \frac{\partial^2 f}{\partial \eta^2} + \sigma_2(\xi) \left(\frac{\partial f}{\partial \xi} \frac{\partial^2 f}{\partial \eta^2} - \frac{\partial f}{\partial \eta} \frac{\partial^2 f}{\partial \xi \partial \eta} \right). \qquad (11.22)$$

令 $q \in [0,1]$ 表示同伦参数, $c_0 \neq 0$ 表示收敛控制参数, $H(\eta) \neq 0$ 表示辅助函数, $\mathcal{L}$ 表示辅助线性算子. 其中 $\mathcal{L}$ 具有如下性质

$$\mathcal{L}(0) = 0, \tag{11.23}$$

$f_0(\eta,\xi)$ 为满足边界条件 (11.18) 的初始猜测解. 注意, 同伦分析方法为我们选择辅助线性算子 $\mathcal{L}$ 和初始猜测解 $f_0(\eta,\xi)$ 提供了极大的自由. 在同伦分析方法的框架下, 我们首先构造一个连续变化 (或形变) $\phi(\eta,\xi;q)$, 随着 q 从 0 增加到 1, $\phi(\eta,\xi;q)$ 从初始猜测解 $f_0(\eta,\xi)$ 变化到 (11.17) 和 (11.18) 的解 $f(\eta,\xi)$. 这种连续变化 (或映射) 由所谓的零阶形变方程

$$(1-q)\mathcal{L}[\phi(\eta,\xi;q) - f_0(\eta,\xi)] = q\, c_0\, H(\eta)\, \mathcal{N}[\phi(\eta,\xi;q)] \tag{11.24}$$

控制, 满足平板上的边界条件

$$\phi(0,\xi;q) = 0, \quad \phi_\eta(0,\xi;q) = \check{U}_w(\xi) \tag{11.25}$$

和无穷远处的边界条件

$$\phi_\eta(+\infty,\xi;q) \to 0. \tag{11.26}$$

注意, 初始猜测解 $f_0(\eta,\xi)$ 满足边界条件 (11.18), 此外 $\mathcal{L}$ 具有性质 (11.23). 因此, 当 $q = 0$ 时, 我们有初始猜测解

$$\phi(\eta,\xi;0) = f_0(\eta,\xi). \tag{11.27}$$

当 $q = 1$ 时, 由于 $c_0 \neq 0$, 零阶形变方程 (11.24) 至 (11.26) 等价于原始方程 (11.17) 和 (11.18), 有

$$\phi(\eta,\xi;1) = f(\eta,\xi). \tag{11.28}$$

因此, 随着嵌入变量 q 从 0 增加到 1, $\phi(\eta,\xi;q)$ 从初始猜测解 $f_0(\eta,\xi)$ 连续地变化到原始方程 (11.17) 和 (11.18) 的精确解 $f(\eta,\xi)$.

然后, 将 $\phi(\eta,\xi;q)$ 展开成关于 q 的 Maclaurin 级数, 运用 (11.27), 我们有如下同伦–Maclaurin 级数

$$\phi(\eta,\xi;q) = f_0(\eta,\xi) + \sum_{m=1}^{+\infty} f_m(\eta,\xi)\, q^m, \tag{11.29}$$

其中

$$f_m(\eta,\xi) = \mathcal{D}_m[\phi(\eta,\xi;q)] = \frac{1}{m!}\frac{\partial^m \phi(\eta,\xi;q)}{\partial q^m}\bigg|_{q=0}$$

为 $\phi(\eta,\xi;q)$ 的 m 阶同伦导数, $\mathcal{D}_m$ 为 m 阶同伦导数算子. 注意, 同伦-Maclaurin 级数 (11.29) 是否收敛取决于初始猜测解 $f_0(\eta,\xi)$, 辅助线性算子 $\mathcal{L}$, 辅助函数 $H(\eta)$ 和收敛控制参数 c_0. 假设所有这些都被恰当地选择, 则同伦-Maclaurin 级数 (11.29) 在 $q=1$ 时绝对收敛, 根据 (11.28) 我们有同伦级数解

$$f(\eta,\xi) = f_0(\eta,\xi) + \sum_{m=1}^{+\infty} f_m(\eta,\xi). \tag{11.30}$$

根据定理 4.15, 我们有 m 阶形变方程

$$\mathcal{L}[f_m(\eta,\xi) - \chi_m f_{m-1}(\eta,\xi)] = c_0 H(\eta) \delta_{m-1}(\eta,\xi), \tag{11.31}$$

满足平板上的边界条件

$$f_m = 0, \quad \frac{\partial f_m}{\partial \eta} = 0, \quad \eta = 0 \tag{11.32}$$

和无穷远处的边界条件

$$\frac{\partial f_m}{\partial \eta} \to 0, \quad \eta \to +\infty, \tag{11.33}$$

其中

$$\chi_m = \begin{cases} 0, & m \leqslant 1, \\ 1, & m > 1, \end{cases} \tag{11.34}$$

以及

$$\delta_n(\eta,\xi) = \frac{\partial^3 f_n}{\partial \eta^3} + \sigma_1(\xi) \sum_{k=0}^{n} f_{n-k} \frac{\partial^2 f_k}{\partial \eta^2}$$
$$+ \sigma_2(\xi) \sum_{k=0}^{n} \left(\frac{\partial f_k}{\partial \xi} \frac{\partial^2 f_{n-k}}{\partial \eta^2} - \frac{\partial f_k}{\partial \eta} \frac{\partial^2 f_{n-k}}{\partial \xi \partial \eta} \right) \tag{11.35}$$

由定理 4.1 获得. 详情请参考第四章.

这里需要强调的是, 高阶形变方程 (11.31) 到 (11.33) 是线性的. 此外, 与摄动方法不同, 我们不需要任何物理小 (大) 参数来获得这些线性微分方程. 而且, 不同于 "局部相似性方法" [209, 217] 和 "局部非相似性方法" [216, 217, 221], 我们

不强制非相似项为零, 也不将变量 ξ 视为常数. 总之, 与之前针对非相似流动的所有其他解析方法不同, 我们的方法不需要任何额外的假设. 更重要的是, 如前所述, 我们有极大的自由来选择 $\mathcal{L}$: 这个自由如此之大, 以至于在同伦分析方法的框架下, 非线性偏微分方程可以转化为无限多个线性常微分方程, 如下所述.

在数学上, 近似非线性微分方程的本质是找到一组合适的基函数来拟合其解. 在物理上, 众所周知, 当 $\eta \to +\infty$ 时, 大多数黏性流动在无穷远处呈指数衰减. 因此, 对于拉伸平板上的非相似边界层流动, 当 $\eta \to +\infty$ 时, 速度 u 和 v 应呈指数衰减. 所以, $f(\eta, \xi)$ 应为如下形式

$$f(\eta, \xi) = \sum_{m=0}^{+\infty} \sum_{n=0}^{+\infty} a_{m,n}(\xi) \, \eta^n \, \exp(-m\,\eta), \tag{11.36}$$

其中 $a_{m,n}(\xi)$ 为 ξ 的多项式. 该表达式被称为 $f(\eta, \xi)$ 的解表达, 它在同伦分析方法的框架下起着重要的作用, 如下所述.

为了满足解表达 (11.36) 和边界条件 (11.18), 我们选择初始猜测解

$$f_0(\eta, \xi) = \check{U}_w(\xi)\left(1 - e^{-\eta}\right), \tag{11.37}$$

其包含了 $\eta \to +\infty$ 时, (11.36) 中最简单但最主要的项. 注意, $f_0(\eta, \xi)$ 满足边界条件 (11.18) 并在无穷远处呈指数衰减.

如前所述, 我们有极大的自由来选择辅助线性算子 $\mathcal{L}$. 然而, 这种自由受到解表达 (11.36) 和边界条件 (11.18) 的限制. 注意, 原始控制方程 (11.17) 是具有变系数的非线性偏微分方程. 因此, 如果我们选择偏微分算子作为 $\mathcal{L}$, 高阶形变方程 (11.31) 是一个偏微分方程. 众所周知, 具有变系数的偏微分方程比具有常系数的常微分方程难求解得多. 所以, 从数学上讲, 如果 $\mathcal{L}$ 是一个线性微分算子, 它只包含关于 η 或 ξ 的微分而且没有任何变系数, 那么求解 (11.31) 会容易得多. 从物理上讲, 对于边界层流动, 垂直流向的速度变化远大于沿流向的速度变化. 因此, 在垂直流动方向的偏导数

$$\frac{\partial f}{\partial \eta}, \frac{\partial^2 f}{\partial \eta^2}, \frac{\partial^3 f}{\partial \eta^3}$$

是相当大的, 所以在物理上比沿流动方向的偏导数

$$\frac{\partial f}{\partial \xi}, \frac{\partial^2 f}{\partial \xi \partial \eta}$$

更重要. 考虑到上述问题, 我们选择辅助线性算子

$$\mathcal{L}(f) = \frac{\partial^3 f}{\partial \eta^3} - \frac{\partial f}{\partial \eta}, \tag{11.38}$$

其与 ξ 无关, 并且不包含任何变系数. 详情请参考廖世俊 [200]. 注意, 辅助线性算子 (11.38) 与 §10.2 中求解一种相似边界层流动所使用的辅助线性算子 (10.16) 完全相同!

使用初始猜测解 (11.37) 和辅助线性算子 (11.38), 很容易求解线性常微分方程 (11.31) 至 (11.33). 方程 (11.31) 的特解为

$$f_m^*(\eta,\xi) = \chi_m\, f_{m-1}(\eta,\xi) + c_0\, \mathcal{L}^{-1}[H(\eta)\, \delta_{m-1}(\eta,\xi)], \tag{11.39}$$

其中 $\mathcal{L}^{-1}$ 表示 $\mathcal{L}$ 的逆算子, 则高阶形变方程 (11.31) 至 (11.33) 的解为

$$f_m(\eta,\xi) = f_m^*(\eta,\xi) + C_0(\xi) + C_1(\xi)\exp(-\eta),$$

其中

$$C_0(\xi) = -f_m^*(0,\xi) - \left.\frac{\partial f_m^*}{\partial \eta}\right|_{\eta=0}, \quad C_1(\xi) = \left.\frac{\partial f_m^*}{\partial \eta}\right|_{\eta=0}$$

由边界条件 (11.32) 和 (11.33) 确定. 通过这样的方式, 很容易求解高阶形变方程 (11.31) 至 (11.33), 特别是借助科学计算软件如 Mathematica. 有关数学公式的详细信息, 请参阅廖世俊 [199].

通过这种方式, 具有变系数的原始非线性偏微分方程被转化为无穷多个具有常系数的线性常微分方程. 这里需要强调的是, 非相似边界层流动的高阶形变方程 (11.31) 与第十章中具有呈指数衰减的解的相似边界层流动的高阶形变方程非常相似: 它们都使用了相同的辅助线性算子. 换句话说, 在同伦分析方法的框架下, 对于非相似边界层流动可以用与求解相似边界层流动类似的方法进行求解. 这极大地简化了对于与非相似边界层流动相关的非线性偏微分方程的求解.

最后, 值得强调的是 $f_m(\eta,\xi)$ 包含收敛控制参数 c_0. 如前所述, 收敛控制参数 c_0 为我们提供了一种简单的方法来保证所有物理变量 (参数) 的同伦级数解的收敛性, 如下所述.

11.3 同伦级数解

不失一般性, 让我们考虑如下的平板拉伸速度

$$U_w(x) = \frac{x}{1+x},$$

其相应的流动是非相似的, 因为它不满足相似性准则 (11.7).

在这种情况下,拉伸速度 $U_w(x)$ 沿着平板从 0 单调增加到 1. 此外,当 $x \to 0$ 时,$U_w \to x$;当 $x \to +\infty$ 时,$U_w \to 1$. 物理上,$x = 0$ 附近的流动应该接近当 $U_w = x$ 时的相似性流动,$x \to +\infty$ 处的流动应该接近当 $U_w = 1$ 时的相似性流动. 众所周知,当 $U_w = x$ 时和当 $U_w = 1$ 时的相似边界层流动对应的相似变量分别为 $y/\sqrt{\nu}$ 和 $y/\sqrt{\nu x}$. 因此,根据 η 的定义 (11.14),我们选择

$$\sigma(x) = \sqrt{1+x}.$$

所以,当 $x \to 0$ 时,η 趋向于对应的相似变量 $y/\sqrt{\nu}$;当 $x \to +\infty$ 时,η 趋向于对应的相似变量 $y/\sqrt{\nu x}$. 此外,我们很自然地定义

$$\xi = \Gamma(x) = \frac{x}{1+x},$$

根据 (11.19),它给出了

$$\check{U}_w(\xi) = \xi, \quad \sigma_1(\xi) = \frac{1}{2}, \quad \sigma_2(\xi) = 1 - \xi.$$

与相似边界层流动一样,非相似流动的高阶形变方程 (11.31) 至 (11.33) 是线性常微分方程. 因此,我们可以直接通过 BVPh 1.0 求解,具体过程见附录 7.1. BVPh 1.0 的相应输入数据见本章附录.

到目前为止,只有辅助函数 $H(\eta)$ 和收敛控制参数 c_0 没有确定. 为了简单起见,我们选择

$$H(\eta) = 1,$$

则高阶形变方程 (11.31) 至 (11.33) 的解 $f_m(\eta, \xi)$ 仅取决于收敛控制参数 c_0,它为我们提供了一种便捷的方法来保证同伦级数解的收敛性,如第二章所述. (11.17) 在 $\eta \in [0, 10]$ 上的 k 阶近似的平均平方残差是通过 BVPh 1.0 中的模块 GetErr[k] 进行计算,然后在区间 $\xi \in [0, 1]$ 上积分获得的. (11.17) 相对于 c_0 的平方残差如图 11.1 所示. 注意,在 10 阶和 15 阶近似处,最优收敛控制参数 c_0 分别约为 -0.8 和 -0.6. 在 20 阶近似处,平方残差的最小值为 3.6×10^{-8},对应的最优收敛控制参数 $c_0^* = -0.5554$. 可以发现,当 $c_0 = -1/2$ 时,同伦级数解 (11.30) 在整个空间区间 $0 \leqslant \xi \leqslant 1$ 和 $0 \leqslant \eta < +\infty$ 上收敛,分别对应 $0 \leqslant x < +\infty$ 和 $0 \leqslant y < +\infty$,如表 11.1 和图 11.2 所示.

11.3 同伦级数解

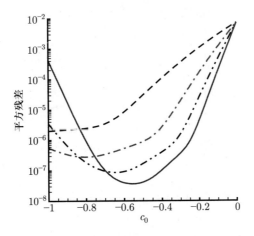

图 11.1 (11.17) 相对于 c_0 的平方残差. 虚线: 5 阶近似; 点划线: 10 阶近似; 双点划线: 15 阶近似; 实线: 20 阶近似

表 11.1 选取 $c_0 = -1/2$ 时, 控制方程 (11.17) 在 $\eta \in [0, 10]$ 和 $\xi \in [0, 1]$ 上的平均平方残差

近似阶数	(11.17) 的平均平方残差
1	2.3×10^{-3}
2	7.4×10^{-4}
4	8.5×10^{-5}
6	1.1×10^{-5}
8	2.3×10^{-6}
10	8.9×10^{-7}
15	1.7×10^{-7}
20	4.2×10^{-8}
25	1.8×10^{-8}
30	1.0×10^{-8}

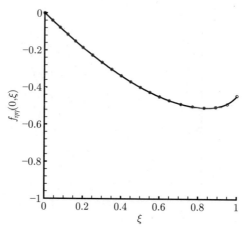

图 11.2 选取 $c_0 = -1/2$ 时, $f_{\eta\eta}(0,\xi)$ 的同伦近似. 实线: 30 阶近似; 空心圆: 20 阶近似

根据图 11.1, 当 $c_0 = -1$ 时同伦级数 (11.30) 发散. 正如廖世俊 [193] 所证明的那样, 一些非摄动方法, 如 Lyapunov 人工小参数法 [207], Adomian 分解法 [172,173], δ-展开法 [184], 是当 $c_0 = -1$ 时同伦分析方法的特例. 此外, 还证明了 [187,212] 所谓的 "同伦摄动方法" [182] 也是当 $c_0 = -1$ 时同伦分析方法的特例. 因此, 如果使用上述任何一种方法, 都无法获得收敛的结果. 这再次说明了收敛控制参数 c_0 的重要性, 以及同伦分析方法对于强非线性问题的有效性.

对于给出非相似边界层流动的精确局部表面摩擦系数是非常重要的, 该系数通过 (11.20) 与 $f_{\eta\eta}(0,\xi)$ 相关. 当 $c_0 = -1/2$ 时, 第 15 阶同伦分析方法近似为

$$\begin{aligned}
f_{\eta\eta}(0,\xi) = & -\xi + 0.357142\,\xi^2 + 0.0745784\,\xi^3 + 0.0329563\,\xi^4 + 0.0187480\,\xi^5 \\
& + 0.0121641\,\xi^6 + 0.00856336\,\xi^7 + 0.00638095\,\xi^8 + 0.00468866\,\xi^9 \\
& + 0.0125286\,\xi^{10} - 0.108975\,\xi^{11} + 0.729854\,\xi^{12} - 2.509780\,\xi^{13} \\
& + 4.702786\,\xi^{14} - 4.475992\,\xi^{15} + 1.690259\,\xi^{16},
\end{aligned} \tag{11.40}$$

这与 20 阶近似非常吻合, 并且在整个区域 $0 \leqslant \xi \leqslant 1$ 中是精确的, 如图 11.2 所示. 然后, 直接通过 (11.20) 计算表面摩擦的局部系数 C_f. 可以发现, 当 $x \to 0$ 时, $C_f \to -2\sqrt{\nu}/x$; 当 $x \to +\infty$ 时, $C_f \to -0.8875\sqrt{\nu/x}$, 如图 11.3 所示. 此外, 对于非相似流动的边界层厚度 $\bar{\delta}(x)$, 当 $x \to 0$ 时, 趋近于 $\sqrt{\nu}$; 当 $x \to +\infty$ 时, 趋近于 $1.61613\sqrt{\nu\,x}$, 如图 11.4 所示. 注意, 当 $U_w = x\ (x \to 0)$ 时, 相应相似流动的边界层厚度仅为 $\sqrt{\nu}$; 当 $U_w = 1\ (x \to +\infty)$ 时, 为 $1.61613\sqrt{\nu\,x}$. 因此, 非相似流动的同伦级数解可分别给出

$$C_f(x) \to -2\sqrt{\nu}/x, \quad \delta(x) \to \sqrt{\nu}, \quad x \to 0$$

和
$$C_f(x) \to -0.8875\sqrt{\nu/x}, \quad \delta(x) \to 1.61613\sqrt{\nu\, x}, \quad x \to +\infty.$$

因此, 从物理上讲, $x \to 0$ 和 $x \to +\infty$ 区域内的非相似边界层流动分别与当 $U_w = x$ 和 $U_w = 1$ 时对应的相似边界层流动非常接近. 然而, 其他区域的流动是非相似的, 分别如图 11.3 和图 11.4 所示. 图 11.5 给出了不同 x 处非相似边界层流动的速度剖面 $u \sim y/\sqrt{\nu}$. 关于同伦级数解物理意义的更多讨论, 请参阅廖世俊 [199].

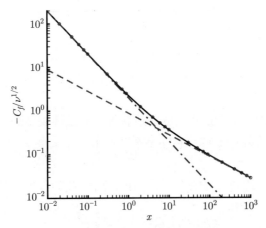

图 11.3 当 $U_w = x/(1+x)$ 时, 非相似流动的局部表面摩擦系数 $C_f(x)/\sqrt{\nu}$. 实线: 30 阶同伦分析方法结果; 空心圆: 20 阶同伦分析方法结果; 虚线: $C_f(x) = -0.8875\sqrt{\nu/x}$; 点划线: $C_f(x) = -2\sqrt{\nu/x}$

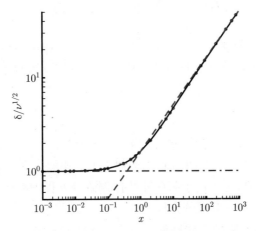

图 11.4 当 $U_w = x/(1+x)$ 时, 非相似流动的边界层厚度 $\delta(x)/\sqrt{\nu}$. 实线: 30 阶同伦分析方法结果; 空心圆: 20 阶同伦分析方法结果; 虚线: $\delta(x) = 1.61613\sqrt{\nu\, x}$; 点划线: $\delta(x) = \sqrt{\nu}$

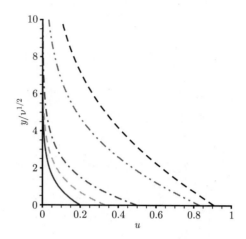

图 11.5 非相似流动在不同 x 处的速度剖面 $u \sim y/\sqrt{\nu}$. 实线: $x = 1/4$; 虚线: $x = 1/2$; 点划线: $x = 1$; 双点划线: $x = 5$; 长虚线: $x = 10$

因此，通过 BVPh 1.0，我们成功地获得了非线性偏微分方程控制的非相似边界层流动的收敛解 (11.17)，它在整个区域 $0 \leqslant x < +\infty$ 和 $0 \leqslant y < +\infty$ 上均有效.

以上所有结果均通过 $H(\eta) = 1$ 给出. 注意，我们有极大的自由来选择辅助函数 $H(\eta)$. 例如，使用

$$H(\eta) = \exp(-\eta),$$

我们获得如下形式的级数解

$$f(\eta, \xi) = \sum_{m=0}^{+\infty} b_m(\xi) \exp(-m\,\eta),$$

其中 $b_m(\xi)$ 为 ξ 的多项式. 在这种情况下，我们可以通过类似的方式选取 $c_0 = -3/2$ 来获得收敛的级数解. 这再次说明了同伦分析方法的灵活性.

11.4 本章小结

本章在同伦分析方法的框架下，描述了一种非相似边界层流动的非线性偏微分方程可以直接通过 BVPh 1.0 以类似于求解相似边界层流动的方式进行求解. 这显示了 BVPh 1.0 对于某些非线性偏微分方程的有效性，特别是对于那些与边界层流动相关的偏微分方程.

需要注意的是，原始非线性偏微分方程 (11.17) 包含 $f(\eta, \xi)$ 关于 η 和 ξ 的导数. 然而，我们在这里选择的辅助线性算子 (11.38) 只包含关于 η 的导数.

从数学上讲，这主要是因为同伦分析方法为我们选取辅助线性算子提供了极大自由，如第二章所述。更重要的是，同伦分析方法还通过选择合适的收敛控制参数 c_0 为我们提供了一种便捷的方式来保证同伦级数解的收敛性：如果不能保证同伦级数解的收敛性，选取辅助线性算子 $\mathcal{L}$ 的自由将没有任何意义。有趣的是，相同的辅助线性算子 (11.38) 也被廖世俊和 Pop [203] 用来求解相似边界层流动。此外，辅助线性算子 (11.38) 与 §10.2 中用来求解一种相似边界层流动所使用的辅助线性算子 (10.16) 完全相同。因此，在同伦分析方法的框架下，对于非相似边界层流动可以用与求解相似边界层流动类似的方式进行求解。从物理上讲，这主要是由于边界层的存在：垂直流向的速度变化远大于沿流向的速度变化。因此，BVPh 1.0 可以采用类似的方式求解其他非相似边界层流动 [185, 186, 222]。

众所周知，非线性偏微分方程比线性常微分方程更难求解。因此，如果可能的话，用一系列线性常微分方程替代非线性偏微分方程是一个很好的办法。因此，这种方法具有普遍意义。例如，如廖世俊 [196, 197] 所示，在同伦分析方法的框架下，一个描述非定常边界层流动的非线性偏微分方程可以转化为类似于 (11.31) 的无穷多个线性常微分方程。此外，如廖世俊，Su 和 Chwang [204] 所示，描述非定常传热的非线性偏微分方程可以被无穷多个线性常微分方程替代。所有这些类型的非线性偏微分方程都可以通过 BVPh 1.0 采用类似的方式进行求解。这显示了 BVPh 1.0 的普遍有效性。

最后，应该指出的是，虽然这种方法具有普遍意义，但它并不是对所有类型的非线性偏微分方程都有效，尤其是与波有关的非线性偏微分方程。正如第七章中提到的，我们的目标是开发一个对尽可能多的非线性边值问题有效的软件包。即便如此，基于同伦分析方法的 Mathematica 软件包 BVPh 1.0 可以作为求解许多非线性偏微分方程的工具，特别是那些与非相似性和 (或) 非定常边界层流动和传热相关的非线性偏微分方程。

参考文献

[172] Adomian, G.: Nonlinear stochastic differential equations. J. Math. Anal. Applic. **55**, 441–452 (1976).

[173] Adomian, G.: Solving Frontier Problems of Physics: The Decomposition Method. Kluwer Academic Publishers, Boston (1994).

[174] Banks,W.H.H.: Similarity solutions of the boundary-layer equations for a stretching wall. Journal de Mecanique theorique et appliquee. **2**, 375–392 (1983).

[175] Blasius, H.: Grenzschichten in Füssigkeiten mit kleiner Reibung. Z. Math. Phys. **56**, 1–37 (1908).

[176] Cheng, W.T., Lin, H.T.: Non-similarity solution and correlation of transient heat transfer in laminar boundary layer flow over a wedge. Int. J. Engineering Science. **40**, 531–548 (2002).

[177] Cimpean, D., Merkin, J.H., Ingham, D.B.: On a free convection problem over a vertical flat surface in a porous medium. Transport in Porous. **64**, 393–411 (2006).

[178] Crane, L.: Flow past a stretching plate. Z. Angew. Math. Phys. **21**, 645–647 (1970).

[179] Duck, P.W., Stow, S.R., Dhanak, M.R.: Non-similarity solutions to the corner boundary-layer equations (and the effects of wall transpiration). J. Fluid Mechanics. **400**, 125–162 (1999).

[180] Gorla, R.S.R., Kumari, M.: Non-similar solutions for mixed convection in non-Newtonian fluids along a vertical plate in a porous medium. Transport in Porous. **33**, 295–307 (1998).

[181] Görtler, H.: Eine neue Reihenentwicklung für laminare Grenzschichten. ZAMM. **32**, 270–271 (1952).

[182] He, J.H.: Homotopy perturbation technique. Comput. Method. Appl. M. **178**, 257–262 (1999).

[183] Howarth, L.: On the solution of the laminar boundary layer equations. Proc. Roy. Soc. London A. **164**, 547–579 (1938).

[184] Karmishin, A.V., Zhukov, A.T., Kolosov, V.G.: Methods of Dynamics Calculation and Testing for Thin-walled Structures (in Russian). Mashinostroyenie, Moscow (1990).

[185] Kousar, N., Liao, S.J.: Series solution of non-similarity boundary flows over a porous wedge. Transport Porous Media. **83**, 397–412 (2010).

[186] Kousar, N., Liao, S.J.: Unsteady non-similarity boundary-layer flows caused by an impulsively stretching flat sheet. Nonlinear Analysis B. **12**, 333–342 (2011).

[187] Liang, S.X., Jeffrey, D.J.: Comparison of homotopy analysis method and homotopy perturbation method through an evalution equation. Commun. Nonlinear Sci. Numer. Simulat. **14**, 4057–4064 (2009).

[188] Liao, S.J.: The Proposed Homotopy Analysis Technique for the Solution of Nonlinear Problems. PhD dissertation, Shanghai Jiao Tong University (1992).

[189] Liao, S.J.: A kind of approximate solution technique which does not depend upon small parameters—(II) An application in fluid mechanics. Int. J. Nonlin. Mech. **32**, 815–822 (1997).

[190] Liao, S.J.: An explicit, totally analytic approximation of Blasius viscous flow problems. Int. J. Nonlin. Mech. **34**, 759–778 (1999a).

[191] Liao, S.J.: A uniformly valid analytic solution of 2D viscous flow past a semi-infinite flat plate. J. Fluid Mech. **385**, 101–128 (1999b).

[192] Liao, S.J.: On the analytic solution of magnetohydrodynamic flows of non-Newtonian fluids over a stretching sheet. J. Fluid Mech. **488**, 189–212 (2003a).

[193] Liao, S.J.: Beyond Perturbation –Introduction to the Homotopy Analysis Method. Chapman & Hall/ CRC Press, Boca Raton (2003b).

[194] Liao, S.J.: On the homotopy analysis method for nonlinear problems. Appl. Math. Comput. **147**, 499–513 (2004).

[195] Liao, S.J.: A new branch of solutions of boundary-layer flows over an impermeable stretched plate. Int. J. Heat Mass Tran. **48**, 2529–2539 (2005).

[196] Liao, S.J.: An analytic solution of unsteady boundary-layer flows caused by an impulsively stretching plate. Communications in Nonlinear Science and Numerical Simulation. **11**, 326–339 (2006a).

[197] Liao, S.J.: Series solutions of unsteady boundary-layer flows over a stretching flat plate. Stud. Appl. Math. **117**, 2529–2539 (2006b).

[198] Liao, S.J.: Notes on the homotopy analysis method –Some definitions and theorems. Commun. Nonlinear Sci. Numer. Simulat. **14**, 983–997 (2009a).

[199] Liao, S.J.: A general approach to get series solution of non-similarity boundary layer flows. Commun. Nonlinear Sci. Numer. Simulat. **14**, 2144–2159 (2009b).

[200] Liao, S.J.: On the relationship between the homotopy analysis method and Euler transform. Commun. Nonlinear Sci. Numer. Simulat. **15**, 1421–1431 (2010a). doi:10.1016/j.cnsns.2009.06.008.

[201] Liao, S.J.: An optimal homotopy-analysis approach for strongly nonlinear differential equations. Commun. Nonlinear Sci. Numer. Simulat. **15**, 2003–2016 (2010b).

[202] Liao, S.J., Campo, A.: Analytic solutions of the temperature distribution in Blasius viscous flow problems. J. Fluid Mech. **453**, 411–425 (2002).

[203] Liao, S.J., Pop, I.: Explicit analytic solution for similarity boundary layer equations. Int. J. Heat Mass Transfer. **47**, 75–85 (2004).

[204] Liao, S.J., Su, J., Chwang, A.T.: Series solutions for a nonlinear model of combined convective and radiative cooling of a spherical body. Int. J. Heat and Mass Transfer. **49**, 2437–2445 (2006).

[205] Liao, S.J., Magyari, E.: Exponentially decaying boundary layers as limiting cases of families of algebraically decaying ones. Z. angew. Math. Phys. **57**, 777–792 (2006).

[206] Liao, S.J., Tan, Y.: A general approach to obtain series solutions of nonlinear differential equations. Stud. Appl. Math. **119**, 297–355 (2007).

[207] Lyapunov, A.M.: General Problem on Stability of Motion (English translation). Taylor & Francis, London (1992).

[208] Magyari, E., Keller, B.: Exact solutions for self-similar boundary-layer flows induced by permeable stretching walls. Eur. J. Mech. B-Fluids. **19**, 109–122 (2000).

[209] Massoudi, M.: Local non-similarity solutions for the flow of a non-Newtonian fluid over a wedge. Int. J. Non-Linear Mech. **36**, 961–976 (2001).

[210] Prandtl, L.: Über Flüssigkeitsbewegungen bei sehr kleiner Reibung. Verhandlg. Int. Math. Kongr. Heidelberg. 484–491 (1904).

[211] Roy, S., Datta, P., Mahanti, N.C.: Non-similar solution of an unsteady mixed convection flow over a vertical cone with suction or injection. Int. J. of Heat and Mass Transfer. **50**, 181–187 (2007).

[212] Sajid, M., Hayat, T.: Comparison of HAM and HPM methods for nonlinear heat conduction and convection equations. Nonlin. Anal. B. **9**, 2296–2301 (2008).

[213] Sahu, A.K., Mathur, M.N., Chaturani, P., Bharatiya, S.S.: Momentum and heat transfer from a continuous moving surface to a power-law fluid. Acta Mechanica. **142**, 119–131 (2000).

[214] Schlichting, H., Gersten, K.: Boundary Layer Theory. Springer, Berlin (2000).

[215] Sobey, I.J.: Introduction to Interactive Boundary Layer Theory. Oxford University Press, Oxford (2000).

[216] Sparrow, E.M., Quack, H.: Local non-similarity boundary-layer solutions. AIAA J. **8**, No.8, 1936–1942 (1970).

[217] Sparrow, E.M., Yu, H.S.: Local non-similarity thermal boundary-layer solutions. J. Heat Transfer Trans. ASME. 328–334 (1971).

[218] Stewartson, C.D., Williams: Viscous flow past a flat plate with uniform injection. Proceedings of Royal Society (A). **284**, 370–396 (1965).

[219] Tani, I.: Histry of boundary-layer theory. Ann. Rev. Fluid Mech. **9**, 87–111 (1977).

[220] Van Dyke, M.: Higher approximations in boundary-layer theory. Part 1: General analysis. Journal of Fluid Mechanics. **14**, 161–177 (1962).

[221] Wanous, K.J., Sparrow, E.M.: Heat transfer for flow longitudinal to a cylinder with surface mass transfer. J. Heat Transfer Trans. Ser. C. **87**, 317–319 (1965).

[222] You, X.C., Xu, H., Liao, S.J.: On the non-similarity boundary-Layer flows of second-order fluid over a stretching sheet. ASME J. App. Mech. **77**: 021002–1 (2010).

附录 11.1 BVPh 的输入数据

```
(* Input Mathematica package BVPh version 1.0 *)
<<BVPh.txt;


(* Define the physical and control parameters *)
TypeEQ      = 1;
```

附录 11.1 BVPh 的输入数据

```
ApproxQ      = 0;
ErrReq       = 10^(-10);
NgetErr      = 100;
zRintegral   = 10;

(* Define the governing equation *)
f[z_,u_,lambda_] := Module[{temp},
temp[1] = D[u,{z,3}] + u*D[u,{z,2}]/2;
temp[2] = D[u,t]*D[u,{z,2}] - D[u,z]*D[u,z,t];
temp[1] + (1-t)* temp[2] // Expand
];

(* Define boundary conditions *)
zR  = infinity;
OrderEQ = 3;
BC[1,z_,u_,lambda_] := u    /. z-> 0 ;
BC[2,z_,u_,lambda_] := D[u,z] - t /.  z -> 0 ;
BC[3,z_,u_,lambda_] := Limit[D[u,z], z -> zR ];

(* Define initial guess *)
u[0]    = t*(1 - Exp[-z]);

(* Define the auxiliary linear operator *)
L[u_] := D[u,{z,3}] - D[u,z];

(* Define output term *)
output[z_,u_,k_]:= Print["output         = ",
                D[u[k],{z,2}]/.{z->0,t->1}//N];

(* Define Getdelta[k] *)
Getdelta[k_]:=Module[{temp,i},
uz[k]    = D[u[k],z]//Expand;
uzz[k]   = D[uz[k],z]//Expand;
uzzz[k]  = D[uzz[k],z]//Expand;
ut[k]    = D[u[k],t]//Expand;
uzt[k]   = D[uz[k],t]//Expand;
uzzu[k]  = Sum[uzz[i]*u[k-i],{i,0,k}]//Expand;
uzuzt[k] = Sum[uz[i]*uzt[k-i],{i,0,k}]//Expand;
uzzut[k] = Sum[uzz[i]*ut[k-i],{i,0,k}]//Expand;
```

```
temp[1] = uzzz[k] + uzzu[k]/2 //Expand;
temp[2] = (1-t)*(uzzut[k] - uzuzt[k]);
delta[k] = temp[1] + temp[2]//Expand;
];


(* Print input and control parameters *)
PrintInput[u[z,t]];

(* Set convergence-control parameter c0 *)
c0 = -1/2;
Print["c0  =  ",c0];

(* Gain 10th-order HAM approximation *)
BVPh[1,10];

(* Calculate the squared residual *)
For[k=2, k<=10, k=k+2,
GetErr[k];
err[k] = Integrate[Err[k],{t,0,1}];
Print[" k  =  ", k, "   Squared residual  =  ",
        err[k]//N];
];
```

需要注意的是，对于在无限区间定义的非线性边值问题，我们必须在 BVPh 1.0 的输入数据中设置 zR = infinity. 此外，当非线性偏微分方程通过 BVPh 1.0 进行求解时，我们必须在输入数据中定义模块 Getdelta，以便计算由 (11.2.14) 定义的项.

第十二章　非定常边界层流动

本章阐述了基于同伦分析方法的 Mathematica 软件包 BVPh 1.0 对于求解与非定常边界层流动相关的非线性偏微分方程的有效性。研究表明，基于 BVPh 1.0，非定常边界层流动可以用与求解由非线性常微分方程控制的定常相似边界层流动以非常类似的方式来进行求解。换句话说，在同伦分析方法的框架下，求解非定常边界层流动和求解定常边界层流动一样简单。这显示了 BVPh 1.0 对于求解某些非线性偏微分方程的有效性，特别是对于那些与边界层流动相关的非线性偏微分方程。

12.1　绪论

第十一章阐述了基于同伦分析方法的 Mathematica 软件包 BVPh 1.0 可用于解决一种在无限区间内由非线性偏微分方程控制的非相似边界层流动。本章将进一步说明 BVPh 1.0 可用于获得由无限时空区间上的非线性偏微分方程控制的非定常相似边界层流动的高精度近似值。

不失一般性，我们在这里考虑由脉冲拉伸平板产生的黏性流体的非定常边界层流动 [223, 224, 241–247]，其由偏微分方程

$$\frac{\partial u}{\partial t} + u\frac{\partial u}{\partial x} + v\frac{\partial u}{\partial y} = \nu\frac{\partial^2 u}{\partial y^2}, \tag{12.1}$$

$$\frac{\partial u}{\partial x} + \frac{\partial v}{\partial y} = 0 \tag{12.2}$$

控制，满足边界条件

$$t > 0 : u = a\,x, \quad v = 0, \quad y = 0, \tag{12.3}$$

$$t > 0 : u \to 0, \quad y \to +\infty \tag{12.4}$$

以及初始条件

$$t = 0 : u = v = 0 \text{ 对于所有点 } (x, y), \tag{12.5}$$

其中 ν 为运动黏度, t 表示时间, (u, v) 分别为沿 x, y 正方向的速度分量. 在这里, 对应于一个拉伸平板, 我们只考虑 $a > 0$ 的情况.

这类流动存在相似变量. 根据 Seshadri 等人 [243] 和 Nazar 等人 [241] 的研究, 我们采用 Williams 和 Rhyne 的相似变换 [248]:

$$\psi = \sqrt{a\nu\xi}\, x\, f(\eta, \xi), \quad \eta = \sqrt{\frac{a}{\nu\xi}}\, y, \quad \xi = 1 - \exp(-\tau), \quad \tau = a\,t, \tag{12.6}$$

其中 ψ 表示流函数. 注意, 新的无量纲时间 ξ 在有限区间

$$0 \leqslant \xi \leqslant 1$$

内有界, 对应于 $0 \leqslant \tau < +\infty$. 此外, 相似变量 η 不仅取决于空间变量 y, 还取决于时间变量 t, 因此自动满足初始条件 (12.5). 因此, 在任何时刻 $\tau \in [0, +\infty)$, 即 $\xi \in [0, 1]$, 它仍然是一个相似边界层流动.

基于上述相似变换, 原始的偏微分方程变为

$$\frac{\partial^3 f}{\partial \eta^3} + \frac{1}{2}(1-\xi)\eta\frac{\partial^2 f}{\partial \eta^2} + \xi\left[f\frac{\partial^2 f}{\partial \eta^2} - \left(\frac{\partial f}{\partial \eta}\right)^2\right] = \xi(1-\xi)\frac{\partial^2 f}{\partial \eta \partial \xi}, \tag{12.7}$$

满足边界条件

$$f(0, \xi) = 0, \quad \left.\frac{\partial f}{\partial \eta}\right|_{\eta=0} = 1, \quad \left.\frac{\partial f}{\partial \eta}\right|_{\eta=+\infty} = 0. \tag{12.8}$$

这仍然是一个非线性偏微分方程. 然而, 由于初始条件 (12.5) 自动满足, 因此这里只存在边界条件 (12.8), 上述非线性偏微分方程比原始的两个耦合偏微分方程更容易求解.

当 $\xi = 0$ 时, 对应于 $\tau = 0$, (12.7) 为 Rayleigh 型方程

$$\frac{\partial^3 f}{\partial \eta^3} + \frac{1}{2}\eta\frac{\partial^2 f}{\partial \eta^2} = 0, \tag{12.9}$$

满足

$$f(0, 0) = 0, \quad \left.\frac{\partial f}{\partial \eta}\right|_{\eta=0, \xi=0} = 1, \quad \left.\frac{\partial f}{\partial \eta}\right|_{\eta=+\infty, \xi=0} = 0. \tag{12.10}$$

上述方程有封闭解

$$f(\eta, 0) = \eta \operatorname{erfc}(\eta/2) + \frac{2}{\sqrt{\pi}} \left[1 - \exp(-\eta^2/4)\right], \tag{12.11}$$

其中 $\operatorname{erfc}(\eta)$ 是互补误差函数, 定义如下

$$\operatorname{erfc}(\eta) = \frac{2}{\sqrt{\pi}} \int_{\eta}^{+\infty} \exp(-z^2) \, dz.$$

当 $\xi = 1$ 时, 对应于 $\tau \to +\infty$, 由 (12.7) 可得

$$\frac{\partial^3 f}{\partial \eta^3} + f \frac{\partial^2 f}{\partial \eta^2} - \left(\frac{\partial f}{\partial \eta}\right)^2 = 0, \tag{12.12}$$

满足

$$f(0,1) = 0, \quad \left.\frac{\partial f}{\partial \eta}\right|_{\eta=0, \xi=1} = 1, \quad \left.\frac{\partial f}{\partial \eta}\right|_{\eta=+\infty, \xi=1} = 0. \tag{12.13}$$

上述方程有封闭解

$$f(\eta, 1) = 1 - \exp(-\eta). \tag{12.14}$$

因此, 随着时间变量 ξ 从 0 增加到 1, $f(\eta, \xi)$ 从初始解 (12.11) 变化到定常解 (12.14). 值得注意的是, 虽然 $f'(+\infty, \xi) \to 0$ 对所有 ξ 都是指数级的, 其中 ' 表示对 η 求导, 但是当 $\eta \to +\infty$ 时, 初始解 (12.11) 比定常解 (12.14) 的衰减速度要快得多. 因此, 从数学角度来讲, 初始解 (12.11) 本质上不同于定常解 (12.14). 这可能就是为什么很难给出对所有时间 $0 \leqslant \tau < +\infty$ 都一致有效的精确解析解.

当 $\xi = 0$ 和 $\xi = 1$ 时, 我们分别有

$$\left.\frac{\partial^2 f}{\partial \eta^2}\right|_{\eta=0, \xi=0} = -\frac{1}{\sqrt{\pi}} \tag{12.15}$$

和

$$\left.\frac{\partial^2 f}{\partial \eta^2}\right|_{\eta=0, \xi=1} = -1. \tag{12.16}$$

表面摩擦系数为

$$c_f^x(x, \xi) = (\xi \operatorname{Re}_x)^{-1/2} f''(0, \xi), \quad 0 \leqslant \xi \leqslant 1, \tag{12.17}$$

其中 $\operatorname{Re}_x = ax^2/\nu$ 为局部雷诺数.

12.2 摄动近似

如 Seshadri 等人 [243] 和 Nazar 等人 [241] 所示,可以把 ξ 看作一个小参数来寻找如下形式的摄动近似

$$f(\eta,\xi) = g_0(\eta) + g_1(\eta)\,\xi + g_2(\eta)\,\xi^2 + \cdots = \sum_{m=0}^{+\infty} g_m(\eta)\,\xi^m, \tag{12.18}$$

将其代入 (12.7) 和 (12.8),并平衡 ξ 的同幂系数,我们有零阶摄动方程

$$g_0'''(\eta) + \frac{\eta}{2} g_0''(\eta) = 0, \quad g_0(0) = 0,\; g_0'(0) = 1,\; g_0'(+\infty) = 0 \tag{12.19}$$

和 m 阶 $(k \geqslant 1)$ 摄动方程

$$g_m'''(\eta) + \frac{\eta}{2} g_m''(\eta) - m\, g_m'(\eta) = \frac{\eta}{2} g_{m-1}''(\eta) - (m-1)g_{m-1}'(\eta)$$
$$- \sum_{i=0}^{m-1} \left[g_i(\eta)\, g_{m-1-i}''(\eta) - g_i'(\eta) g_{m-1-i}'(\eta) \right], \tag{12.20}$$

满足边界条件

$$g_m(0) = 0, \quad g_m'(0) = 0, \quad g_m'(+\infty) = 0. \tag{12.21}$$

上述所有摄动方程都仅仅是关于 η 的线性常微分方程.

零阶摄动方程 (12.19) 的解为

$$g_0(\eta) = f(\eta,0) = \eta\,\mathrm{erfc}(\eta/2) + \frac{2}{\sqrt{\pi}} \left[1 - \exp(-\eta^2/4)\right], \tag{12.22}$$

其中 $\mathrm{erfc}(\eta/2)$ 是互补误差函数,$\exp(-\eta^2/4)$ 是高斯分布函数. 将上述表达式代入 (12.20) 和 (12.21) 得到一阶摄动方程

$$g_1'''(\eta) + \frac{\eta}{2} g_1''(\eta) - g_1'(\eta) = \left[\frac{2}{\pi} - \frac{\eta}{2\sqrt{\pi}} + \frac{\eta}{\sqrt{\pi}}\mathrm{erfc}\left(\frac{\eta}{2}\right)\right] \exp\left(-\frac{\eta^2}{4}\right)$$
$$- \frac{2}{\pi} \exp\left(-\frac{\eta^2}{2}\right) + \left[\mathrm{erfc}\left(\frac{\eta}{2}\right)\right]^2, \tag{12.23}$$

满足边界条件

$$g_1(0) = 0, \quad g_1'(0) = 0, \quad g_1'(+\infty) = 0. \tag{12.24}$$

上述方程虽然是线性常微分方程，但求解难度较大. 首先, 即使是齐次方程

$$g_1'''(\eta) + \frac{\eta}{2}g_1''(\eta) - g_1'(\eta) = 0$$

也存在一个相当复杂的特解

$$g_1^*(\eta) = \left(1 + \frac{\eta^2}{4}\right)\exp\left(-\frac{\eta^2}{4}\right) - \frac{3\sqrt{\pi}}{4}\left(\eta + \frac{\eta^3}{6}\right)\operatorname{erfc}\left(\frac{\eta}{2}\right),$$

虽然其他两个特解

$$g_1^*(\eta) = 1, \quad g_1^*(\eta) = \eta + \frac{\eta^3}{6}$$

比较简单. 因此, 摄动近似似乎不可能避免互补误差函数 $\operatorname{erfc}(\eta/2)$ 和高斯分布函数 $\exp(-\eta^2/4)$. 其次, 一阶摄动方程 (12.23) 的右端项包含互补误差函数 $\operatorname{erfc}(\eta/2)$ 和高斯分布函数 $\exp(-\eta^2/4)$ 以及它们的组合形式, 如

$$\eta \operatorname{erfc}\left(\frac{\eta}{2}\right)\exp\left(-\frac{\eta^2}{4}\right).$$

这也可能是 Seshadri 等人 [243] 和 Nazar 等人 [241] 仅给出如下一阶摄动近似的原因

$$\begin{aligned}g_1(\eta) =& \left(\frac{1}{2} - \frac{2}{3\pi}\right)\left[\left(1 + \frac{\eta^2}{2}\right)\operatorname{erfc}(\eta/2) - \frac{\eta}{\sqrt{\pi}}e^{-\eta^2/4}\right] \\ & - \frac{1}{2}\left(1 - \frac{\eta^2}{2}\right)\operatorname{erfc}^2(\eta/2) - \frac{3\eta}{2\sqrt{\pi}}\,e^{-\eta^2/4}\operatorname{erfc}(\eta/2) \\ & - \frac{1}{\sqrt{\pi}}\left(\frac{4}{3\sqrt{\pi}} - \frac{\eta}{4}\right)e^{-\eta^2/4} + \frac{2}{\pi}\,e^{-\eta^2/2}. \end{aligned} \quad (12.25)$$

值得注意的是, 随着近似阶数的增加, 高阶摄动方程 (12.20) 的右端项变得越来越复杂. 所以, 虽然我们很幸运地获得了一阶摄动方程 (12.23) 和 (12.24) 的解 (12.25), 但是求解高阶问题变得越来越困难. 更重要的是, 即使我们可以有效地求解所有高阶摄动方程, 我们仍然不能保证摄动级数 (12.18) 的收敛性.

因此, 虽然原始非线性偏微分方程 (12.7) 通过将 ξ 看作一个物理小参数并将 $f(\eta,\xi)$ 展开为摄动级数 (12.18) 而后转化为无穷多个线性常微分方程 (12.19) 和 (12.20), 但我们仍不能有效地获得高阶摄动近似值, 这是因为相应的线性摄动方程 (12.20) 变得越来越难以求解. 注意, 摄动方程 (12.19) 和 (12.20) 完全取决于摄动量 ξ 和原始非线性偏微分方程 (12.7). 所以我们没有选择比 (12.22) 更好的初始解, 或者比下式更好的线性算子

$$\mathcal{L}_p(f) = f''' + \frac{\eta}{2}\,f'' - m\,f' \quad (12.26)$$

的自由，其中 ′ 表示对 η 求导.

一阶摄动近似给出的表面摩擦系数为

$$C_f^x(x,\xi) \approx -\frac{1}{\sqrt{\pi\,\xi\,\mathrm{Re}_x}}\left[1+\left(\frac{5}{4}-\frac{4}{3\pi}\right)\xi\right], \tag{12.27}$$

其在整个时间区间内并不是一个很好的近似值 (见后图 12.5). 因此摄动方法对于这种非定常的边界层流动，无法在整个时空区间内获得足够精确的近似值.

12.3 同伦级数解

12.3.1 简明数学公式

如廖世俊 [235,236] 所示，一些非定常相似边界层流动可以通过同伦分析方法 [225-240,249] 进行求解. 这里，我们说明了由非线性偏微分方程 (12.7) 和边界条件 (12.8) 控制的非定常边界层流动可以通过 BVPh 1.0 与求解定常相似边界层流动类似的方式进行求解.

根据 (12.7)，我们定义非线性算子

$$\mathcal{N}(\phi) = \frac{\partial^3 \phi}{\partial \eta^3} + \frac{1}{2}(1-\xi)\,\eta\,\frac{\partial^2 \phi}{\partial \eta^2} + \xi\left[\phi\,\frac{\partial^2 \phi}{\partial \eta^2}-\left(\frac{\partial \phi}{\partial \eta}\right)^2\right]$$
$$-\xi(1-\xi)\,\frac{\partial^2 \phi}{\partial \eta \partial \xi}. \tag{12.28}$$

令 $f_0(\eta,\xi)$ 表示 $f(\eta,\xi)$ 的初始猜测解，$q \in [0,1]$ 表示嵌入变量. 在同伦分析方法的框架下，我们先构造一个连续变化 (或形变) $\phi(\eta,\xi;q)$，随着嵌入变量 q 从 0 增加到 1，$\phi(\eta,\xi;q)$ 从初始猜测解 $f_0(\eta,\xi)$ 连续变化 (或形变) 到解 $f(\eta,\xi)$. 这种连续变化 $\phi(\eta,\xi;q)$ 由零阶形变方程

$$(1-q)\,\mathcal{L}\left[\phi(\eta,\xi;q)-f_0(\eta,\xi)\right] = q\,c_0\,H(\eta)\,\mathcal{N}\left[\phi(\eta,\xi;q)\right] \tag{12.29}$$

控制，满足边界条件

$$\phi(0,\xi;q)=0,\quad \left.\frac{\partial \phi(\eta,\xi;q)}{\partial \eta}\right|_{\eta=0}=1,\quad \left.\frac{\partial \phi(\eta,\xi;q)}{\partial \eta}\right|_{\eta=+\infty}=0, \tag{12.30}$$

其中 $\mathcal{L}$ 为具有性质 $\mathcal{L}[0]=0$ 的辅助线性算子，$c_0 \neq 0$ 表示收敛控制参数，$H(\eta) \neq 0$ 表示辅助函数. 注意，与摄动方法不同，我们有极大的自由来选择 $\mathcal{L}$，c_0 和 $H(\eta)$. 显然，当 $q=0$ 和 $q=1$ 时，我们分别有

$$\phi(\eta,\xi;0) = f_0(\eta,\xi) \tag{12.31}$$

和
$$\phi(\eta,\xi;1) = f(\eta,\xi). \tag{12.32}$$

因此, 随着嵌入变量 q 从 0 增加到 1, $\phi(\eta,\xi;q)$ 从初始猜测解 $f_0(\eta,\xi)$ 连续变化到原始方程 (12.7) 和 (12.8) 的解 $f(\eta,\xi)$.

将 $\phi(\eta,\xi;q)$ 展开成关于 q 的 Maclaurin 级数, 运用 (12.31), 我们有如下同伦 – Maclaurin 级数

$$\phi(\eta,\xi;q) = f_0(\eta,\xi) + \sum_{n=1}^{+\infty} f_n(\eta,\xi)\, q^n, \tag{12.33}$$

其中

$$f_n(\eta,\xi) = \mathcal{D}_n\left[\phi(\eta,\xi;q)\right] = \frac{1}{n!}\left.\frac{\partial^n \phi(\eta,\xi;q)}{\partial q^n}\right|_{q=0} \tag{12.34}$$

以及 $\mathcal{D}_n$ 为 n 阶同伦导数算子.

如上所述, 我们有极大的自由选择初始猜测解 $f_0(\eta,\xi)$, 辅助线性算子 $\mathcal{L}$, 尤其是收敛控制参数 c_0. 假设所有这些都被恰当地选取, 则同伦 – Maclaurin 级数 (12.33) 在 $q=1$ 时绝对收敛, 由 (12.32) 我们有同伦级数解

$$f(\eta,\xi) = f_0(\eta,\xi) + \sum_{n=1}^{+\infty} f_n(\eta,\xi). \tag{12.35}$$

根据定理 4.15, 我们有 m 阶形变方程

$$\mathcal{L}[f_m(\eta,\xi) - \chi_m\, f_{m-1}(\eta,\xi)] = c_0\, H(\eta)\, \delta_{m-1}(\eta,\xi), \tag{12.36}$$

满足边界条件

$$f_m(0,\xi) = 0, \quad \left.\frac{\partial f_m(\eta,\xi)}{\partial \eta}\right|_{\eta=0} = 0, \quad \left.\frac{\partial f_m(\eta,\xi)}{\partial \eta}\right|_{\eta=+\infty} = 0, \tag{12.37}$$

其中

$$\chi_n = \begin{cases} 1, & n > 1, \\ 0, & n = 1, \end{cases} \tag{12.38}$$

以及

$$\delta_k(\eta,\xi) = \mathcal{D}_k\left\{\mathcal{N}\left[\phi(\eta,\xi;q)\right]\right\}$$

$$= \frac{\partial^3 f_k}{\partial \eta^3} + \frac{1}{2}(1-\xi)\eta \frac{\partial^2 f_k}{\partial \eta^2} - \xi(1-\xi)\frac{\partial^2 f_k}{\partial \eta \partial \xi}$$
$$+ \xi \sum_{n=0}^{k} \left[f_{k-n} \frac{\partial^2 f_n}{\partial \eta^2} - \frac{\partial f_{k-n}}{\partial \eta} \frac{\partial f_n}{\partial \eta} \right] \tag{12.39}$$

由定理 4.1 获得. 详情请参考第四章和廖世俊 [236].

由于高阶形变方程 (12.36) 和 (12.37) 是线性的, 因此原始非线性问题可被转化为无穷多个线性子问题. 然而, 与上面提到的摄动方法不同, 这种转化不需要依赖任何小的摄动量. 更重要的是, 我们有极大的自由选择初始猜测解 $f_0(\eta,\xi)$, 辅助线性算子 $\mathcal{L}$, 辅助函数 $H(\eta)$ 和收敛控制参数 c_0: 正是由于这种自由, 我们可以获得整个时空区间内有效的精确解析近似值, 如下所述.

一般来说, 一个连续函数可以用不同的基函数来近似. 在数学上, 根据上面关于摄动方法失败的讨论, $f(\eta,\xi)$ 不应该包含互补误差函数 $\mathrm{erfc}(\eta/2)$ 和高斯分布函数 $\exp(-\eta^2/4)$, 否则高阶形变方程 (12.36) 和 (12.37) 将很难求解. 物理上, 在任意时间 $\tau \in [0,+\infty)$, 对应于 $\xi \in [0,1]$, (12.7) 和 (12.8) 描述了一个相似边界层流动, 它沿 $\eta \to +\infty$ 呈指数形式衰减. 因此, 可以合理地假设 $f(\eta,\xi)$ 可以由基函数

$$\left\{ \xi^k \, \eta^m \, \exp(-n\eta) \mid k \geqslant 0, m \geqslant 0, n \geqslant 1 \right\} \tag{12.40}$$

表示为以下形式

$$f(\eta,\xi) = a_{0,0}(\xi) + \sum_{m=0}^{+\infty}\sum_{n=1}^{+\infty} a_{m,n}(\xi)\, \eta^m \, \exp(-n\eta), \tag{12.41}$$

其中 $a_{m,n}(\xi)$ 为待定的 ξ 多项式, 它为我们提供了 $f(\eta,\xi)$ 所谓的解表达, 其在同伦分析方法的框架下起着重要作用, 如下所述.

与摄动方法不同, 我们现在有极大的自由来选择零阶形变方程中的初始猜测解 $f_0(\eta,\xi)$, 辅助线性算子 $\mathcal{L}$ 和辅助函数 $H(\eta)$: 所有的选择都应该使高阶形变方程 (12.36) 和 (12.37) 易于求解, 并且保证同伦级数解 (12.35) 在整个时空区间内收敛.

初始猜测解 $f_0(\eta,\xi)$ 应满足边界条件 (12.8) 并符合解表达 (12.41) 的要求. 因此, 我们选择初始猜测解

$$f_0(\eta,\xi) = 1 - \exp(-\eta). \tag{12.42}$$

注意, 该初始猜测解恰好满足 $\xi=1$, 对应于 $\tau \to +\infty$ 处的控制方程 (12.7), 虽然它在 $\xi=0$ 处并不是 $f(\eta,\xi)$ 的一个好的近似值.

需要注意的是，控制方程 (12.7) 包含线性算子

$$\mathcal{L}_0(f) = \frac{\partial^3 f}{\partial \eta^3} + \frac{1}{2}(1-\xi)\eta \frac{\partial^2 f}{\partial \eta^2} - \xi(1-\xi)\frac{\partial^2 f}{\partial \eta \partial \xi}.$$

但是，如果我们选择上述线性算子作为辅助线性算子 $\mathcal{L}$，则高阶形变方程 (12.36) 就变成了包含以下变系数的偏微分方程

$$\frac{1}{2}(1-\xi)\eta, \quad -\xi(1-\xi),$$

因此难以求解。如果我们选择 (12.26) 定义的 $\mathcal{L}_p$ 作为辅助线性算子，高阶形变方程为一个常微分方程，但它的解包含互补误差函数 $\text{erfc}(\eta/2)$ 和高斯分布函数 $\exp(-\eta^2/4)$，这不符合 $f(\eta,\xi)$ 的解表达 (12.41)。所以，$\mathcal{L}_0$ 和 $\mathcal{L}_p$ 都不是好的选择。幸运的是，同伦分析方法为我们提供了极大的自由来选择辅助线性算子 $\mathcal{L}$。注意，当 $\xi=1$ 时，对应的定常相似边界层流动在 §10.2 中通过以下辅助线性算子进行求解

$$\mathcal{L}(u) = \frac{\partial^3 u}{\partial \eta^3} - \frac{\partial u}{\partial \eta}, \qquad (12.43)$$

具有性质

$$\mathcal{L}\left[C_1 + C_2 \exp(-\eta) + C_3 \exp(\eta)\right] = 0. \qquad (12.44)$$

此外，对于 ξ 的任意多项式 $b(\xi)$，线性常微分方程

$$\frac{\partial^3 u}{\partial \eta^3} - \frac{\partial u}{\partial \eta} = b(\xi)\,\eta^m\,\exp(-n\eta)$$

可以通过像 Mathematica 这样的科学计算软件快速解决。所以，如果我们选择 (12.43) 作为辅助线性算子，高阶形变方程 (12.36) 可以很容易地进行求解：有趣的是，即使通过这样一个简单的辅助线性算子，我们也可以得到整个时空区间内有效的高精度近似值，如下所述。

因此，基于同伦分析方法提供的这种自由，我们只需选择 (12.43) 作为辅助线性算子 $\mathcal{L}$。令

$$f_m^*(\eta,\xi) = \chi_m\,f_{m-1}(\eta,\xi) + c_0\,\mathcal{L}^{-1}\left[H(\eta)\,\delta_{m-1}(\eta,\xi)\right]$$

表示 (12.36) 的特解，其中 $\mathcal{L}^{-1}$ 是 (12.43) 定义的辅助线性算子 $\mathcal{L}$ 的逆算子。根据性质 (12.44)，其通解为

$$f_m(\eta,\xi) = f_m^*(\eta,\xi) + C_1 + C_2 \exp(-\eta) + C_3 \exp(\eta),$$

其中系数 C_1, C_2 和 C_3 由边界条件 (12.37) 唯一确定.

值得注意的是, 辅助线性算子 (12.43) 与时间变量 ξ 无关. 此外, m 阶形变方程 (12.36) 的右端项 $\delta_{m-1}(\eta, \xi)$ 总是已知的. 在数学上, m 阶形变方程 (12.36) 是线性常微分方程, 因此可以用与 §10.2 中描述的定常相似边界层流动类似的方式进行求解, 因为时间变量 ξ 可以看作一个常数. 唯一的区别是, 当通过 (12.39) 计算 $\delta_{m-1}(\eta, \xi)$ 时, ξ 必须被看作一个变量. 通过这样的方式, 连续求解线性高阶形变方程 (12.36) 和 (12.37) 是相当容易的, 尤其是通过科学计算软件 Mathematica.

本质上, 上述方法将一个非线性偏微分方程转化为无穷多个线性常微分方程. 更重要的是, 与摄动方法不同, 同伦分析方法为我们提供了很大的自由来选择初始猜测解 $f_0(\eta, \xi)$ 和辅助线性算子 $\mathcal{L}$, 使得高阶近似很容易通过科学计算软件如 Mathematica 获得. 这样, 我们大大简化了原始非线性偏微分方程的求解, 从而可以获得在整个时空区间内有效的精确解析近似.

12.3.2 同伦近似

如上所述, 在同伦分析方法的框架下, 与 §10.2 中考虑的定常边界层流动类似, 描述非定常边界层流动的原始非线性偏微分方程 (12.7) 可转化为无穷多个线性常微分方程. 值得注意的是, 非定常流动的高阶形变方程 (12.36) 与 §10.2 中定常流动的高阶形变方程 (10.11) 非常类似: 非定常流动所采用的辅助线性算子 (12.43) 与定常流动所采用的辅助线性算子 (10.16) 完全相同! 因此, 与第十一章中提到的非相似边界层流动一样, 非线性偏微分方程 (12.7) 控制的非定常边界层流动也可以通过 BVPh 1.0 来进行求解, 具体过程见第七章, 相应输入数据见本章附录.

需要注意的是, 我们有很大的自由来选择辅助函数 $H(\eta)$. 为简单起见, 我们选择

$$H(\eta) = 1,$$

则 $f_m(\eta, \xi)$ 只包含所谓的收敛控制参数 c_0. 虽然 c_0 没有任何物理意义, 但为我们提供了一种便捷的方式来保证同伦级数解 (12.35) 的收敛性. 控制方程 (12.7) 的 m 阶同伦近似对应的平均平方残差 E_m 首先是通过 Mathematica 软件包 BVPh 1.0 中的模块 `GetErr[m]` 在区间 $\eta \in [0, 10]$ 上对 (12.7) 的平方残差进行积分, 然后在区间 $\xi \in [0, 1]$ 上进一步积分得到的. 平均平方残差 E_m 相对于 c_0 的曲线如图 12.1 所示. 这表明, 20 阶同伦近似对应的最优收敛控制参数 c_0^* 约为 -0.3. 还可以发现, 当 $c_0 = -1/4$ 时, 对应的 25 阶同伦近似已足

够精确,此时平均平方残差为 3.5×10^{-7}, 如表 12.1 所示.

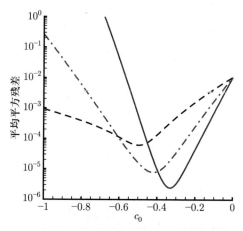

图 12.1 选取辅助函数 $H(\eta) = 1$ 和初始猜测解 (12.42),控制方程的 m 阶同伦近似对应的 c_0 的平均平方残差. 虚线: $m = 5$; 点划线: $m = 10$; 实线: $m = 20$

表 12.1 选取 $c_0 = -1/4$ 和 $H(\eta) = 1$ 时,控制方程 (12.7) 在 $\eta \in [0, 10]$ 和 $\xi \in [0, 1]$ 上的平均平方残差

近似阶数	(12.7) 的平均平方残差
1	6.5×10^{-3}
3	2.5×10^{-3}
5	9.5×10^{-4}
10	9.6×10^{-5}
15	1.1×10^{-5}
20	1.4×10^{-6}
25	3.5×10^{-7}

研究发现,当 $c_0 = -1/4$ 时,$f''(0, 0)$ 的同伦近似与精确结果 $f''(0, 0) = -1/\sqrt{\pi} \approx -0.56419$ 吻合良好,如表 12.2 所示,其中 $'$ 表示对 η 求导. 此外,通过所谓的同伦 – 帕德近似方法 [232], 极大地修正了同伦近似 $f''(0, 0)$ 的精度,如表 12.3 所示. 另外,速度剖面 $f'(\eta, 0)$ 的 20 阶同伦近似及其 [3, 3] 阶同伦 – 帕德近似与精确解 (12.11) 在整个区间 $0 \leqslant \eta < +\infty$ 内吻合良好,如图 12.2 所示. 这表明包含互补误差函数 $\text{erfc}(\eta/2)$ 和高斯分布函数 $\exp(-\eta^2/4)$ 的初始猜测解 (12.11) 可以用基函数 (12.40) 很好地近似.

表 12.2　选取 $c_0 = -1/4$ 和 $H(\eta) = 1$ 时, $f''(0,0)$ 的同伦近似

近似阶数	$f''(0,0)$
5	-0.69303
10	-0.60114
15	-0.57440
20	-0.56693
25	-0.56491
30	-0.56438
35	-0.56424
40	-0.56420
45	-0.56419
50	-0.56419

表 12.3　选取 $H(\eta) = 1$ 时, $f''(0,0)$ 的 $[m,m]$ 阶同伦–帕德近似

m	$f''(0,0)$
5	-0.56415
10	-0.56418
15	-0.56419
20	-0.56419
25	-0.56419
30	-0.56419

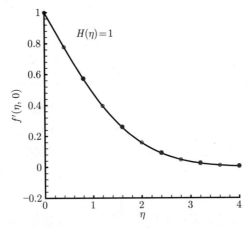

图 12.2　精确解 (12.11) 与选取 $c_0 = -1/4$ 和 $H(\eta) = 1$ 时在 $\xi = 0$ 处的 20 阶同伦近似以及与其 $[3,3]$ 阶同伦–帕德近似的比较. 实线: 精确解 (12.11); 空心圆: $\xi = 0$ 处的 20 阶同伦近似; 实心圆: $[3,3]$ 阶同伦–帕德近似

值得注意的是，$f''(0,\xi)$ 与表面摩擦有关．研究发现，当 $c_0 = -1/4$ 时，整个区间 $\xi \in [0,1]$ 上，$f''(0,\xi)$ 的 20 阶同伦近似与 30 阶同伦近似吻合良好，如图 12.3 所示．此外，对应的速度剖面在整个区间 $\xi \in [0,1]$ 和 $0 \leqslant \eta < +\infty$ 上也都是精确的，如图 12.4 所示．值得注意的是，随着 τ 从 0 增加到 $+\infty$，速度剖面变化平缓.

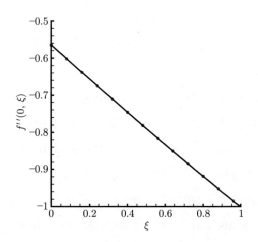

图 12.3 选取 $c_0 = -1/4$ 和 $H(\eta) = 1$ 时，$f''(0,\xi)$ 的同伦近似．实线：30 阶近似；空心圆：20 阶近似

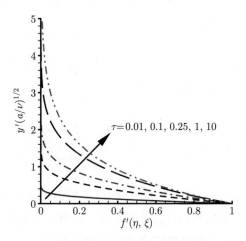

图 12.4 选取 $c_0 = -1/4$ 和 $H(\eta) = 1$ 时，不同的无量纲时间 $\tau = at$ 对应的速度剖面．实线：$\tau = 0.01$；虚线：$\tau = 0.1$；点划线：$\tau = 0.25$；长虚线：$\tau = 1$；双点划线：$\tau = 10$

局部表面摩擦与 $f''(0,\xi)$ 有关. $f''(0,\xi)$ 的 30 阶同伦近似为

$$\begin{aligned}
& f''(0,\xi) \\
= & -0.5643747892 - 0.4653303619\,\xi + 2.998049008 \times 10^{-2}\,\xi^2 \\
& - 2.518392990 \times 10^{-3}\,\xi^3 - 2.561860658 \times 10^{-5}\,\xi^4 \\
& - 2.531901893 \times 10^{-5}\,\xi^5 + 3.073353805 \times 10^{-5}\,\xi^6 \\
& + 5.063224875 \times 10^{-5}\,\xi^7 + 5.780083670 \times 10^{-5}\,\xi^8 \\
& - 3.019750875 \times 10^{-4}\,\xi^9 + 0.2746188078\,\xi^{10} - 48.017634463\,\xi^{11} \\
& + 3.4358227736 \times 10^{3}\,\xi^{12} - 1.2769915485 \times 10^{5}\,\xi^{13} \\
& + 2.8257114566 \times 10^{6}\,\xi^{14} - 4.0636817506 \times 10^{7}\,\xi^{15} \\
& + 4.0325552732 \times 10^{8}\,\xi^{16} - 2.8809318182 \times 10^{9}\,\xi^{17} \\
& + 1.5277371296 \times 10^{10}\,\xi^{18} - 6.1471279622 \times 10^{10}\,\xi^{19} \\
& + 1.9058081506 \times 10^{11}\,\xi^{20} - 4.5980815420 \times 10^{11}\,\xi^{21} \\
& + 8.6766946108 \times 10^{11}\,\xi^{22} - 1.2809186178 \times 10^{12}\,\xi^{23} \\
& + 1.4720293398 \times 10^{12}\,\xi^{24} - 1.3019682527 \times 10^{12}\,\xi^{25} \\
& + 8.6859786813 \times 10^{11}\,\xi^{26} - 4.2255442112 \times 10^{11}\,\xi^{27} \\
& + 1.4139635601 \times 10^{11}\,\xi^{28} - 2.9087215325 \times 10^{10}\,\xi^{29} \\
& + 2.7723411925 \times 10^{9}\,\xi^{30}. \tag{12.45}
\end{aligned}$$

无量纲时间 $\tau \in [0,+\infty)$ 对应的局部表面摩擦, 如图 12.5 所示. 利用 (12.45) 的前四项, 我们有简化的局部表面摩擦公式

$$C_f^x(x,\xi) = (\xi\,\mathrm{Re}_x)^{-1/2}\,\left(-0.5643747892 - 0.4653303619\,\xi \right. \\ \left. + 2.998049008 \times 10^{-2}\,\xi^2 - 2.518392990 \times 10^{-3}\,\xi^3\right), \tag{12.46}$$

这与整个时间区间 $0 \leqslant \tau < +\infty$ 上的精确 30 阶同伦近似非常吻合, 如图 12.5 所示. 因此, 通过选取适当的收敛控制参数 c_0, 我们可以获得由非线性偏微分方程 (12.7) 控制的非定常边界层流动的精确解析近似, 其不仅在整个时间区间 $0 \leqslant \tau < +\infty$ 上有效, 而且在整个空间区间 $0 \leqslant \eta < +\infty$ 上也是有效的. 据我们所知, 对于非定常边界层流动局部摩擦的这种简单而精确的解析近似从未被报道过. 这再次验证了同伦分析方法对于复杂非线性问题的有效性.

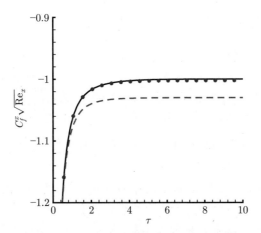

图 12.5 $C_f^x\sqrt{\text{Re}_x}$ 的四项近似 (12.46) 与精确的 30 阶同伦近似 (12.45) 以及与摄动近似 (12.27) 的比较. 实线: 精确的 30 阶同伦近似 (12.45); 虚线: 摄动近似 (12.27); 实心圆: 简化的同伦近似 (12.46)

需要注意的是，上述所有同伦近似都是通过选取最简单的辅助函数 $H(\eta) = 1$ 获得的. 然而, 在同伦分析方法的框架下, 我们有极大的自由来选取辅助函数 $H(\eta)$. 例如, 当我们选取 $H(\eta) = \exp(-\eta)$ 时, 控制方程 (12.7) 的 20 阶近似对应的平均平方残差有最小值 1.0×10^{-4}, 对应的最优收敛控制参数 $c_0^* = -0.44$, 如图 12.6 所示. 实际上, 使用工具 BVPh 1.0, 选取 $c_0 = -2/5$ 和 $H(\eta) = \exp(-\eta)$, 我们也可以获得精确解析近似. 这再次验证了 BVPh 1.0 的有效性和灵活性.

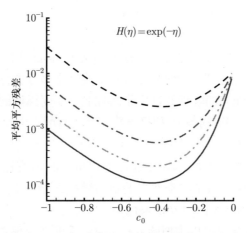

图 12.6 选取 $c_0 = -2/5$ 和 $H(\eta) = \exp(-\eta)$ 时, 控制方程 (12.7) 在 $\eta \in [0, 10]$ 和 $\xi \in [0, 1]$ 上的平均平方残差. 虚线: $m = 5$; 点划线: $m = 10$; 双点划线: $m = 15$; 实线: $m = 20$

12.4 本章小结

本章说明了在同伦分析方法的框架下,描述非定常边界层流动的非线性偏微分方程 (12.7) 可以转化为无穷多个线性常微分方程. 因此,可直接通过 BVPh 1.0 以与求解定常相似方程类似的方式进行求解. 换句话说,通过同伦分析方法,求解非定常边界层流动和定常边界层流动一样简单. 这验证了 BVPh 1.0 对于某些非线性偏微分方程的有效性,特别是对于那些与边界层流动相关的偏微分方程.

值得注意的是,尽管原始的非线性偏微分方程 (12.7) 也可以通过将 ξ 看作一个小的摄动量来转换为无穷多个线性常微分方程,但是通过摄动方法无法获得精确的近似解. 这主要是因为,基于摄动方法,我们在高阶摄动方程中没有选择初始近似 (12.22) 和相应的线性算子 (12.26) 的自由,因此这些高阶摄动方程很难求解. 然而,不同于摄动方法,同伦分析方法为我们选择初始猜测解 $f_0(\eta,\xi)$ 和辅助线性算子 $\mathcal{L}$ 提供了极大的自由,因此我们可以选取简单的初始猜测解 (12.42) 和辅助线性算子 (12.43) 使得相应的高阶形变方程 (12.36) 可以被轻松求解. 更重要的是,在同伦分析方法的框架下,可以通过收敛控制参数 c_0 来保证高阶同伦近似的准确性. 因此,同伦分析方法适用于更加复杂的非线性问题.

如第十一章所述,在同伦分析方法的框架下,描述定常非相似边界层流动的非线性偏微分方程可以转化为由如下辅助线性算子控制的无穷多个线性常微分方程

$$\mathcal{L}(f) = \frac{\partial^3 f}{\partial \eta^3} - \frac{\partial f}{\partial \eta}. \tag{12.47}$$

这里,我们进一步表明,在同伦分析方法的框架下,描述非定常相似边界层流动的非线性偏微分方程也可以转化为无穷多个线性常微分方程,这些线性常微分方程由与如上完全相同的辅助线性算子所控制! 需要强调的是,上述辅助线性算子也正是 §10.2 中用来求解一种呈指数衰减的解的定常相似边界层流动所使用的辅助线性算子 (10.16)! 注意,这三章中考虑的边界层流动不仅在物理上而且在数学上都是非常不同的. 然而,通过 BVPh 1.0,所有这些问题都可以通过在同伦分析方法框架下选取完全相同的辅助线性算子 (12.47) 来解决! 这些显示了同伦分析方法和 BVPh 1.0 的通用性.

因此,BVPh 1.0 可以以类似的方式求解许多非线性偏微分方程,尤其是那些与非定常和 (或) 非相似边界层流动和传热相关的非线性偏微分方程,但是我们应该注意的是,它并不适用于所有非线性偏微分方程.

事实上，开发一种适用于所有非线性边值问题的解析方法是不可能的．然而，正如 Rabindranth Tagore (1861—1941) 所指出的那样，"如果你对所有错误关上大门，真理就会被拒之门外"．因此，我们的策略是开发一种对尽可能多的非线性边值问题有效的解析方法．尽管基于同伦分析方法的 Mathematica 软件包 BVPh 的开发还需要进一步改进和应用，但是 BVPh 1.0 及其在第二部分中对于某些非线性常微分方程和偏微分方程的成功应用表明了这种策略具有良好的前景．

无论怎样，不管我们能走多远，我们都要踏上充满希望的征程！

参考文献

[223] Dennis, S.C.R.: The motion of a viscous fluid past an impulsively started semi-infinite flat plate. J. Inst. Math. Its Appl. **10**, 105–117 (1972)

[224] Hall, M.G.: The boundary layer over an impulsively started flat plate. Proc. R. Soc. A. **310**, 401–414 (1969)

[225] Li, Y.J., Nohara, B.T., Liao, S.J.: Series solutions of coupled van der Pol equation by means of homotopy analysis method. J. Mathematical Physics **51**, 063517 (2010)

[226] Liao, S.J.: The proposed Homotopy Analysis Technique for the Solution of Nonlinear Problems. PhD dissertation, Shanghai Jiao Tong University (1992)

[227] Liao, S.J.: A kind of approximate solution technique which does not depend upon small parameters—(II) an application in fluid mechanics. Int. J. Nonlin. Mech. **32**, 815–822 (1997)

[228] Liao, S.J.: An explicit, totally analytic approximation of Blasius viscous flow problems. Int. J. Nonlin. Mech. **34**, 759–778 (1999a)

[229] Liao, S.J.: A uniformly valid analytic solution of 2D viscous flow past a semi-infinite flat plate. J. Fluid Mech. **385**, 101–128 (1999b)

[230] Liao, S.J., Campo, A.: Analytic solutions of the temperature distribution in Blasius viscous flow problems. J. Fluid Mech. **453**, 411–425 (2002)

[231] Liao, S.J.: On the analytic solution of magnetohydrodynamic flows of non-Newtonian fluids over a stretching sheet. J. Fluid Mech. **488**, 189–212 (2003a)

[232] Liao, S.J.: Beyond Perturbation – Introduction to the Homotopy Analysis Method. Chapman & Hall/ CRC Press, Boca Raton (2003b)

[233] Liao, S.J.: On the homotopy analysis method for nonlinear problems. Appl. Math. Comput. **147**, 499–513 (2004)

[234] Liao, S.J.: A new branch of solutions of boundary-layer flows over an impermeable stretched plate. Int. J. Heat Mass Tran. **48**, 2529–2539 (2005)

[235] Liao, S.J.: Series solutions of unsteady boundary-layer flows over a stretching flat plate. Stud. Appl. Math. **117**, 2529–2539 (2006a)

[236] Liao, S.J.: An analytic solution of unsteady boundary-layer flows caused by an impulsively stretching plate. Communications in Nonlinear Science and Numerical Simulation. **11**, 326–339 (2006b)

[237] Liao, S.J.: Notes on the homotopy analysis method - some definitions and theorems. Commun. Nonlinear Sci. Numer. Simulat. **14**, 983–997 (2009)

[238] Liao, S.J.: On the relationship between the homotopy analysis method and Euler transform. Commun. Nonlinear Sci. Numer. Simulat. **15**, 1421–1431 (2010a)

[239] Liao, S.J.: An optimal homotopy-analysis approach for strongly nonlinear differential equations. Commun. Nonlinear Sci. Numer. Simulat. **15**, 2003–2016 (2010b).

[240] Liao, S.J., Tan, Y.: A general approach to obtain series solutions of nonlinear differential equations. Stud. Appl. Math. **119**, 297–355 (2007)

[241] Nazar, N., Amin, N., Pop, I.: Unsteady boundary layer flow due to stretching surface in a rotating fluid. Mechanics Research Communications. **31**, 121–128 (2004)

[242] Pop, I., Na, T.Y.: Unsteady flow past a stretching sheet. Mech. Research Comm. **23**, 413 (1996)

[243] Seshadri, R., Sreeshylan, N., Nath, G.: Unsteady mixed convection flow in the stagnation region of a heated vertical plate due to impulsive motion. Int. J. Heat and Mass Transfer. **45**, 1345–1352 (2002)

[244] Stewartson, K.: On the impulsive motion of a flat plate in a viscous fluid (Part I). Quart. J. Mech. Appl. Math. **4**, 182–198 (1951)

[245] Stewartson, K.: On the impulsive motion of a flat plate in a viscous fluid (Part II). Quart. J. Mech. Appl. Math. **22**, 143–152 (1973)

[246] Wang, C.Y., Du, G., Miklavcic, M., Chang, C.C.: Impulsive stretching of a surface in a viscous fluid. SIAM J. Appl. Math. **57**, 1 (1997)

[247] Watkins, C.B.: Heat transfer in the boundary layer over an impulsively started flat plate. J. Heat Transfer. **97**, 482–484 (1975)

[248] Williams, J.C., Rhyne, T.H.: Boundary layer development on a wedge impulsively set into motion. SIAM J. Appl. Math. **38**, 215–224 (1980)

[249] Xu, H., Lin, Z.L., Liao, S.J., Wu, J.Z., Majdalani, J.: Homotopy-based solutions of the Navier-Stokes equations for a porous channel with orthogonally moving walls. Physics of Fluids. **22**, 053601 (2010)

附录 12.1 BVPh 的输入数据

```
(* Input Mathematica package BVPh version 1.0 *)
<<BVPh.txt;
```

附录 12.1 BVPh 的输入数据

```
(* Define the physical and control parameters *)
TypeEQ     = 1;
ApproxQ    = 0;
ErrReq     = 10^(-10);
NgetErr    = 100;
zRintegral = 10;

(* Define the governing equation *)
f[z_,u_,lambda_] := Module[{temp},
temp[1] = D[u,{z,3}] + (1-t)*z/2*D[u,{z,2}]
            - t*(1-t)*D[u,z,t];
temp[2] = u*D[u,{z,2}] - D[u,z]^2;
temp[1] + t* temp[2] // Expand
];

(* Define boundary conditions *)
zR = infinity;
OrderEQ = 3;
BC[1,z_,u_,lambda_] := Limit[u,   z -> 0];
BC[2,z_,u_,lambda_] := Limit[D[u,z] - 1, z -> 0];
BC[3,z_,u_,lambda_] := Limit[D[u,z] , z -> zR ];

(* Define initial guess *)
u[0]  = 1 - Exp[-z];

(* Define the auxiliary linear operator *)
L[u_] := D[u,{z,3}] - D[u,z];

(* Define output term *)
output[z_,u_,k_]:= Print["output            = ",
               D[u[k],{z,2}]/.{z->0,t->0}//N];

(* Define Getdelta[k]  *)
Getdelta[k_]:=Module[{temp,i},
uz[k]   = D[u[k],z]//Expand;
uzz[k]  = D[uz[k],z]//Expand;
uzzz[k] = D[uzz[k],z]//Expand;
uzuz[k] = Sum[uz[i]*uz[k-i],{i,0,k}]//Expand;
```

```
uzzu[k] = Sum[uzz[i]*u[k-i],{i,0,k}]//Expand;
uzt[k]  = D[uz[k],t]//Expand;
temp[1] = uzzz[k] + (1-t)*z/2*uzz[k]
              - t*(1-t)*uzt[k]//Expand;
temp[2] = t*(uzzu[k] - uzuz[k]);
delta[k] = temp[1] + temp[2]//Expand;
];


(* Print input and control parameters *)
PrintInput[u[z,t]];

(* Set convergence-control parameter c0 *)
c0  = -1/4;
Print["c0  =  ",c0];

(* Gain 10th-order HAM approximation *)
Print[" c0  = ",c0];
BVPh[1,10];

(* Calculate the squared residual *)
For[k=2, k<=10, k=k+2,
GetErr[k];
err[k] = Integrate[Err[k],{t,0,1}];
Print[" k  =  ", k, "   Squared residual =  ",
         err[k]//N];
];
```

需要注意的是，对于在无限区间上定义的非线性边值问题，我们必须在 BVPh 1.0 的输入数据中设置 zR = infinity. 此外，当非线性偏微分方程通过 BVPh 1.0 进行求解时，必须在输入数据中定义模块 Getdelta.

第十三章 金融领域的应用：美式认沽期权

本章将同伦分析方法与 Laplace 变换相结合成功地解决了金融界著名的美式认沽期权方程. 与通常有效期仅为几天或几周的渐近和 (或) 摄动公式不同, 最优行权边界 $B(\tau)$ 的同伦近似解, 以 $\sqrt{\tau}$ 级数展开到 $o(\tau^M)$, 只要 M 足够大, 能在很长时间内都有效, 甚至半个世纪. 我们发现, $B(\tau)$ 的同伦近似解以 $\sqrt{\tau}$ 级数展开到 $o(\tau^{48})$, 这个结果与长期的最优行权价格已经非常接近了. 因此, 这两者的结合就可以看作这个问题在整个时域 $0 \leqslant \tau < +\infty$ 上的解析公式. 附录 13.3 中提供了实用的 Mathematica 程序 APOh, 可供读者在数秒内通过笔记本电脑获得足够准确的美式认沽期权在很长时间内有效的最优行权价格.

13.1 数学模型

金融领域涉及很多非线性偏微分方程. 传统上, 这些非线性偏微分方程的解析近似是通过渐近方法给出的. 然而, 大多数渐近近似仅对短时间和 (或) 一些小数量级有效. 本章说明, 基于同伦分析方法 [276–290, 293], 我们可以获得金融领域中某些非线性偏微分方程更好的解析近似.

让我们考虑一个著名的金融问题: 美式认沽期权, 其行权价格为 X, 到期时间为 t. 令 $V(S,t)$ 表示美式认沽期权价格, S 表示标的资产价格, t 表示时间. 此外, 令 σ 表示标的波动率, r 表示无风险利率, 两者均为常数. 在到期日前的任何时刻, 美式认沽期权都存在一个最优行权边界 $B(t)$, 使得当 S 等于或

低于 $B(t)$ 时行权认沽期权是最优的. 因此, 当 $S \leqslant B(t)$ 时, 认沽期权的价格为

$$V(S,t) = X - S, \tag{13.1}$$

其中 X 为行权价格. 当 $S > B(t)$ 时, $V(S,t)$ 满足著名的 Black-Scholes 方程

$$\frac{\partial V}{\partial t} + \frac{1}{2}\sigma^2 S^2 \frac{\partial^2 V}{\partial S^2} + rS\frac{\partial V}{\partial S} - rV = 0, \tag{13.2}$$

行权边界 $B(t)$ 处的光滑条件

$$\lim_{S \to B(t)} V(S,t) = X - B(t), \quad \lim_{S \to B(t)} \frac{\partial V}{\partial S} = -1, \tag{13.3}$$

上边界条件

$$\lim_{S \to +\infty} V(S,t) \to 0, \tag{13.4}$$

以及终止条件

$$\lim_{t \to T} V(S,t) = \max\{X - S, 0\}. \tag{13.5}$$

定义变量 $\tau \equiv T - t$. 当 $r > 0$ 时, 终止条件 (13.5) 可以进一步简化为

$$\lim_{\tau \to 0} V(S,\tau) = 0, \tag{13.6}$$

其在

$$\Sigma_1 = \{(S,\tau)|\ B(\tau) \leqslant S < +\infty,\ 0 \leqslant \tau \leqslant T\}$$

范围内成立. Kim [272] 和 Carr 等人 [258] 推导出期权价格公式

$$V(S,\tau) = V_E(S,\tau) + X\int_0^\tau r\exp(-r\xi)N(-d_{\xi,2})\mathrm{d}\xi, \tag{13.7}$$

其中

$$V_E(S,\tau) = X\exp(-r\tau)N(-d_2) + SN(-d_1)$$

是具有如下定义的欧式认沽期权价格

$$d_1 = \frac{\ln(S/X) + (r + \sigma^2/2)\tau}{\sigma\sqrt{\tau}},$$

$$d_2 = d_1 - \sigma\sqrt{\tau},$$

$$d_{\xi,1} = \frac{\ln[S/B(\tau-\xi)] + (r + \sigma^2/2)\xi}{\sigma\sqrt{\xi}},$$

$$d_{\xi,2} = d_{\xi,1} - \sigma\sqrt{\xi},$$

$N(x)$ 是标准正态分布的累计概率分布函数, 定义如下

$$N(x) = \frac{1}{\sqrt{2\pi}} \int_{-\infty}^{x} \exp\left(-\frac{w^2}{2}\right) dw. \tag{13.8}$$

因此, 从 (13.7) 不难看出, 只要最优行权边界 $B(\tau)$ 已知, 很容易得到期权价格 $V(S,\tau)$. 所以, 最优行权边界 $B(\tau)$ 的求解是该问题的关键.

尽管控制方程 (13.2) 是线性的, 但由于未知移动边界 $B(\tau)$ 的存在, 所以该问题本质上是非线性的. 上述具有未知移动边界的偏微分方程可以通过数值方法求解, 例如二项式 (三项式) 方法 [255, 264], 蒙特卡罗模拟 [268], 最小二乘法 [291], 变分不等式 [265, 271], 以及其他基于求解偏微分方程的技术 [250, 254, 256, 260, 269, 292]. 然而, 由于大多数市场从业人员不熟悉数值方法, 因此解析近似在实际计算和理论上都非常有价值.

传统的解析方法大多基于摄动法或渐近法, 例如 Barles 等人 [252], Kuske 和 Keller [274], Alobaidi 和 Mallier [251], Evans, Kuske 和 Keller [266], Zhang 和 Li [294] 以及 Knessl [273]. 根据 Chen 等人 [260] 以及 Chen 和 Chadam [261] 的论述, 所有这些近似解的有效期都很短, 通常是几天或几周.

2006 年, 诸颂平 [295] 首次将同伦分析方法应用于美式认沽期权. 利用 Landau 变换 [275] 和同伦分析方法, 诸颂平 [295] 给出了一个包含二重积分的无穷递推级数形式的解. 诸颂平 [295] 通过数值积分方法求得了 30 阶的近似解, 并在数值上证明了其结果的收敛性, 这对在不涉及额外参数的情况下找到解析公式做了重大的贡献. 此外, 诸颂平 [296] 还应用同伦分析方法获得了具有恒定股息收益率的可转换债券估值的解析解. 2010 年, 成钧、诸颂平和廖世俊 [263] 进一步应用同伦分析方法, 给出了最优行权边界 $B(\tau)$ 以 $\sqrt{\tau}$ 级数展开到 $o(\tau^6)$ 的显式解析近似解, 其在到期前的更长时间内有效, 通常以年为单位, 且和许多数值结果一样精确. 他们的方法基于 Laplace 变换, 与 Landau 变换无关 [275]. 基于同伦分析方法, 成钧 [262] 给出了 $B(\tau)$ 以 $\sqrt{\tau}$ 级数展开到 $o(\tau^7)$ 的解析近似解, 该研究显示了同伦分析方法在求解以美式期权为代表的金融问题上的潜力和有效性.

本章进一步修正了成钧、诸颂平和廖世俊 [263] 基于同伦分析方法的求解方式, 并给出了更精确的最优行权边界 $B(\tau)$ 的显式表达式. 特别地, 我们研究了对 $B(\tau)$ 按 $\sqrt{\tau}$ 作多项式, 展开到 M 阶后对解的有效性的影响. $B(\tau)$ 要对更长时间的 $B(\tau)$ 有效, 就要对 τ 展开到更高阶. 因此, 当 M 足够大时, 同伦分析方法给出的 $B(\tau)$ 的有效期甚至可以长达半个世纪, 这比摄动或渐近方法

给出的时间要长约 1000 倍！基于这种显式近似，本章附录中给出了一个简短的 Mathematica 程序 `APOh`，读者可通过笔记本电脑在数秒内获得美式认沽期权在相当长的有效期内的精确最优行权价格.

13.2 简明数学公式

当 $\sigma \neq 0$ 时，我们引入如下无量纲变换

$$V^* = \frac{V}{X}, \quad S^* = \frac{S}{X}, \quad \tau^* = \frac{\sigma^2}{2}\tau, \quad \gamma = \frac{2r}{\sigma^2}. \tag{13.9}$$

在后文中，我们将省略星号，无量纲控制方程变为

$$-\frac{\partial V}{\partial \tau} + S^2 \frac{\partial^2 V}{\partial S^2} + \gamma S \frac{\partial V}{\partial S} - \gamma V = 0, \tag{13.10}$$

满足如下边界条件

$$V(B(\tau), \tau) = 1 - B(\tau), \tag{13.11}$$

$$\frac{\partial V}{\partial S}(B(\tau), \tau) = -1, \tag{13.12}$$

$$V(S, \tau) \to 0, \quad S \to +\infty, \tag{13.13}$$

$$V(S, 0) = 0. \tag{13.14}$$

令 $V_0(S, \tau)$ 和 $B_0(\tau)$ 分别表示 $V(S, \tau)$ 和最优行权边界 $B(\tau)$ 的初始近似值. 令 $q \in [0, 1]$ 表示嵌入变量. 在同伦分析方法的框架下，我们首先构造两个连续变化 (或形变) $\phi(S, \tau; q)$ 和 $\Lambda(\tau; q)$，随着嵌入变量 q 从 0 增加到 1，$\phi(S, \tau; q)$ 从初始猜测解 $V_0(S, \tau)$ 连续变化到解 $V(S, \tau)$，$\Lambda(\tau; q)$ 从初始猜测解 $B_0(\tau)$ 连续变化到最优行权边界 $B(\tau)$. 这种连续变化由零阶形变方程

$$-\frac{\partial \phi(S, \tau; q)}{\partial \tau} + S^2 \frac{\partial^2 \phi(S, \tau; q)}{\partial S^2} + \gamma S \frac{\partial \phi(S, \tau; q)}{\partial S} - \gamma \phi(S, \tau; q) = 0 \tag{13.15}$$

控制，其定义域为

$$\Lambda(\tau; q) \leqslant S < +\infty, \quad 0 \leqslant \tau \leqslant \tau_{\exp},$$

满足初始 (边界) 条件

$$\phi(S, 0; q) = 0, \tag{13.16}$$

$$\frac{\partial \phi(S, \tau; q)}{\partial S} = -1, \quad S = \Lambda(\tau; q), \tag{13.17}$$

$$\phi(+\infty, \tau; q) \to 0, \tag{13.18}$$

以及

$$(1-q)\left[\Lambda(\tau;q) - B_0(\tau)\right] = c_0\, q\, \{\Lambda(\tau;q) + \phi[\Lambda(\tau;q), \tau; q] - 1\}, \tag{13.19}$$

其中 $c_0 \neq 0$ 为收敛控制参数,

$$\tau_{\exp} = \frac{1}{2}\sigma^2\, T$$

为无量纲到期时间.

当 $q=1$ 时, 零阶形变方程 (13.15) 至 (13.19) 等价于原始方程 (13.10) 至 (13.14), 因此

$$\phi(S,\tau;1) = V(S,\tau), \quad \Lambda(\tau;1) = B(\tau). \tag{13.20}$$

当 $q=0$ 时, 由 (13.19) 可得

$$\Lambda(\tau;0) = B_0(\tau), \tag{13.21}$$

由 (13.15) 至 (13.18), 可进一步得到控制方程

$$-\frac{\partial V_0(S,\tau)}{\partial \tau} + S^2 \frac{\partial^2 V_0(S,\tau)}{\partial S^2} + \gamma S \frac{\partial V_0(S,\tau)}{\partial S} - \gamma V_0(S,\tau) = 0, \tag{13.22}$$

满足初始 (边界) 条件

$$V_0(S,0) = 0, \tag{13.23}$$

$$\frac{\partial V_0(S,\tau)}{\partial S} = -1, \quad S = B_0(\tau), \tag{13.24}$$

$$V_0(+\infty, \tau) \to 0, \tag{13.25}$$

其中

$$V_0(S,\tau) = \phi(S,\tau;0). \tag{13.26}$$

为了简单起见, 我们选择

$$B_0(\tau) = 1$$

作为最优行权边界 $B(\tau)$ 的初始猜测解. 然后, 初始猜测解 $V_0(S,\tau)$ 由线性偏微分方程 (13.22) 控制, 满足线性初始条件 (13.23), 固定边界 $S=1$ 处的线性边界

条件 (13.24) 和无穷远处的线性边界条件 (13.25). 因此, 零阶形变方程 (13.15) 至 (13.19) 确实构造了两个连续变化 (或形变) $\phi(S,\tau;q)$ 和 $\Lambda(\tau;q)$, 这是拓扑中的两个同伦

$$\phi(S,\tau;q): V_0(S,\tau) \sim V(S,\tau), \quad \Lambda(\tau;q): B_0(\tau) \sim B(\tau).$$

将 $\phi(S,\tau;q)$ 和 $\Lambda(\tau;q)$ 展开成关于 $q \in [0,1]$ 的 Maclaurin 级数, 我们有如下同伦 – Maclaurin 级数

$$\phi(S,\tau;q) = V_0(S,\tau) + \sum_{n=1}^{+\infty} V_n(S,\tau)\, q^n, \qquad (13.27)$$

$$\Lambda(\tau;q) = B_0(\tau) + \sum_{n=1}^{+\infty} B_n(\tau)\, q^n, \qquad (13.28)$$

其中

$$V_n(S,\tau) = \frac{1}{n!} \left.\frac{\partial^n \phi(S,\tau;q)}{\partial q^n}\right|_{q=0} = \mathcal{D}_n\left[\phi(S,\tau;q)\right],$$

$$B_n(\tau) = \frac{1}{n!} \left.\frac{\partial^n \Lambda(\tau;q)}{\partial q^n}\right|_{q=0} = \mathcal{D}_n\left[\Lambda(\tau;q)\right] \qquad (13.29)$$

分别为 $\phi(S,\tau;q)$ 和 $\Lambda(q)$ 的 n 阶同伦导数, $\mathcal{D}_n$ 为 n 阶同伦导数算子. 注意, 零阶形变方程包含收敛控制参数 c_0. 假设 c_0 选择恰当, 则上述级数在 $q=1$ 时绝对收敛, 我们有同伦级数解

$$V(S,\tau) = V_0(S,\tau) + \sum_{n=1}^{+\infty} V_n(S,\tau), \qquad (13.30)$$

$$B(\tau) = B_0(\tau) + \sum_{n=1}^{+\infty} B_n(\tau). \qquad (13.31)$$

$V_n(S,\tau)$ 和 $B_n(\tau)$ 的方程可以直接由零阶形变方程 (13.15) 至 (13.19) 推导获得. 将级数 (13.27) 和 (13.28) 代入 (13.15), (13.16) 和 (13.18), 然后将 q 的相同次幂取等式, 可以得到 n 阶形变方程 ($n \geq 1$)

$$-\frac{\partial V_n(S,\tau)}{\partial \tau} + S^2 \frac{\partial^2 V_n(S,\tau)}{\partial S^2} + \gamma S \frac{\partial V_n(S,\tau)}{\partial S} - \gamma V_n(S,\tau) = 0, \qquad (13.32)$$

满足初始 (边界) 条件

$$V_n(S,0) = 0, \qquad (13.33)$$

$$V_n(+\infty, \tau) \to 0. \tag{13.34}$$

以上公式详见成钧、诸颂平和廖世俊 [263].

需要注意的是, (13.17) 和 (13.19) 定义在依赖于 q 的移动边界 $S = \Lambda(\tau; q)$ 上, 因此 (13.27) 对它们无效. 成钧、诸颂平和廖世俊 [263] 开发了一个 Mathematica 程序来获得 $\phi(S, \tau; q)$ 在关于 q 的移动边界 $S = \Lambda(\tau; q)$ 上的 Maclaurin 级数. 与成钧、诸颂平和廖世俊 [263] 不同, 我们在本章中将给出其显式表达式.

令 $'$ 表示对 S 求导. 在移动边界 $S = \Lambda(\tau; q)$ 上, 将 $\phi(S, \tau; q)$, $\phi'(S, \tau; q)$ 分别展开成关于 q 的 Maclaurin 级数, 运用 (13.28), 我们有

$$\phi(S, \tau; q) = V_0(B_0, \tau) + \sum_{n=1}^{+\infty} [V_n(B_0, \tau) + f_n(\tau)] q^n \tag{13.35}$$

和

$$\frac{\partial \phi(S, \tau; q)}{\partial S} = V_0'(B_0, \tau) + \sum_{n=1}^{+\infty} [V_n'(B_0, \tau) + g_n(\tau)] q^n, \tag{13.36}$$

其中

$$f_n(\tau) = \sum_{j=0}^{n-1} \alpha_{j,n-j}(\tau), \quad g_n(\tau) = \sum_{j=0}^{n-1} \beta_{j,n-j}(\tau), \tag{13.37}$$

且有定义

$$\alpha_{n,i}(\tau) = \sum_{m=1}^{i} \psi_{n,m}(\tau)\, \mu_{m,i}(\tau), \quad i \geqslant 1, \tag{13.38}$$

$$\beta_{n,i}(\tau) = \sum_{m=1}^{i} (m+1)\psi_{n,m+1}(\tau)\, \mu_{m,i}(\tau), \quad i \geqslant 1, \tag{13.39}$$

$$\psi_{n,0}(\tau) = V_n(B_0, \tau), \quad \psi_{n,m}(\tau) = \frac{1}{m!} \frac{\partial^m V_n(S, \tau)}{\partial S^m}\bigg|_{S=1} \tag{13.40}$$

和递推公式

$$\mu_{1,n}(\tau) = B_n(\tau), \quad \mu_{m+1,n}(\tau) - \sum_{i=m}^{n-1} \mu_{m,i}(\tau)\, B_{n-i}(\tau). \tag{13.41}$$

上述显式表达式的详细推导过程在附录 13.1 中给出. 这些显式表达式极大地提高了计算效率. 更重要的是, 通过这些显式公式, 可以更容易地研究 $o(\tau^M)$ 的阶对 $\sqrt{\tau}$ 多项式表示的最优行权边界 $B(\tau)$ 的影响, 如下所述.

然后, 将 (13.36) 代入 (13.17), 并将 q 的相同次幂取等式, 有边界条件

$$\frac{\partial V_n(S,\tau)}{\partial S} = -g_n(\tau), \quad S = B_0(\tau) = 1, \ n \geqslant 1. \tag{13.42}$$

类似地, 将 (13.35) 代入 (13.19) 并将 q 的相同次幂取等式, 有

$$B_n(\tau) = \begin{cases} c_0\, V_0(B_0,\tau), & n = 1, \\ B_{n-1}(\tau) + c_0\left[B_{n-1}(\tau) + V_{n-1}(B_0,\tau) + f_{n-1}(\tau)\right], & n > 1, \end{cases} \tag{13.43}$$

其中 $B_0 = 1$.

为了简单起见, 定义 $g_0(\tau) = 1$. 然后, 未知的初始猜测解 $V_0(S,\tau)$ 的 (13.22) 至 (13.25) 分别具有与 $V_n(S,\tau)$ 的 n 阶形变方程 (13.32) 至 (13.34) 和 (13.42) 相同的形式. 由上述显式递推公式, 我们有

$$g_0(\tau) = 1,$$
$$g_1(\tau) = 2\psi_{0,2}(\tau)\, B_1(\tau),$$
$$g_2(\tau) = 2\psi_{0,2}(\tau)\, B_2(\tau) + 3\psi_{0,3}(\tau)\, B_1^2(\tau) + 2\psi_{1,2}(\tau)\, B_1(\tau),$$
$$g_3(\tau) = 2\psi_{0,2}(\tau)\, B_3(\tau) + 6\psi_{0,3}(\tau)\, B_1(\tau)\, B_2(\tau) + 4\psi_{0,4}(\tau)\, B_1^3(\tau)$$
$$\qquad + 2\psi_{1,2}(\tau)\, B_2(\tau) + 3\psi_{1,3}(\tau)\, B_1^2(\tau) + 2\psi_{2,2}(\tau)\, B_1(\tau),$$
$$\cdots$$

和

$$f_0(\tau) = 0,$$
$$f_1(\tau) = \psi_{0,1}(\tau)\, B_1(\tau),$$
$$f_2(\tau) = \psi_{0,1}(\tau)\, B_2(\tau) + \psi_{0,2}(\tau)\, B_1^2(\tau) + \psi_{1,1}(\tau)\, B_1(\tau),$$
$$f_3(\tau) = \psi_{0,1}(\tau)\, B_3(\tau) + 2\psi_{0,2}(\tau)\, B_1(\tau)\, B_2(\tau) + \psi_{0,3}(\tau)\, B_1^3(\tau)$$
$$\qquad + \psi_{1,1}(\tau)\, B_2(\tau) + \psi_{1,2}(\tau)\, B_1^2(\tau) + \psi_{2,1}(\tau)\, B_1(\tau),$$
$$\cdots$$

需要注意的是, $g_n(\tau)$ 仅取决于 $V_m(S,\tau)$ 和 $B_m(\tau)$, 其中 $m = 0,1,2,\cdots,n-1$. 所以, 对于 n 阶形变方程, $g_n(\tau)$ 总是已知的. 因此, 当 $n \geqslant 0$ 时, $V_n(S,\tau)$ 由线性偏微分方程 (13.32) 控制, 满足线性初始条件 (13.33), 无穷远处的线性边界条件 (13.34), 以及固定边界 $S = B_0(\tau) = 1$ 处的线性条件 (13.42). 只要 $S = B_0(\tau) = 1$ 处的 $V_n(S,\tau)$ 已知, 通过上面提到的递推公式的显式公式 (13.37) 就可以得到 $f_n(\tau)$, 然后通过 (13.43) 得到 $B_{n+1}(\tau)$.

然而, 线性偏微分方程 (13.32) 包含变系数, 因此不容易求解. 成钧、诸颂平和廖世俊 [263] 通过 Laplace 变换解决了这种具有变系数的线性偏微分方程系统. 令

$$\hat{V}_n(S,\zeta) = \mathcal{L}_T[V_n(S,\tau)]$$

和

$$\hat{g}_n(\zeta) = \mathcal{L}_T[g_n(\tau)]$$

分别表示 $V_n(S,\tau)$ 和 $g_n(\tau)$ 关于 τ 的 Laplace 变换, 其中 $\mathcal{L}_T$ 是 Laplace 变换算子, ζ 为 τ 的复数. 通过 Laplace 变换并基于初始条件 (13.23), 当 $m \geqslant 0$ 时, 有

$$\mathcal{L}_T\left[\frac{\partial V_n(S,\tau)}{\partial \tau}\right] = \zeta\, \mathcal{L}_T[V_n(S,\tau)] - V_n(S,0) = \zeta\, \hat{V}_n(S,\zeta)$$

和

$$\mathcal{L}_T\left[\frac{\partial^m V_n(S,\tau)}{\partial S^m}\right] = \frac{\partial^m}{\partial S^m}\{\mathcal{L}_T[V_n(S,\tau)]\} = \frac{\partial^m \hat{V}_n(S,\zeta)}{\partial S^m}.$$

因此, 通过 Laplace 变换, 满足初始 (边界) 条件 (13.33), (13.34) 和 (13.42) 的高阶形变方程 (13.32) 转化为线性常微分方程

$$S^2 \frac{\partial^2 \hat{V}_n}{\partial S^2} + \gamma S \frac{\partial \hat{V}_n}{\partial S} - (\gamma + \zeta)\hat{V}_n = 0, \qquad (13.44)$$

满足边界条件

$$\frac{\partial \hat{V}_n}{\partial S} = -\hat{g}_n(\zeta), \quad S = 1, \qquad (13.45)$$

$$\hat{V}_n(S,\zeta) \to 0, \quad S \to +\infty. \qquad (13.46)$$

上述线性常微分方程具有封闭解

$$\hat{V}_n(S,\zeta) = \hat{K}(S,\zeta)\,\hat{g}_n(\zeta), \qquad (13.47)$$

其中

$$\hat{K}(S,\zeta) = -\frac{S^\lambda}{\lambda}, \quad \lambda = \frac{1-\gamma-\sqrt{4\zeta+(1+\gamma)^2}}{2}. \qquad (13.48)$$

需要注意的是, $\hat{g}_n(\zeta) = \mathcal{L}_T[g_n(\tau)]$ 是 $g_n(\tau)$ 的 Laplace 变换. 然后, 利用 Laplace 逆变换, 我们有

$$V_n(S,\tau) = \mathcal{L}_T^{-1}\left[\hat{K}(S,\zeta)\,\hat{g}_n(\zeta)\right] \qquad (13.49)$$

及其导数
$$V_n^{(m)}(S,\tau) = \frac{\partial^m V_n(S,\tau)}{\partial S^m} = \mathcal{L}_T^{-1}\left[\hat{K}^{(m)}(S,\zeta)\,\hat{g}_n(\zeta)\right], \qquad (13.50)$$

其中
$$\hat{K}^{(m)}(S,\zeta) = \frac{\partial^m \hat{K}(S,\zeta)}{\partial S^m},$$

$\mathcal{L}_T^{-1}$ 表示 Laplace 变换的逆算子.

根据 (13.40), (13.42), (13.43) 和相关的递推公式, 当 $0 \leqslant m \leqslant n-1$ 时, $g_n(\tau)$, $f_n(\tau)$ 和 $B_n(\tau)$ 仅仅依赖于 $V_n^{(m)}(1,\tau)$, 其中上标 (m) 表示 S 的 m 阶导数. 因此, 除了 $S = B_0(\tau) = 1$, 不需要对所有的 S 都得到一般表达式 $V_n^{(m)}(S,\tau)$. 所以, 关键是要有效地获得 $V_n^{(m)}(1,\tau)$. 由 (13.50) 可得

$$V_n^{(m)}(1,\tau) = \mathcal{L}_T^{-1}\left[\hat{K}^{(m)}(1,\zeta)\,\hat{g}_n(\zeta)\right]. \qquad (13.51)$$

根据 (13.48), 可以得到
$$\hat{K}(1,\zeta) = -\frac{2}{(1-\gamma) - \sqrt{4\zeta + (1+\gamma)^2}},$$
$$\hat{K}^{(1)}(1,\zeta) = \left.\frac{\partial \hat{K}(S,\zeta)}{\partial S}\right|_{S=1} = -1,$$
$$\hat{K}^{(2)}(1,\zeta) = \frac{1+\gamma}{2} + \frac{1}{2}\sqrt{4\zeta + (1+\gamma)^2},$$
$$\cdots$$

它们的 Laplace 逆变换为
$$K(1,\tau) = \frac{1}{\sqrt{\pi\tau}}\exp\left[-\frac{1}{4}(1+\gamma)^2\tau\right] + \frac{(1-\gamma)}{2}\exp(-\gamma\tau)\,\text{Erfc}\left[-\frac{(1-\gamma)}{2}\sqrt{\tau}\right],$$
$$K^{(1)}(1,\tau) = -\delta(\tau),$$
$$K^{(2)}(1,\tau) = -\frac{1}{2\sqrt{\pi\tau^3}}\exp\left[-\frac{1}{4}(1+\gamma)^2\tau\right] + \frac{1}{2}(1+\gamma)\delta(\tau),$$
$$\cdots$$

其中 $\text{Erfc}(x)$ 是互补误差函数, $\delta(\tau)$ 是狄拉克 δ 函数.

如上所述, 我们只需要获得 $V_n^{(m)}(1,\tau)$ 即可获得最优行权边界 $B(\tau)$. 该过程始于初始猜测解 $B_0(\tau) = 1$. 因为 $g_0(\tau) = 1$, 所以我们有其 Laplace 变换

$$\hat{g}_0(\zeta) = \mathcal{L}_T[g_0(\tau)] = \frac{1}{\zeta}.$$

然后, $V_0(1,\tau)$ 及其导数由 Laplace 逆变换 (13.51) 给出, $B_1(\tau)$ 由 (13.43) 获得, $g_1(\tau)$ 和 $f_1(\tau)$ 由具有显式公式 (13.38) 至 (13.41) 的 (13.37) 获得. 类似地, $\hat{g}_1(\zeta)$ 由 Laplace 变换给出, 然后 $V_1(1,\tau)$ 及其导数由 Laplace 逆变换 (13.51) 给出, $B_2(\tau)$ 由 (13.43) 获得, $g_2(\tau)$ 和 $f_2(\tau)$ 由具有显式公式 (13.38) 到 (13.41) 的 (13.37) 获得. 理论上, 我们可以通过这种方式依次获得 $B_m(\tau)$, 其中 $m = 1, 2, 3, \cdots$.

需要注意的是, Laplace 逆变换 (13.51) 可以用 $K^{(m)}(1,\tau)$ 和 $g_n(\tau)$ 的卷积积分来表示, 比如说,

$$V_n^{(m)}(1,\tau) = \int_0^\tau K^{(m)}(1,\tau-t)\, g_n(t)\, dt. \tag{13.52}$$

因此, 由于 $K(1,\tau)$ 包含互补误差函数, $V_n^{(m)}(1,\tau)$, $B_n(\tau)$, $g_n(\tau)$ 和 $f_n(\tau)$ 的表达式, 随着 n 的增加而变得越来越复杂. 所以, 精确计算 Laplace 变换 $\hat{g}_n(\zeta) = \mathcal{L}_T[g_n(\tau)]$, 尤其是 Laplace 逆变换 (13.51) 变得越来越困难. 因此, 精确求解高阶形变方程变得越来越困难.

为了获得高阶近似, 诸颂平 [295] 采用数值方法计算与 (13.52) 类似的相关积分. 与诸颂平 [295] 不同, 成钧、诸颂平和廖世俊 [263] 得到了 $B(\tau)$ 在有效期 $\tau = 0$ 时以 $\sqrt{\tau}$ 级数展开的解析近似公式. 注意, $K(1,\tau)$ 由指数函数和互补误差函数表示, 其在 $\tau = 0$ 处的泰勒级数的收敛半径无穷大. 此外, 无量纲有效期 τ 在实际中通常很小. 因此, 只要使用了足够多的项, 例如 $o(\tau^M)$ 的阶足够高, $B(\tau)$ 就可以很好地由 $\sqrt{\tau}$ 的多项式近似表示.

成钧、诸颂平和廖世俊 [263] 运用了如下策略. 首先, 将 $K^{(m)}(1,\tau)$ 和 $g_n(\tau)$ 在 $\tau = 0$ 时展开到 $o(\tau^M)$, 分别用 $\bar{K}^{(m)}(1,\tau)$ 和 $\bar{g}_n(\tau)$ 表示. 它们可以表示为

$$\bar{K}^{(m)}(1,\tau) = \sum_{n=0}^{2M} a_{m,n} \left(\sqrt{\tau}\right)^n \tag{13.53}$$

和

$$\bar{g}_n(\tau) = \sum_{n=0}^{2M} d_n \left(\sqrt{\tau}\right)^n, \tag{13.54}$$

其在时间上是指数为 $1/2$ 的 Hölder 连续, 这与 Blanchet [253] 的证明非常一致. 其次, 很容易得到它们的 Laplace 变换

$$\check{K}^{(m)}(1,\zeta) = \mathcal{L}_T[\bar{K}^{(m)}(1,\tau)], \quad \check{g}_m(\zeta) = \mathcal{L}_T[\bar{g}_m(\tau)].$$

更重要的是，获得它们的 Laplace 逆变换是相当方便的

$$V_n^{(m)}(1,\tau) = \mathcal{L}_T^{-1}\left[\check{K}^{(m)}(1,\zeta)\,\check{g}_m(\zeta)\right],$$

因为 $\check{K}^{(m)}(1,\zeta)$ 和 $\check{g}_m(\zeta)$ 都以 $\zeta^{-1/2}$ 的多项式表示.

无量纲 $B(\tau)$ 以 $\sqrt{\tau}$ 级数展开到 $o(\tau^M)$ 的 N 阶同伦近似由以下公式显式给出

$$B(\tau) \approx \sum_{m=0}^{N} B_m(\tau) = \sum_{k=0}^{2M} b_k \left(\sqrt{\tau}\right)^k, \qquad (13.55)$$

其中系数 b_k 取决于 $\gamma = 2r/\sigma^2$ 和收敛控制参数 c_0. 为了进一步修正上述对 $B(\tau)$ 的近似，我们先写出 $z = \sqrt{\tau}$，然后计算 $B = \sum_{n=0}^{2M} b_n z^n$ 以 $z = 0$ 为中心的 $[M,M]$ 阶帕德近似. 然后，用 $\sqrt{\tau}$ 替换 z，可以得到 $o(\tau^M)$ 阶的 $B(\tau)$ 的 $[M,M]$ 阶帕德近似.

13.3 显式同伦近似的有效性

为了证明上述同伦分析方法给出的最优行权边界 $B(\tau)$ 的显式表达式 (13.55) 的精确性和有效性，我们将其与一些已发表的分析结果进行比较. 正如成钧、诸颂平和廖世俊 [263] 所指出的那样，所有已发表的 $B(\tau)$ 的显式近似值都对 $\tau \ll \tau_{\text{exp}}$ 有效. 例如，

$$B(\tau) = \exp(-2\sqrt{\alpha\tau}), \quad \tau \ll \tau_{\text{exp}}, \qquad (13.56)$$

其中

$$\alpha = -\frac{1}{2}\ln\left(9\pi\gamma^2\tau\right), \quad \text{Kushe 和 Keller [274],} \qquad (13.57)$$

$$\alpha = -\frac{1}{2}\ln\left(\frac{4e\gamma^2\tau}{2-B_p^2}\right), \quad \text{Bunch 和 Johnson [257],} \qquad (13.58)$$

具有长期最优行权价格

$$B_p = \frac{\gamma}{1+\gamma}. \qquad (13.59)$$

此外，Knessl [273] 给出了以下渐近公式

$$\ln[B(\tau)] = -\sqrt{2\tau|\ln(4\pi\gamma^2\tau)|}\left[1 + \left|\ln(4\pi\gamma^2\tau)\right|^{-2}\right], \quad \tau \ll \tau_{\text{exp}}. \qquad (13.60)$$

通过两个例子, 将这些渐近或摄动公式与上述同伦分析方法给出的显式公式 (13.55) 进行比较. 下文给出的所有结果均通过附录 13.2 中的 Mathematica 程序获得. 此外, 下面给出的所有结果都是有单位的, 也就是说, 它们不是无量纲的.

例子 13.1

让我们首先考虑 Wu 和 Kwok [292], Carr 和 Faguet [259], 诸颂平 [295], 成钧、诸颂平和廖世俊 [263] 以及成钧 [262] 所讨论的例子, 参数如下:
- 行权价格 $X = 100$ (美元),
- 无风险利率 $r = 0.1$,
- 波动率 $\sigma = 0.3$,
- 有效期 $T = 1$ (年).

需要注意的是, 零阶形变方程包含收敛控制参数 c_0, 该参数在同伦分析方法框架中可以保证同伦级数的收敛性. 为了简单起见, 我们在本小节例子中选取 $c_0 = -1$.

我们发现, 不管是 Kuske 和 Keller [274] 给出的 (13.57), Knessl [273] 给出的 (13.60), 还是 Bunch 和 Johnson [257] 给出的 (13.58) 的有效期在到期前的几周内, 如图 13.1 所示.

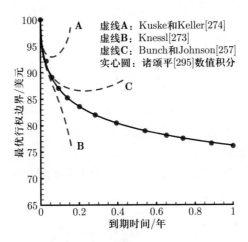

图 13.1 例子 13.1 的最优行权边界的近似值: $X = 100$ (美元), $r = 0.1$, $\sigma = 0.3$ 以及 $T = 1$ (年). 实线: 选取 $c_0 = -1$ 时, 以 $\sqrt{\tau}$ 级数展开到 $o(\tau^8)$ 的 12 阶同伦近似; 实心圆: 诸颂平 [295] 基于数值积分给出的结果; 虚线 A: Kuske 和 Keller [274] 给出的 (13.57); 虚线 B: Knessl [273] 给出的 (13.60); 虚线 C: Bunch 和 Johnson [257] 给出的 (13.58)

基于同伦分析方法，最优行权边界 $B(\tau)$ 以 $\sqrt{\tau}$ 级数展开到 $o(\tau^8)$ 的近似解，其与诸颂平 [295] 在整个时间范围内的数值积分给出的结果非常吻合，如图 13.1 所示. 对于有效期 $T = 1$（年），Wu 和 Kwok [292] 给出的数值解的最优行权边界 $B(\tau)$ 为 76.25 美元，诸颂平 [295] 通过数值积分给出的最优行权边界 $B(\tau)$ 为 76.11 美元，由同伦分析方法给出的最优行权边界 $B(\tau)$ 展开到 $o(\tau^8)$ 的 20 阶同伦近似为 76.17 美元，如表 13.1 所示. 这显示了上述基于同伦分析方法的有效性. 因此，与上述所有渐近和摄动近似（通常有效期仅为几天或几周）不同的是，基于同伦分析方法给出的最优行权边界的显式多项式近似是精确的且有效期可长达几年. 注意，即使 $B(\tau)$ 以 $\sqrt{\tau}$ 级数展开到 $o(\tau^8)$ 的 12 阶同伦近似也是足够准确的，如图 13.1 所示. 从数学上讲，所有渐近和摄动公式仅对 $\tau \ll T$ 有效，但 $B(\tau)$ 以 $\sqrt{\tau}$ 级数展开到 $o(\tau^8)$ 的同伦近似即使在 $\tau = T$ 时，即到期日，也是有效的！

表 13.1 对于例子 $X = 100$（美元），$r = 0.1$，$\sigma = 0.3$ 和 $T = 1$（年），选取 $c_0 = -1$ 时，最优行权边界 $B(\tau)$ 以 $\sqrt{\tau}$ 级数展开到 $o(\tau^8)$ 的 m 阶同伦近似在不同时间的值

同伦近似的阶 m	3 月	6 月	9 月	12 月
4	84.02	79.86	77.44	75.84
8	82.86	79.22	77.24	75.96
12	82.63	79.27	77.41	76.15
16	82.60	79.35	77.49	76.18
18	82.62	79.38	77.50	76.18
20	82.63	79.40	77.50	76.17

研究发现，选取 $c_0 = -1$ 时，最优行权边界 $B(\tau)$ 以 $\sqrt{\tau}$ 级数展开到 $o(\tau^8)$ 的 10 阶同伦近似在到期前 3 年内足够精确，如图 13.2 所示，这比上述提到的其他渐近和（或）摄动公式给出的有效期长约 60 倍！此外，可以使用帕德技术进一步扩大 $B(\tau)$ 的有效时间：令 $z = \sqrt{\tau}$，然后使用帕德技术，可以获得以 $z = 0$ 为中心的 $B(\tau)$ 的 [8,8] 阶帕德近似值，即

$$B(\tau) \approx 100 \left[\frac{1 + \check{b}_1(\tau)}{1 + \check{b}_2(\tau)} \right], \tag{13.61}$$

其中

$$\check{b}_1(\tau) = -0.758595\tau^{1/2} + 0.8748811\tau - 0.474758\tau^{3/2} + 0.209634\tau^2$$
$$- 0.0551746\tau^{5/2} + 0.01333\tau^3 - 0.00210274\tau^{7/2} + 2.15592 \times 10^{-4}\tau^4,$$

$$\check{b}_2(\tau) = -0.228157\tau^{1/2} + 0.297781\tau - 0.0406677\tau^{3/2} + 0.0324789\tau^2$$
$$- 0.0018480\tau^{5/2} + 0.0016651\tau^3 + 3.545 \times 10^{-6}\tau^{7/2} + 4.245 \times 10^{-5}\tau^4.$$

如图 13.2 所示,$B(\tau)$ 的 [8,8] 阶帕德近似值已足够精确, 甚至在到期前 10 年内足够精确, 比上述其他渐近和 (或) 摄动公式给出的有效期长约 200 倍!

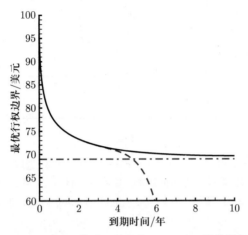

图 13.2 对于例子 13.1: $X = 100$ (美元), $r = 0.1$, $\sigma = 0.3$, 最优行权边界 $B(\tau)$ 的 [8,8] 阶帕德近似. 虚线: 选取 $c_0 = -1$ 时, $B(\tau)$ 以 $\sqrt{\tau}$ 级数展开到 $o(\tau^8)$ 的 10 阶同伦近似; 实线: 对应的 $B(\tau)$ 的 [8,8] 阶帕德近似; 点划线: 长期最优行权价格

对于同样的问题, 成钧、诸颂平和廖世俊 [263] 以及成钧 [262] 基于同伦分析方法分别给出了 $B(\tau)$ 展开到 $o(\tau^6)$ 和 $o(\tau^7)$ 的同伦近似. 即使在 $\tau = T = 1$ 年, 即到期日, 它们都是有效的. 本章获得了 $B(\tau)$ 展开到 $o(\tau^8)$ 的 10 阶同伦近似, 该近似值在到期前 3 年是精确的. 理论上, 对于任意正整数 $M \geqslant 1$, $B(\tau)$ 可以按 $\sqrt{\tau}$ 级数展开到 $o(\tau^M)$. 那么, $B(\tau)$ 展开到 $o(\tau^M)$ 的同伦近似的最长有效期是否强烈依赖于 M?

使用附录 13.1 中的显式公式, 我们可以方便地研究 $o(\tau^M)$ 的阶对于 $B(\tau)$ 的影响. 我们发现, $o(\tau^M)$ 的阶越高, $B(\tau)$ 的 10 阶同伦近似在到期前的最长有效时间越长, 如图 13.3 所示. 注意, $B(\tau)$ 以 $\sqrt{\tau}$ 级数展开到 $o(\tau^{48})$ 的 10 阶同伦近似对于有效期为 20 年已足够精确, 这比上面提到的渐近和 (或) 摄动公式长约 400 倍! 如图 13.3 所示, 当 $M = 48$ 时, 最优行权价格与长期最优行权价格 B_p 非常接近, 所以 $B(\tau)$ 展开到 $o(\tau^{48})$ 的 10 阶同伦近似结合已知的长期最优行权价格 $B_p = \gamma/(1 + \gamma)$ 甚至在到期前的整个时域 ($0 \leqslant \tau < +\infty$) 上也可以给出足够精确的最优行权价格! 使用附录 13.2 中给出的 Mathematica 程序, 通过笔记本电脑 (配备 2.8 GHz 和 4 GB MHz DDR3 的 MacBook Pro) 仅在

102 s 内就可以获得精确的解析近似值.

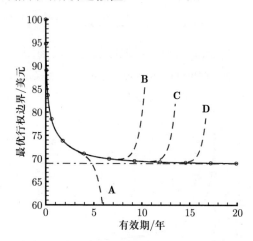

图 13.3 对于例子 13.1: $X = 100$ (美元), $r = 0.1$, $\sigma = 0.3$, 选取 $c_0 = -1$ 时, $o(\tau^M)$ 的阶对最优行权边界的 10 阶同伦近似的影响. 虚线 A: $M = 8$; 虚线 B: $M = 16$; 虚线 C: $M = 24$; 虚线 D: $M = 32$; 实线: $M = 48$; 空心圆: [48,48] 阶帕德近似; 点划线: 长期最优行权价格

如图 13.4 和表 13.2 所示, $B(\tau)$ 以 $\sqrt{\tau}$ 级数展开到 $o(\tau^{128})$ 的 10 阶同伦近似的有效期甚至为半个世纪, 这比上述渐近和 (或) 摄动公式给出的有效期长约 1000 倍!

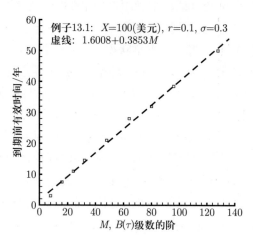

图 13.4 对于例子 13.1: $X = 100$ (美元), $r = 0.1$, $\sigma = 0.3$, 选取 $c_0 = -1$ 时, $o(\tau^M)$ 的阶对最优行权边界的 10 阶同伦近似的最长有效时间 (年) 的影响. 虚线: 最小二乘法拟合给出的公式 $1.6008 + 0.3853M$; 空心方块: 最优行权边界 $B(\tau)$ 以 $\sqrt{\tau}$ 级数展开到 $o(\tau^8)$ 的到期前的最长有效时间

表 13.2 对于例子 13.1: $X = 100$ (美元), $r = 0.1$, $\sigma = 0.3$, 选取 $c_0 = -1$ 时, $B(\tau)$ 以 $\sqrt{\tau}$ 级数展开到 $o(\tau^M)$ 的 10 阶同伦近似的到期前的最长有效时间 (年)

$B(\tau)$ 展开到 $o(\tau^M)$ 的阶	到期前的最长有效时间 (年)
$M = 8$	3
$M = 16$	7.5
$M = 24$	11
$M = 32$	14.5
$M = 48$	21
$M = 64$	28
$M = 80$	32
$M = 96$	38.5
$M = 128$	50

特别地, 我们发现, $B(\tau)$ 以 $\sqrt{\tau}$ 级数展开到 $o(\tau^M)$ 的 10 阶同伦近似的最长有效时间与 M 近似成正比, 如图 13.4 所示. 这表明, 给定到期前的任意时间 T, 只要 $B(\tau)$ 以 $\sqrt{\tau}$ 级数展开到 $o(\tau^M)$ 的 N 阶同伦近似有足够大的 M 和相当高的阶 N, 我们总能获得足够精确的最优行权价格. 这个例子很好地说明了同伦分析方法可以为某些非线性问题提供比渐近和 (或) 摄动方法更好的显式解析近似!

事实上, 一种全新的方法总会带来一些新的和 (或) 不同的东西. 这个例子再次展示了同伦分析方法的原创性和巨大潜力.

例子 13.2

第二个例子是 Chen 和 Chadam 等人 [260] 考虑的长期期权, 参数如下:
- 行权价格 $X = 1$ (美元),
- 无风险利率 $r = 0.08$,
- 波动率 $\sigma = 0.4$,
- 有效期 $T = 3$ (年).

如图 13.5 所示, 选取 $c_0 = -1$ 时, $B(\tau)$ 以 $\sqrt{\tau}$ 级数展开到 $o(\tau^{48})$ 的 10 阶同伦近似在到期前 20 年内有效, 而不管是 Kuske 和 Keller [274] 给出的 (13.57),

Knessl [273] 给出的 (13.60), 还是 Bunch 和 Johnson [257] 给出的 (13.58) 的有效期均不超过到期前一个月! 值得注意的是, 我们的同伦近似与其 [48,48] 阶帕德近似在整个时域 $0 \leqslant \tau \leqslant 20$ (年) 内很好地拟合. 这种精确的解析近似可通过笔记本电脑 (配备 2.8 GHz 和 4 GB MHz DDR3 的 MacBook Pro) 在 57 s 内获得. 有效期为 $T = 20$ 年时, 10 阶同伦近似和对应的 [48,48] 阶帕德近似给出了相同的最优行权价格 0.5029 美元, 非常接近于长期最优行权价格 0.5 美元, 即 $B(\tau) = 0.5$ 是 $\tau > 20$ 足够精确的近似值. 因此, 将 $B(\tau)$ 以 $\sqrt{\tau}$ 级数展开到 $o(\tau^{48})$ 的 10 阶同伦近似与 $\tau > 20$ (年) 时的长期最优行权价格 $B(\tau) = 0.5$ 相结合, 我们在整个时域 $0 \leqslant \tau < +\infty$, 即任意有效期内都有一个足够精确的解析近似值!

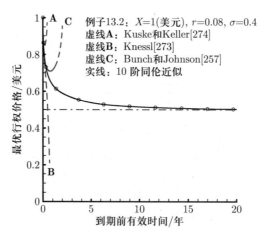

图 13.5 对于例子 13.2: $X = 1$ (美元), $r = 0.08$ 以及 $\sigma = 0.4$ 的最优行权边界 $B(\tau)$. 虚线 A: Kuske 和 Keller [274] 给出的 (13.57); 虚线 B: Knessl [273] 给出的 (13.60); 虚线 C: Bunch 和 Johnson [257] 给出的 (13.58); 实线: 选取 $c_0 = -1$ 时, 以 $\sqrt{\tau}$ 级数展开到 $o(\tau^{48})$ 的 10 阶同伦近似; 空心圆: 帕德方法给出的 $B(\tau)$ 的 10 阶同伦近似; 点划线: 长期最优行权价格 0.5 美元

与成钧、诸颂平和廖世俊 [263] 以及成钧 [262] 一样, 我们通过选择收敛控制参数 $c_0 = -1$ 获得了上述所有同伦近似. 然而, 正如第四章和本书中所证明的那样, 收敛控制参数 c_0 为我们提供了一种保证同伦级数收敛的便捷方法. 如图 13.6 所示, 选取 $c_0 = -1$ 时, $B(\tau)$ 以 $\sqrt{\tau}$ 级数展开到 $o(\tau^8)$ 的 10 阶同伦近似的有效期约为到期前 6 年. 然而, 选取 $c_0 = -1/2$ 时, 其有效期约为到期前 10 年: $B(\tau)$ 的最长有效时间增加了约 66%. 因此, 收敛控制参数 c_0 也为我们提供了另一种方法来扩大 $B(\tau)$ 的同伦近似在到期前的最长有效时间.

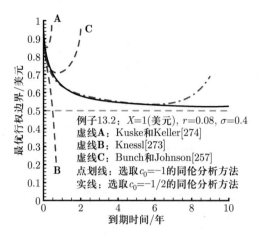

图 13.6 对于例子 13.2: $X = 1$ (美元), $r = 0.08$ 以及 $\sigma = 0.4$ 的最优行权边界. 虚线 A: Kuske 和 Keller [274] 给出的 (13.57); 虚线 B: Knessl [273] 给出的 (13.60); 虚线 C: Bunch 和 Johnson [257] 给出的 (13.58); 点划线: 选取 $c_0 = -1$ 时, 以 $\sqrt{\tau}$ 级数展开到 $o(\tau^8)$ 的 10 阶同伦近似; 实线: 选取 $c_0 = -1/2$ 时, 以 $\sqrt{\tau}$ 级数展开到 $o(\tau^8)$ 的 10 阶同伦近似; 空心圆: 帕德方法给出的 $B(\tau)$ 的 10 阶同伦近似; 长虚线: 长期最优行权价格 0.5 美元

这两个例子说明, 与其他有效期通常仅为几天或几周的渐近和 (或) 摄动公式不同的是, 同伦分析方法为我们提供了最优行权边界 $B(\tau)$ 以 $\sqrt{\tau}$ 级数展开到 $o(\tau^M)$ 的高精度近似值, 只要 M 和近似阶数相当大, 并适当选择收敛控制参数 c_0, 有效期通常为几十年甚至半个世纪.

正如 Kim [272] 和 Carr 等人 [258] 所指出的那样, 当最优行权边界 $B(\tau)$ 已知时, 通过 (13.7) 很容易获得美式认沽期权的价格 $V(S, \tau)$. 因此, 通过本章提到的基于同伦分析方法给出的 $B(\tau)$ 以 $\sqrt{\tau}$ 级数展开的显式解析近似 (13.55), 可以很方便地获得价格 $V(S, \tau)$ 的高精度近似值, 该价格的有效期可能为半个世纪!

13.4 可供商业使用的实用程序

成钧、诸颂平和廖世俊 [263] 以及成钧 [262] 考虑的例子通过 $B(\tau)$ 以 $\sqrt{\tau}$ 级数展开到 $o(\tau^{48})$ 的 10 阶同伦近似在更长的时间间隔内 (到期前) 获得了精确的最优行权价格. 此外, 我们发现当选取 $c_0 = -1$ 时, $B(\tau)$ 以 $\sqrt{\tau}$ 级数展开到 $o(\tau^{48})$ 的 10 阶同伦近似与长期的最优行权价格已经非常接近. 因此, 实际上, 对于任意长的有效期 $0 < T < +\infty$, 我们都有精确的最优行权价格 $B(\tau)$ 和期权价格 $V(S, \tau)$. 这表明 $B(\tau)$ 以 $\sqrt{\tau}$ 级数展开到 $o(\tau^{48})$ 的同伦近似是普遍

有效的, 因此可以广泛应用于美式认沽期权相关业务.

如上所述, 通过笔记本电脑 (配备 2.8 GHz 和 4 GB MHz DDR3 的 MacBook Pro) 计算例子 13.1 和例子 13.2, 选取 $c_0 = -1$ 时, 只需 102 s 和 57 s 的 CPU 时间即可获得 $B(\tau)$ 以 $\sqrt{\tau}$ 级数展开到 $o(\tau^{48})$ 的 10 阶同伦近似. 这对于学者来说已经足够快了, 但对于商人来说则不然. 为了方便起见, 附录 13.3 中提供了一个实用的 Mathematica 程序, 它可以在数秒内给出有效期 T 相当长的美式认沽期权的足够精确的最优行权价格!

需要注意的是, 无量纲方程 (13.10) 至 (13.14) 仅包含一个无量纲参数 $\gamma = 2r/\sigma^2$. 因此, 无量纲 $B(\tau)$ 仅依赖于 γ. 使用附录 13.2 中的 Mathematica 程序和高性能计算机, 对于未知参数 γ, 当选取 $c_0 = -1$ 时, 我们获得了无量纲 $B(\tau)$ 以 $\sqrt{\tau}$ 级数展开到 $o(\tau^M)$ 的 10 阶同伦近似, 其中 M 是一个相当大的整数, 例如 $M = 24, 36, 48$ 等. 然后, 这些冗长的无量纲 $B(\tau)$ 的同伦近似被保存在不同名称的数据文件中, 例如 `APO-48-10.txt`, 对应于展开到 $o(\tau^{48})$ 的 10 阶同伦近似. 在附录 13.3 中给出了一个简短的 Mathematica 程序, 即 `APOh`, 它首先读取保存在数据文件中的 $B(\tau)$ 的所有无量纲同伦近似值, 然后根据给定的行权价格 X, 无风险利率 r, 波动率 σ 和有效期 T (年) 计算有量纲的最优行权价格. 使用笔记本电脑, 通常可以在数秒内获得足够精确且有效期 T 相当长的美式认沽期权的行权价格!

还要注意的是, $B(\tau)$ 以 $\sqrt{\tau}$ 级数展开表示的这种显式同伦近似是解析的, 因为它们对于任意的 $0 < \tau < +\infty$ 都是连续的, 其各阶导数 $B'(\tau), B''(\tau)$ 等对所有的 $\tau > 0$ 都是存在的. 因此, 根本不需要任何插值. 的确, 同伦分析方法给出的这种解析近似非常冗长, 如果打印出来可能有数百页. 从传统的角度来看, 这种显式的、冗长的解析公式可能是无用的. 幸运的是, 我们现在已经进入了计算机时代: 笔记本电脑可以在数秒内读取并计算出这种显式的、冗长的解析公式, 这比手动计算一个简单的半页长度的解析公式还要快! 所以, 如果我们把笔记本电脑的键盘看成笔, 把它的硬盘看成纸, 把它的中央处理器 (CPU) 看成人类的大脑, 那么在计算机时代, 一个显式的解析公式就可以很冗长, 如本节所示. 注意, 我们的传统概念 "解析" 出现在数百年前, 当时我们的计算工具效率非常低. 实用的 Mathematica 程序 `APOh` 很好地说明了我们的传统概念 "解析" 是过时的, 因此在计算机和互联网时代该传统概念应进行修正. 另外, 这也显示了同伦分析方法的原创性.

事实上, 一种全新的方法总会带来一些新的和 (或) 不同的东西.

13.5 本章小结

美式认沽期权由具有变系数的线性偏微分方程控制, 满足一定的线性初始条件和边界条件. 但是, 由于两个边界条件都在未知的移动边界 $B(\tau)$ 上, 所以其本质是一个非线性问题. 不管是 Kuske 和 Keller [274] 给出的 (13.57), Bunch 和 Johnson [257] 给出的 (13.58), 还是 Knessl [273] 给出的 (13.60), 这些渐近和 (或) 摄动公式给出的有效期都仅为几天或几周, 这对于商业中的实际使用来说太短了.

成钧、诸颂平和廖世俊 [263] 首次将同伦分析方法与 Laplace 变换结合, 得到了最优行权边界 $B(\tau)$ 以 $\sqrt{\tau}$ 级数展开到 $o(\tau^6)$ 的同伦近似. 成钧 [262] 进一步给出了 $B(\tau)$ 展开到 $o(\tau^7)$ 的同伦近似. 与渐近和 (或) 摄动公式不同的是, 这些 $B(\tau)$ 的同伦近似的有效期为几年, 比上述所有渐近和 (或) 摄动公式得到的结果要好得多.

本章通过导出与未知移动边界相关的显式公式 (13.35) 至 (13.41), 进一步修正了成钧、诸颂平和廖世俊 [263] 以及成钧 [262] 基于同伦分析方法的求解过程. 特别地, 我们首次研究了 $B(\tau)$ 多项式解析近似解中 $o(\tau^M)$ 的阶对其最长有效时间的影响. 研究发现, $B(\tau)$ 以 $\sqrt{\tau}$ 级数展开到 $o(\tau^M)$ 的最长有效时间与 M 成正比. 因此, 只要 M 足够大, $B(\tau)$ 的显式同伦近似的有效期甚至可以长达半个世纪, 比本文提到的渐近和 (或) 摄动公式得到的有效期长约 1000 倍, 如例子 13.1 所示. 此外, 我们发现, 当选取 $c_0 = -1$ 时, $B(\tau)$ 以 $\sqrt{\tau}$ 级数展开到 $o(\tau^{48})$ 的 10 阶同伦近似通常在到期前的很多年都有效, 以至于理论上的长期最优行权价格

$$B_p = \left(\frac{\gamma}{1+\gamma}\right) X$$

在此后足够精确. 因此, 将 $B(\tau)$ 以 $\sqrt{\tau}$ 级数展开到 $o(\tau^{48})$ 的同伦近似与长期最优行权价格 B_p 结合, 我们在整个时域 $0 \leqslant \tau < +\infty$ 内都有一个足够精确的解析近似值! 基于同伦分析方法, 附录 13.2 中给出了 Mathematica 程序. 在高性能计算机上使用此程序, 对于未知无量纲参数 $\gamma = 2r/\sigma^2$, 我们得到了无量纲 $B(\tau)$ 以 $\sqrt{\tau}$ 级数展开到 $o(\tau^{48})$ 的 10 阶同伦近似. 然后, 将其保存在名为 APO-48-10.txt 的文件中, 可直接通过简短实用的 Mathematica 程序 APOh 在数秒内获得足够精确的美式认沽期权的最优行权价格. 实用程序 APOh 和数据文件 APO-48-10.txt 都可以在网上免费获得. 显然, APOh 为商人提供了一个方便的工具.

此外, 我们还首次研究了收敛控制参数 c_0 对 $B(\tau)$ 以 $\sqrt{\tau}$ 级数展开表示的

同伦近似的最长有效时间 (到期前) 的影响. 如图 13.6 所示, $B(\tau)$ 以 $\sqrt{\tau}$ 级数展开到 $o(\tau^8)$ 的 10 阶同伦近似的最长有效时间在 $c_0 = -1/2$ 时比 $c_0 = -1$ 时增加了约 66%. 这说明了收敛控制参数 c_0 也为我们提供了一种便捷的方式来增加 $B(\tau)$ 的最长有效时间.

需要注意的是, 此处未使用 Landau 变换 [275]. 因此, 这种基于同伦分析方法和 Laplace 变换的解析方法具有相当普遍的意义, 因此可以广泛应用于解决金融领域的类似问题, 例如具有股息收益率的标的资产的美式期权的最优行权边界 [296] 等.

最后, 我们强调, 与上述通常有效期仅为几天或几周的渐近和 (或) 摄动公式不同的是, 最优行权边界 $B(\tau)$ 以 $\sqrt{\tau}$ 级数展开表示的同伦近似的有效期可能是几十年, 甚至半个世纪! 此外, 虽然保存在数据文件 APO-48-10.txt 中的这些 $B(\tau)$ 的精确同伦近似值在打印出来时可能长达数百页, 但是通过笔记本电脑中简短实用的 Mathematica 程序 APOh, 我们在数秒内就可以得到精确的美式认沽期权的最优行权价格! 这个例子很好地表明了传统概念 "解析" 已经过时, 因此应该在计算机和互联网时代进行修正.

致谢 附录 13.2 中的 Laplace 逆变换公式由成钧博士提供.

参考文献

[250] Allegretto,W., Lin, Y., Yang, H.: Simulation and the early-exercise option problem. Discr. Contin. Dyn. Syst. B. **8**, 127–136 (2001).

[251] Alobaidi, G., Mallier, R.: On the optimal exercise boundary for an American put option. Journal of Applied Mathematics. **1**, No. 1, 39–45 (2001).

[252] Barles, G., Burdeau, J., Romano, M., Samsoen, N.: Critical stock price near expiration. Mathematical Finance. **5**, No. 2, 77–95 (1995).

[253] Blanchet, A.: On the regularity of the free boundary in the parabolic obstacle problem - Application to American options. Nonlinear Analysis. **65**, 1362–1378 (2006).

[254] Brennan, M., Schwartz, E.: The valuation of American put options. Journal of Finance. **32**, 449–462 (1977).

[255] Broadie, M., Detemple, J.: American option valuation: new bounds, approximations, and a comparison of existing methods. Review of Financial Studies. **9**, No. 4, 1211–1250 (1996).

[256] Broadie, M., Detemple, J.: Recent advances in numerical methods for pricing derivative securities. In Numerical Methods in Finance, 43–66, edited by Rogers, L.C.G. and Talay, D., Cambridge University Press, England (1997).

[257] Bunch, D.S., Johnson, H.: The American put option and its critical stock price. Journal of Finance. **5**, 2333–2356 (2000).

[258] Carr, P., Jarrow, R., Myneni, R.: Alternative characterizations of American put options. Mathematical Finance. **2**, 87–106 (1992).

[259] Carr, P., Faguet, D.: Fast accurate valuation of American options. Working paper, Cornell University (1994).

[260] Chen, X.F., Chadam, J., Stamicar R.: The optimal exercise boundary for American put options: analytic and numerical approximations. Working paper (http://www.math.pitt.edu/-xfc/Option/CCSFinal.ps. Accessed 15 April 2011), University of Pittsburgh (2000).

[261] Chen, X.F., Chadam, J.: A mathematical analysis for the optimal exercise boundary American put option. Working paper (http://www.pitt.edu/-chadam/ papers/2CC9-30-05.pdf. Accessed 15 April 2011), University of Pittsburgh (2005).

[262] Cheng, J.: Application of the Homotopy Analysis Method in the Nonlinear Mechanics and Finance (in Chinese). PhD Dissertation, Shanghai Jiao Tong University (2008).

[263] Cheng, J., Zhu, S.P., Liao, S.J.: An explicit series approximation to the optimal exercise boundary of American put options. Communications in Nonlinear Science and Numerical Simulation. **15**, 1148–1158 (2010).

[264] Cox, J., Ross, S., Rubinstein, M.: Option pricing: a simplified approach. Journal of Financial Economics. **7**, 229–263 (1979).

[265] Dempster, M.: Fast numerical valation of American, exotic and complex options. Department of Mathematics Research Report, University of Essex, Colchester, England (1994).

[266] Evans, J.D., Kuske, R., Keller, J.B.: American options on asserts with dividends near expiry. Mathematical Finance. **12**, No. 3, 219–237 (2002).

[267] Geske, R., Johnson, H.E.: The American put option valued analytically. Journal of Finance. **5**, 1511–1523 (1984).

[268] Grant, D., Vora, G., Weeks, D.: Simulation and the early-exercise option problem. Journal of Financial Engineering. **5**, 211–227 (1996).

[269] Hon, Y.C., Mao, X.Z.: A radial basis function method for solving options pricing model. Journal of Financial Engineering. **8**, 31–49 (1997).

[270] Huang, J.Z., Marti, G.S., Yu, G.G.: Pricing and Hedging American Options: A Recursive Integration Method. Review of Financial Studies. **9**, 277–300 (1996).

[271] Jaillet, P., Lamberton, D., Lapeyre, B.: Variational inequalities and the pricing of American options. Acta Applicandae Math. **21**, 263–289 (1990).

[272] Kim, I.J.: The analytic valuation of American options. Review of Financial Studies. **3**, 547–572 (1990).

[273] Knessl, C.: A note on a moving boundary problem arising in the American put option. Studies in Applied Mathematics. **107**, 157–183 (2001).

[274] Kuske, R.A., Keller, J.B.: Optional exercise boundary for an American put option. Applied Mathematical Finance. **5**, 107–116 (1998).

[275] Landau, H.G.: Heat conduction in melting solid. Quarterly of Applied Mathematics. **8**, 81–94 (1950).

[276] Li, Y.J., Nohara, B.T., Liao, S.J.: Series solutions of coupled van der Pol equation by means of homotopy analysis method. J. Mathematical Physics **51**, 063517 (2010). doi:10.1063/1.3445770.

[277] Liao, S.J.: The Proposed Homotopy Analysis Technique for the Solution of Nonlinear Problems. PhD dissertation, Shanghai Jiao Tong University (1992).

[278] Liao, S.J.: A kind of approximate solution technique which does not depend upon small parameters—(II) An application in fluid mechanics. Int. J. Nonlin. Mech. **32**, 815–822 (1997).

[279] Liao, S.J.: An explicit, totally analytic approximation of Blasius viscous flow problems. Int. J. Nonlin. Mech. **34**, 759–778 (1999a).

[280] Liao, S.J.: A uniformly valid analytic solution of 2D viscous flow past a semi-infinite flat plate. J. Fluid Mech. **385**, 101–128 (1999b).

[281] Liao, S.J., Campo, A.: Analytic solutions of the temperature distribution in Blasius viscous flow problems. J. Fluid Mech. **453**, 411–425 (2002).

[282] Liao, S.J.: On the analytic solution of magnetohydrodynamic flows of non-Newtonian fluids over a stretching sheet. J. Fluid Mech. **488**, 189–212 (2003a).

[283] Liao, S.J.: Beyond Perturbation - Introduction to the Homotopy Analysis Method. Chapman & Hall/ CRC Press, Boca Raton (2003b).

[284] Liao, S.J.: On the homotopy analysis method for nonlinear problems. Appl. Math. Comput. **147**, 499–513 (2004).

[285] Liao, S.J.: A new branch of solutions of boundary-layer flows over an impermeable stretched plate. Int. J. Heat Mass Tran. **48**, 2529–2539 (2005).

[286] Liao, S.J.: Series solutions of unsteady boundary-layer flows over a stretching flat plate. Stud. Appl. Math. **117**, 2529–2539 (2006).

[287] Liao, S.J.: Notes on the homotopy analysis method - Some definitions and theorems. Commun. Nonlinear Sci. Numer. Simulat. **14**, 983–997 (2009).

[288] Liao, S.J.: On the relationship between the homotopy analysis method and Euler transform. Commun. Nonlinear Sci. Numer. Simulat. **15**, 1421–1431 (2010a).

[289] Liao, S.J.: An optimal homotopy-analysis approach for strongly nonlinear differential equations. Commun. Nonlinear Sci. Numer. Simulat. **15**, 2003–2016 (2010b).

[290] Liao, S.J., Tan, Y.: A general approach to obtain series solutions of nonlinear differential equations. Stud. Appl. Math. **119**, 297–355 (2007).

[291] Longstaff, F., Schwartz, E.S.: A radial basis function method for solving options pricing model. Review of Finanial Studies. **14**, 113–147 (2001).

[292] Wu, L., Kwok, Y.K.: A front-fixing finite difference method for the valuation of American options. Journal of Financial Engineering. **6**, 83–97 (1997).

[293] Xu, H., Lin, Z.L., Liao, S.J., Wu, J.Z., Majdalani, J.: Homotopy-based solutions of the Navier-Stokes equations for a porous channel with orthogonally moving walls. Physics of Fluids. **22**, 053601 (2010) doi:10.1063/1.3392770.

[294] Zhang, J.E., Li, T.C.: Pricing and hedging American options analytically: a perturbation method.Working paper, University of Hong Kong (2006).

[295] Zhu, S.P.: An exact and explicit solution for the valuation of American put options. Quant. Financ. **6**, 229–242 (2006a).

[296] Zhu, S.P.: A closed-form analytical solution for the valuation of convertible bonds with constant dividend yield. ANZIAM J. **47**, 477–494 (2006b).

附录 13.1　$f_n(\tau)$ 和 $g_n(\tau)$ 的详细推导

根据 (13.28), 定义

$$[\Lambda(\tau;q) - B_0(\tau)]^m = \left[\sum_{i=1}^{+\infty} B_i(\tau) q^i\right]^m = \sum_{n=m}^{+\infty} \mu_{m,n}(\tau)\, q^n, \tag{13.62}$$

其中

$$\mu_{1,n}(\tau) = B_n(\tau), \quad n \geqslant 1. \tag{13.63}$$

则有

$$[\Lambda(\tau;q) - B_0(\tau)]^{m+1} = \sum_{n=m+1}^{+\infty} \mu_{m+1,n}(\tau)\, q^n$$

$$= \left[\sum_{n=m}^{+\infty} \mu_{m,n}(\tau)\, q^n\right] \left[\sum_{i=1}^{+\infty} B_i(\tau) q^i\right] = \sum_{n=m+1}^{+\infty} q^n \left[\sum_{i=m}^{n-1} \mu_{m,i}(\tau)\, B_{n-i}(\tau)\right] \tag{13.64}$$

给出如下递推公式

$$\mu_{m+1,n}(\tau) = \sum_{i=m}^{n-1} \mu_{m,i}(\tau)\, B_{n-i}(\tau). \tag{13.65}$$

为了简单起见, 定义

$$\psi_{n,0}(\tau) = V_n(B_0, \tau), \quad \psi_{n,m}(\tau) = \frac{1}{m!} \frac{\partial^m V_n(S,\tau)}{\partial S^m}\bigg|_{S=B_0(\tau)}. \tag{13.66}$$

然后，在移动边界 $S = \Lambda(\tau;q)$ 上，我们通过在 $S = B_0(\tau)$ 处的泰勒展开，有

$$V_n(S,\tau) = \psi_{n,0}(\tau) + \sum_{m=1}^{+\infty} \psi_{n,m}(\tau) \left[\Lambda(\tau;q) - B_0(\tau)\right]^m$$

$$= \psi_{n,0}(\tau) + \sum_{m=1}^{+\infty} \psi_{n,m}(\tau) \left[\sum_{i=m}^{+\infty} \mu_{m,i}(\tau) \, q^i\right]$$

$$= \psi_{n,0}(\tau) + \sum_{i=1}^{+\infty} q^i \left[\sum_{m=1}^{i} \psi_{n,m}(\tau)\mu_{m,i}(\tau)\right]$$

$$= V_n(B_0,\tau) + \sum_{i=1}^{+\infty} \alpha_{n,i}(\tau) \, q^i, \tag{13.67}$$

其中

$$\alpha_{n,i}(\tau) = \sum_{m=1}^{i} \psi_{n,m}(\tau) \, \mu_{m,i}(\tau), \quad i \geqslant 1. \tag{13.68}$$

因此，在移动边界 $S = \Lambda(\tau;q)$ 上，我们有

$$\phi(S,\tau;q) = \sum_{n=0}^{+\infty} V_n(S,\tau) \, q^n = \sum_{n=0}^{+\infty} q^n \left[V_n(B_0,\tau) + \sum_{i=1}^{+\infty} \alpha_{n,i}(\tau) \, q^i\right]$$

$$= \sum_{n=0}^{+\infty} q^n \left[V_n(B_0,\tau) + \sum_{j=0}^{n-1} \alpha_{j,n-j}(\tau)\right]$$

$$= \sum_{n=0}^{+\infty} \left[V_n(B_0,\tau) + f_n(\tau)\right] q^n, \tag{13.69}$$

其中

$$f_n(\tau) = \sum_{j=0}^{n-1} \alpha_{j,n-j}(\tau). \tag{13.70}$$

当 $q = 0$ 时，在 $S = \Lambda(\tau;0) = B_0(\tau)$ 上有

$$\phi(\Lambda,\tau;q) = \phi(B_0,\tau;0) = V_0(B_0,\tau).$$

所以，我们有 $f_0(\tau) = 0$，因此在移动边界 $S = \Lambda(\tau;q)$ 上有

$$\phi(S,\tau;q) = V_0(B_0,\tau) + \sum_{n=1}^{+\infty} \left[V_n(B_0,\tau) + f_n(\tau)\right] q^n. \tag{13.71}$$

相似地, 我们在移动边界 $S = \Lambda(\tau;q)$ 上有

$$\frac{\partial V_n(S,\tau)}{\partial S} = \psi_{n,1}(\tau) + \sum_{m=1}^{+\infty}(m+1)\psi_{n,m+1}(\tau)\left[\Lambda(\tau;q) - B_0(\tau)\right]^m$$

$$= V_n'(B_0,\tau) + \sum_{i=1}^{+\infty}\beta_{n,i}(\tau)\,q^i, \qquad (13.72)$$

其中 ′ 表示对 S 求导, 且

$$\beta_{n,i}(\tau) = \sum_{m=1}^{i}(m+1)\psi_{n,m+1}(\tau)\,\mu_{m,i}(\tau), \quad i \geqslant 1. \qquad (13.73)$$

此外

$$\frac{\partial \phi(S,\tau;q)}{\partial S} = V_0'(B_0,\tau) + \sum_{n=1}^{+\infty}\left[V_n'(B_0,\tau) + g_n(\tau)\right]q^n \qquad (13.74)$$

在移动边界 $S = \Lambda(\tau;q)$ 上成立, 其中

$$g_n(\tau) = \sum_{j=0}^{n-1}\beta_{j,n-j}(\tau), \quad n \geqslant 1. \qquad (13.75)$$

附录 13.2　美式认沽期权的 Mathematica 程序

基于同伦分析方法和 Laplace 变换求解美式认沽期权方程.

版 权 声 明

版权所有 ©2011, 上海交通大学, 程序开发者. 保留所有权利.

如果满足以下条件, 则允许以源程序和二进制形式重新发布和使用, 无论是否修改:

- 源程序的重新发布必须保留上述版权声明, 此条件列表和以下免责声明.
- 以二进制形式重新发布必须在随发布提供的文档和 (或) 其他材料中复制上述版权声明, 此条件列表和以下免责声明.
- 未经程序开发人员书面同意, 不得出于盈利目的以源程序和二进制形式重新发布, 无论是否修改.

本软件由版权所有人和贡献者 "按原样" 提供, 不提供任何明示或暗示的保证. 在任何情况下, 版权所有人或贡献者均不对任何损害承担责任.

美式认沽期权的 Mathematica 程序

廖世俊编写
上海交通大学
2010 年 12 月

```
<<Calculus`Pade`;
<<Graphics`Graphics`;

(* Define approx[f] for Taylor expansion of f *)
approx[f_] := Module[{temp},
temp[0] = Series[f, {t, 0, OrderTaylor}]//Normal;
temp[1] = temp[0]/.t^(n_.)*Derivative[j_][DiracDelta][0]->0;
temp[2] = temp[1]/.t^(n_.)*DiracDelta[0]->0;
temp[3] = temp[2]/.DiracDelta[0]->0;
temp[4] = temp[3]/.Derivative[j_][DiracDelta][0]->0;
temp[5] = N[temp[4],60]//Expand;
If[KeyCutOff == 1, temp[5] = temp[5]//Chop];
temp[5]
];

(* Define approx2[f] for Taylor expansion of f *)
approx2[f_] := Module[{temp},
temp[0] = Expand[f];
temp[1] = temp[0] /. Derivative[n_][DiracDelta][t] -> dd[n];
temp[2] = temp[1] /. DiracDelta[t] -> dd[0];
temp[3] = Series[temp[2],{t, 0, OrderTaylor}]//Normal;
temp[4] = temp[3] /. dd[0] -> DiracDelta[t];
temp[5]=temp[4]/.dd[n_]->Derivative[n][DiracDelta][t];
temp[6] = N[temp[5],60]//Expand;
If[KeyCutOff == 1, temp[6] = temp[6]//Chop];
temp[6]
];

(* Define GetLK[n] *)
lamda      := (1 - gamma)/2 - 1/2*Sqrt[(4 p + (1 + gamma)^2)];
```

```
kernel[s_] := -s^lamda/lamda;
lK0[0]     = -1/lamda;
lK0[i_]    := D[kernel[s], {s, i}] /. s -> 1 // Expand;

GetLK[m0_,m1_,Nappr_]:= Module[{temp,K1,K2,lK1,lK2},
For[i = Max[m0,0], i <= m1, i++,
   K[i]    = invl[lK0[i]];
   K1[i]   = K[i]/.{Derivative[_][DiracDelta][t_]->0,
                    DiracDelta[t_]->0};
   K2[i]   = Collect[K[i]-K1[i],{DiracDelta[t],
                    Derivative[Blank[]][DiracDelta][t]}];
   temp    = Series[K1[i],{t,0,Nappr}]//Normal;
   lK1[i]  = LaplaceTransform[temp, t, p];
   lK2[i]  = LaplaceTransform[K2[i],t, p];
   LK[i]   = Collect[lK1[i] + lK2[i], p];
  ];
];


(* Define Getf[n] and Getg[n] *)
mu[m_,n_]:=If[m ==1,b[n],Sum[mu[m-1,i]*b[n-i],{i,m-1,n-1}]];
psi[n_,m_]   := dV[n,m]/m!;
alpha[n_,i_] := Sum[psi[n,m]*mu[m,i],{m,1,i}];
beta[n_,i_]  := Sum[(m+1)*psi[n,m+1]*mu[m,i],{m,1,i}];
f[0]       := 0;
g[0]       := 1;
Getf[n_]   := Sum[alpha[j,n-j],{j,0,n-1}];
Getg[n_]   := Sum[beta[j,n-j]  ,{j,0,n-1}];

(* Define Getb[n] *)
b[0]  := 1;
BB[0] := 1;
B[0]  := X;
Getb[n_] := Module[{temp},
If[n == 1,
   b[1] = c0*dV[0, 0] // Expand,
   temp = b[n - 1] + c0*(b[n -1]+dV[n-1,0]+f[n-1])//Expand;
   b[n] = approx[temp];
   ];
];
```

```
(* Define GetDV[m,n] *)
GetDV[m_, n_] := Module[{temp},
If[n == 1, DV[m, 1] = -g[m],
   temp[1]  = Expand[LK[n]*Lg[m]];
   DV[m, n] = invl[temp[1]];
   ];
DV[m, n] = approx[DV[m, n]];
];


(* Define dV[m,n] *)
dV[m_,n_] := Module[{temp},
If[NumberQ[flag[m,n]],
   Goto[100],
   GetDV[m,n];
   flag[m,n] = 1
   ];
Label[100];
DV[m,n]
];


(* Define hp[f_,m_,n_] *)
hp[f_,m_,n_]:= Module[{k,i,df,res,q},
df[0] = f[0];
For[k = 1, k <= m+n, k++, df[k] = f[k] - f[k-1]//Expand ];
res = df[0] + Sum[df[i]*q^i,{i,1,m+n}];
Pade[res,{q,0,m,n}]/.q->1
];



(* Get [m,n] Pade approximant of B *)
pade[order_]:= Module[{temp,s,i,j},
temp[0] = BB[order] /. t^i_. -> s^(2*i);
temp[1] = Pade[temp[0],{s,0,OrderTaylor,OrderTaylor}];
If[KeyCutOff == 1, temp[1] = temp[1]//Chop];
BBpade[order] = temp[1] /. s^j_. -> t^(j/2);
Bpade[order]  = X*BBpade[order]/. t -> (sigma^2*t/2);
];
```

```
(* Define inverse Laplace transformation *)
invl[Sqrt[p]] := -1/(2*Sqrt[Pi]*t^(3/2));

invl[p^n_] := Module[{temp, nInt},
nInt = IntegerPart[n];
If[n > 1/2 && n > nInt,
   Goto[100],
   temp[2] = InverseLaplaceTransform[p^n, p, t];
   Goto[200];
   ];
Label[100];
temp[1] = -1/2/Sqrt[Pi]/t^(3/2);
temp[2] = D[temp[1], {t, nInt}];
Label[200];
temp[2]//Expand
];

invl[d_./(c_. + a_.*Sqrt[4p + b_.])] := Module[{temp},
temp[1] = d/(4a)*Exp[-b*t/4];
temp[2] = 2/Sqrt[Pi*t];
temp[3] = c/a*Exp[c^2*t/(4a^2)]*Erfc[c*Sqrt[t]/(2a)];
temp[1]*(temp[2]-temp[3])//Expand
];

invl[d_./(p*(c_. + a_.*Sqrt[4p + b_.]))]:= Module[{temp},
temp[1] = Sqrt[b]*Erf[Sqrt[b*t]/2];
temp[2] = c/a*Exp[-(b-(c/a)^2)*t/4]*Erfc[c*Sqrt[t]/(2a)];
temp[3] = -1/(b - (c/a)^2)*d/a*(c/a-temp[1]-temp[2]);
temp[3]//Expand
];

invl[p^i_.*Sqrt[c_.*p + a_.]] := Module[{temp},
temp = D[-Exp[-a*t/c]/(2*c*Sqrt[Pi]*(t/c)^(3/2)),{t, i}];
temp//Expand
];

invl[Sqrt[c_.*p+a_.]]:=-Exp[-a*t/c]/(2*c*Sqrt[Pi]*(t/c)^(3/2));
invl[f_]      := InverseLaplaceTransform[f, p, t] // Expand;
invl[p_Plus] := Map[invl, p];
```

```
invl[c_*f_]  := c*invl[f] /; FreeQ[c, p];

(* Main code *)
ham[m0_, m1_] := Module[{temp, k, n},
If[m0 == 1,
Print[" Strike price = ?"];
temp[0] = Input[];
If[!NumberQ[temp[0]],Goto[100]];
X = IntegerPart[temp[0]*10^10]/10^10;
Print[" Risk-free interest rate = ?"];
temp[0] = Input[];
If[!NumberQ[temp[0]],Goto[100]];
r = IntegerPart[temp[0]*10^10]/10^10;
Print[" Volatility = ?"];
temp[0] = Input[];
If[!NumberQ[temp[0]],Goto[100]];
sigma = IntegerPart[temp[0]*10^10]/10^10;
Print[" Time to expiry = ?"];
temp[0] = Input[];
If[!NumberQ[temp[0]],Goto[100]];
T = IntegerPart[temp[0]*10^10]/10^10;
gamma   = 2*r/sigma^2;
texp    = sigma^2*T/2;
Bp      = X*gamma/(1 + gamma);
Label[100];
If[!NumberQ[gamma],
    X     = .;
    r     = .;
    sigma = .;
    gamma = .;
    T     = .;
    ];
Print["--------------------------------------------------"];
Print[" INPUT PARAMETERS: "];
Print["    Strike price              (X) = ",X," ($) "];
Print["    Risk-free interest rate (r) = ",r];
Print["    Volatility         (sigma) = ",sigma];
Print["    Time to expiry           (T) = ",T," (year)"];
Print["--------------------------------------------------"];
```

```
Print[" CORRESPONDING PARAMETERS: "];
Print["   gamma      = ",gamma];
Print["   dimensionless time to expiry   (texp)=",texp//N];
Print["   perpetual optimal exercise price (Bp)=",Bp//N,"($)"];
Print["----------------------------------------------------"];
Print[" CONTROL PARAMETERS: "];
Print["   OrderTaylor = ",OrderTaylor];
Print["   c0          = ",c0];
Print["----------------------------------------------------"];
KeyCutOff = If[OrderTaylor < 80  &&  NumberQ[gamma], 1, 0];
If[KeyCutOff == 1,
   Print["Command Chop is used to simplify the result"],
   Print["Command Chop is NOT used "]
   ];
If[NumberQ[gamma],
   Print["Pade technique is used"],
   Print["Pade technique is NOT used"]
   ];
Clear[flag,DV];
];
For[k = Max[1, m0], k <= m1, k++,
    Print[" k = ", k];
    If[k == 1, GetKK[]; GetBJ[]; GetKn[]];
    If[k == 1, Lg[0] = LaplaceTransform[g[0], t, p]];
    If[k == 1, GetLK[0,2,OrderTaylor],
               GetLK[k+1,k+1,OrderTaylor]];
    Getb[k];
    BB[k] = Collect[BB[k - 1] + b[k], t];
    temp[0] = X*BB[k] /. t-> (sigma^2*t/2)//Expand;
    B[k] = Collect[temp[0],t];
    If[NumberQ[gamma],pade[k]];
    temp[1] = Getg[k];
    temp[2] = Getf[k];
    g[k] = approx2[temp[1]];
    f[k] = approx2[temp[2]];
    Lg[k] = LaplaceTransform[g[k], t, p];
    If[NumberQ[gamma] && NumberQ[sigma] && NumberQ[X],
        Print[" Optimal exercise price at the time
                  to expiration = ", B[k]/.t->T//N];
```

```
            Print[" Modified result given
                       by Pade technique = ",Bpade[k]/.t->T//N];
               ];
         ];
   Print[" Well done !"];
   If[NumberQ[gamma] && NumberQ[sigma] && NumberQ[X],
      Plot[{Bp,B[m1],Bpade[m1]},{t,0,1.25*T},
             PlotRange -> {0.8*Bp,X}, PlotStyle ->
                 {RGBColor[1,0,0],RGBColor[0,1,0],RGBColor[0,0,1]}];
      Print[" Order of homotopy-approximation : ",m1];
      Print[" Green line : optimal exercise boundary B
                                in polynomial "];
      Print[" Blue line  : optimal exercise boundary B
                                by Pade method "];
      Print[" Red line   : perpetual optimal exercise price "];
      ];
   ];


(* Dimensionless analytic formula given by Kuske and Keller *)
GetKK[] := Module[{alpha},
alpha = -Log[9*Pi*gamma^2*t]/2;
KK0 = Exp[-2*Sqrt[alpha*t]];
KK  = X*KK0 /. t->sigma^2/2*t;
];


(* Dimensionless formula given by Bunch and Johnson *)
GetBJ[] := Module[{alpha},
Bp0   = gamma/(1+gamma);
alpha = -Log[4*E*gamma^2*t/(2 - Bp0^2)]/2;
BJ0 = Exp[-2*Sqrt[alpha*t]];
BJ  = X*BJ0 /. t->sigma^2/2*t;
];


(* Dimensionless formula given by Knessl *)
GetKn[] := Module[{z},
z   = Abs[Log[4*Pi*gamma^2*t]];
Kn0 = Exp[-Sqrt[2*t*z]*(1+1/z^2)];
Kn  = X*Kn0 /. t->sigma^2/2*t;
];
```

```
(* Define the order of Taylor's series expansion *)
OrderTaylor  = 8;

(* Assign the convergence-control parameter *)
c0 = -1;

(* Get 8th-order homotopy-approximation of B *)
ham[1,8];

(* Get 10th-order homotopy-approximation of B *)
ham[8,10]
```

附录 13.3　可供商人使用的 Mathematica 程序 APOh

　　Mathematica 程序 APOh 能在数秒内给出到期前给定时间的最优行权价格.

　　该实用程序首先读取数据文件 APO-48-10.txt 中的无量纲结果, 对于未知的无量纲参数 $\gamma = 2r/\sigma^2$, 该数据文件由廖世俊使用附录 13.2 中的 Mathematica 程序和高性能计算机获得, 然后计算有效期内的无量纲最优行权价格. 与通常有效期仅为几天或几周的其他渐近和 (或) 摄动公式不同, 此程序能够提供精确的近似值, 且通常在几十年内有效.

> **版 权 声 明**
>
> 版权所有 ©2011, 上海交通大学和 APOh 开发者. 保留所有权利.
>
> 　　如果满足以下条件, 则允许以源程序和二进制形式重新发布和使用, 无论是否修改:
>
> - 源程序的重新发布必须保留上述版权声明, 此条件列表和以下免责声明.
> - 以二进制形式重新发布必须在随发布提供的文档和 (或) 其他材料中复制上述版权声明, 此条件列表和以下免责声明.
> - 未经程序 AOPh 开发人员书面同意, 不得出于盈利目的以源程序和二进制形式重新发布, 无论是否修改.
>
> 　　本软件由版权所有人和贡献者 "按原样" 提供, 不提供任何明示或暗示的保证. 在任何情况下, 版权所有人或贡献者均不对任何损害承担责任.

使用指南

APOh[order_] 该模块给出了有量纲的最优行权价格 $B(\tau)$ 的同伦近似, 其中 order 表示同伦近似的阶. 该程序首先读取数据文件 APO-48-10.txt, 然后要求用户输入行权价格 X, 无风险利率 r, 波动率 σ 和有效期 T (年), 最后列出了 $B(\tau)$ 在不同近似阶下的结果, 以及它们通过帕德方法得到的修正近似, 并绘制了在时域 $0 \leqslant \tau \leqslant 1.25T$ 上 $B(\tau)$ 的曲线, 该曲线理论上具有长期最优行权价格 $B_p = X\gamma/(1+\gamma)$. 如果想要开始一个新案例, 只需再次运行程序 APOh 并输入新的参数.

B[n] 具有量纲 (美元) 的最优行权价格以 $\sqrt{\tau}$ 级数展开到 $o(\tau^M)$ 的 n 阶同伦近似, 其中 $M =$ OrderTaylor.

Bpade[n] $B(\tau)$ 以 $\sqrt{\tau}$ 级数展开到 $o(\tau^M)$ 的 n 阶同伦近似的 $[M, M]$ 阶帕德近似, 其中 $M =$ OrderTaylor.

OrderHAM 数据文件 APO-48-10.txt 中无量纲结果的最高阶同伦近似值, 默认值为 10.

OrderTaylor $B(\tau)$ 以 $\sqrt{\tau}$ 级数展开的阶. 它在数据文件 APO-48-10.txt 中的默认值为 48.

Bp 长期最优行权价格 $B_p = X\gamma/(1+\gamma)$.

可供商人使用的 Mathematica 程序 **APOh**

廖世俊编写
上海交通大学
2010 年 12 月

```
<<Calculus`Pade`;
<<Graphics`Graphics`;

(* Input dimensionless results given by means of the HAM *)
<<APO-48-10.txt;
Print["----------------------------------------------------"];
Print["OrderTaylor = ",OrderTaylor];
Print["OrderHAM    = ",OrderHAM];
Print["----------------------------------------------------"];
```

附录 13.3 可供商人使用的 Mathematica 程序 APOh

```
(* Define APOh[Order_] *)
APOh[Order_]:=Module[{temp,n,i,j,s},
Print[" OrderTaylor   = ",OrderTaylor];
Print[" strike price = ?"];
temp[0] = Input[];
X = IntegerPart[temp[0]*10^10]/10^10;
Print[" risk-free interest rate = ?"];
temp[0] = Input[];
r = IntegerPart[temp[0]*10^10]/10^10;
Print[" volatility = ?"];
temp[0] = Input[];
sigma = IntegerPart[temp[0]*10^10]/10^10;
Print[" time to expiry (year)   = ?"];
temp[0] = Input[];
T = IntegerPart[temp[0]*10^10]/10^10;
gamma    = 2*r/sigma^2;
texp     = sigma^2*T/2;
Bp       = X*gamma/(1 + gamma);
Print["----------------------------------------------------"];
Print[" INPUT PARAMETERS: "];
Print["    Strike price              (X) = ",X," ($) "];
Print["    Risk-free interest rate (r) = ",r];
Print["    Volatility            (sigma) = ",sigma];
Print["    Time to expiry            (T) = ",T," (year)"];
Print["----------------------------------------------------"];
Print[" CORRESPONDING PARAMETERS: "];
Print["    gamma   = ",gamma];
Print["    dimensionless time to expiry   (texp)=",texp//N];
Print["    perpetual optimal exercise price (Bp)=",Bp//N,"($)"];
Print["----------------------------------------------------"];
Print[" CONTROL PARAMETERS: "];
Print["    OrderTaylor   = ",OrderTaylor];
Print["    c0            = ",c0];
Print["----------------------------------------------------"];
For[n = 1,n <= Min[Order,OrderHAM], n++,
    Print[" n = ",n];
    temp[0] = X*BB[n] /. t-> (sigma^2*t/2)//Expand;
    B[n] = Collect[temp[0],t];
```

```
        If[NumberQ[gamma],
            temp[0] = BB[n] /. t^i_. -> s^(2*i);
            temp[1] = Pade[temp[0],{s,0,OrderTaylor,OrderTaylor}];
            BBpade[n] = temp[1] /. s^j_. -> t^(j/2);
            Bpade[n]  = X*BBpade[n]/. t -> (sigma^2*t/2);
            ];
        If[NumberQ[gamma] && NumberQ[sigma] && NumberQ[X],
            Print["Optimal exercise price at the time to expiration
                    = ", B[n]/.t->T//N];
            Print["Modified result given by Pade technique = ",
                        Bpade[n]/.t->T//N];
            ];
        ];
Print["Well done"];
If[NumberQ[gamma] && NumberQ[sigma] && NumberQ[X],
    n = Min[Order,OrderHAM];
    Plot[{Bp, B[n], Bpade[n]}, {t, 0, 1.25*T},
            PlotRange -> {0.8*Bp, X}, PlotStyle ->
            {RGBColor[1,0,0],RGBColor[0,1,0],RGBColor[0,0,1]}];
    Print["  Order of homotopy-approximation : ",n];
    Print["  Green line : optimal exercise boundary B
                            in polynomial "];
    Print["  Blue line  : optimal exercise boundary B
                            by Pade method "];
    Print["  Red line   : perpetual optimal exercise price "];
    ];
];
```

第十四章　二维和三维 Gelfand 方程

本章以二维和三维 Gelfand 方程为例, 说明了同伦分析方法可以将二阶非线性偏微分方程转化为无穷多个四阶或六阶线性偏微分方程, 从而以一种相当简单的方式求解该方程. 这主要是因为同伦分析方法为我们选择辅助线性算子提供了极大的自由, 同时也为保证级数解的收敛性提供了一种便捷的方法. 据我们所知, 这种变换从未被其他解析 (数值) 方法使用过. 这说明了同伦分析方法对于强非线性问题的原创性和极大的灵活性. 这还表明了, 我们必须保持开放的心态, 因为我们在解决非线性问题上的自由度可能比我们传统上认为的要大得多.

14.1　绪论

第二章说明了同伦分析方法 [308–321, 328] 为我们选择辅助线性算子提供了极大的自由度: 描述周期振动的二阶非线性常微分方程可以转化为无穷多个 2κ 阶线性常微分方程, 其中 $\kappa \geqslant 1$ 为任意正整数, 且通过所谓的收敛控制参数 c_0 可以保证级数解的收敛性. 此外, 正是由于这种极大的自由度, 描述非相似边界层流动 (第十一章) 和非定常边界层流动 (第十二章) 的非线性偏微分方程可以转化为无穷多个线性常微分方程. 因此, 该方程可以通过基于同伦分析方法的 Mathematica 软件包 BVPh 1.0 求解. 本章将进一步说明这种选择辅助线性算子 $\mathcal{L}$ 的自由可以大大简化某些高维非线性偏微分方程的求解过程.

例如, 考虑高维 Gelfand 方程 [301, 322]

$$\begin{cases} \Delta u + \lambda e^u = 0, & \mathbf{x} \in \Omega \subset \mathbf{R}^N, \\ u = 0, & \mathbf{x} \in \partial\Omega, \end{cases} \tag{14.1}$$

其中 Δ 表示拉普拉斯算子, λ 为特征值, u 为特征函数, $\mathbf{x}$ 为空间变量的向量, $N = 1, 2, 3$ 表示维度, Ω 为域, $\partial\Omega$ 为边界. 从物理上讲, Gelfand 方程出现在多种背景下, 如化学反应器理论, 非线性热传导问题的定常方程, 宇宙膨胀的几何学和相对论问题等. 从数学上讲, 因为其包含指数项 $\exp(u)$, 所以控制方程具有很强的非线性.

一般来说, 强非线性的高维偏微分方程的特征值和特征函数很难得到精确的解析近似. 关于 Gelfand 问题 [301, 322] 的研究由来已久. Liouville [322] 给出了一维 Gelfand 方程特征值的封闭形式表达式. 基于 Chebyshev 函数, Boyd [301] 提出了在 $[-1, 1] \times [-1, 1]$ 方形区域内求解二维 Gelfand 方程的解析方法和数值方法, 并给出了一个单点解析近似

$$\lambda = 3.2\, A\, e^{-0.64\,A} \tag{14.2}$$

和一个三点解析近似

$$\lambda = (2.667\, A + 4.830\, B + 0.127\, C)\, e^{-0.381A - 0.254B - 0.018C}, \tag{14.3}$$

其中 $A = u(0, 0)$ 以及

$$B = A \left(0.829 - 0.566 e^{0.463A} - 0.0787 e^{-0.209A}\right)/G,$$
$$C = A \left(-1.934 + 0.514 e^{0.463A} + 1.975 e^{-0.209A}\right)/G,$$
$$G = 0.2763 + e^{0.463A} + 0.0483 e^{-0.209A},$$

该问题也引起了当前研究人员的关注 [304, 323].

本章基于同伦分析方法将具有强非线性的二阶二维 (或三维) Gelfand 方程转化为无穷多个四阶 (或六阶) 线性二维 (或三维) 偏微分方程, 从而以一种相当简单的方式求解该方程.

14.2 二维 Gelfand 方程的同伦近似

14.2.1 简明数学公式

根据 Boyd [301] 的研究, 考虑二维 Gelfand 方程

$$\Delta u + \lambda\, e^u = 0, \quad -1 < x < 1, \quad -1 < y < 1, \tag{14.4}$$

满足四个壁面处的边界条件

$$u(x, \pm 1) = 0, \quad u(\pm 1, y) = 0. \tag{14.5}$$

上述非线性特征方程有无数个特征值和特征函数. 显然, 不同的特征函数在原点 $(0,0)$ 处有不同的值. 所以, 我们可以利用

$$A = u(0,0) \tag{14.6}$$

的不同值来区分不同的特征函数和对应的特征值, 写为

$$u(x,y) = A + w(x,y), \tag{14.7}$$

其中 A 为给定的常数, 原始的二维 Gelfand 方程变为

$$\Delta w + \lambda\, e^A\, e^w = 0, \quad -1 < x < 1, \quad -1 < y < 1, \tag{14.8}$$

满足四个壁面处的边界条件

$$w(x, \pm 1) = -A, \quad w(\pm 1, y) = -A, \tag{14.9}$$

且有限制条件

$$w(0,0) = 0. \tag{14.10}$$

令 $w_0(x,y)$ 和 λ_0 分别表示特征函数 $w(x,y)$ 和特征值 λ 的初始猜测解. 注意, 初始猜测解 $w_0(x,y)$ 不需要满足边界条件 (14.9) 和限制条件 (14.10). 令 $q \in [0,1]$ 表示嵌入变量. 在同伦分析方法的框架下, 我们首先构造两个连续变化 (或形变) $\phi(x,y;q)$ 和 $\Lambda(q)$, 随着嵌入变量 q 从 0 增加到 1, $\phi(x,y;q)$ 从初始猜测解 $w_0(x,y)$ 连续变化到特征函数 $w(x,y)$, 与此同时, $\Lambda(q)$ 从初始猜测解 λ_0 连续变化到特征值 λ. 这两种连续变化由零阶形变方程

$$(1-q)\mathcal{L}\left[\phi(x,y;q) - w_0(x,y)\right] = c_0\, q\, \mathcal{N}\left[\phi(x,y;q), \Lambda(q)\right] \tag{14.11}$$

控制, 在方形区域 $(x,y) \in [-1,1] \times [-1,1]$ 上, 满足四个壁面处的边界条件

$$(1-q)\left[\phi(\pm 1,y;q) - w_0(\pm 1,y)\right] = c_0\, q\, \left[\phi(\pm 1,y;q) + A\right], \qquad (14.12)$$

$$(1-q)\left[\phi(x,\pm 1;q) - w_0(x,\pm 1)\right] = c_0\, q\, \left[\phi(x,\pm 1;q) + A\right], \qquad (14.13)$$

在原点有额外的限制条件

$$(1-q)\left[\phi(0,0;q) - w_0(0,0)\right] = c_0\, q\, \phi(0,0;q), \qquad (14.14)$$

其中

$$\begin{aligned}&\mathcal{N}[\phi(x,y;q), \Lambda(q)]\\ &= \frac{\partial^2 \phi(x,y;q)}{\partial x^2} + \frac{\partial^2 \phi(x,y;q)}{\partial y^2} + e^A\, \Lambda(q)\, \exp[\phi(x,y;q)]\end{aligned} \qquad (14.15)$$

是二维 Gelfand 方程 (14.8) 的非线性算子, $\mathcal{L}$ 为具有性质 $\mathcal{L}[0] = 0$ 的辅助线性算子, $c_0 \neq 0$ 为收敛控制参数. 这里需要强调的是, 我们有极大的自由来选择辅助线性算子 $\mathcal{L}$ 和收敛控制参数 c_0, 如下所述.

当 $q = 0$ 时, 因为 $\mathcal{L}[0] = 0$, 零阶形变方程 (14.11) 至 (14.14) 有解

$$\phi(x,y;0) = w_0(x,y). \qquad (14.16)$$

当 $q = 1$ 时, 因为 $c_0 \neq 0$, 零阶形变方程 (14.11) 至 (14.14) 等价于原始偏微分方程 (14.8) 至 (14.10), 有

$$\phi(x,y;1) = w(x,y), \quad \Lambda(1) = \lambda. \qquad (14.17)$$

因此, 随着嵌入变量 $q \in [0,1]$ 从 0 增加到 1, $\phi(x,y;q)$ 确实从初始猜测解 $w_0(x,y)$ 连续变化到特征函数 $w(x,y)$, $\Lambda(q)$ 也一样地从初始猜测解 λ_0 连续变化到特征值 λ. 所以, 在数学上, 零阶形变方程 (14.11) 至 (14.14) 构造了两个连续的同伦

$$\phi(x,y;q): w_0(x,y) \sim w(x,y), \quad \Lambda(q): \lambda_0 \sim \lambda.$$

将 $\phi(x,y;q)$ 和 $\Lambda(q)$ 展开成关于嵌入变量 $q \in [0,1]$ 的 Maclaurin 级数, 并利用 (14.16), 我们有如下同伦 – Maclaurin 级数

$$\phi(x,y;q) = w_0(x,y) + \sum_{n=1}^{+\infty} w_n(x,y)\, q^n, \qquad (14.18)$$

$$\Lambda(q) = \lambda_0 + \sum_{n=1}^{+\infty} \lambda_n\, q^n, \qquad (14.19)$$

其中

$$w_n(x,y) = \frac{1}{n!} \left.\frac{\partial^n \phi(x,y;q)}{\partial q^n}\right|_{q=0} = \mathcal{D}_n\left[\phi(x,y;q)\right], \quad (14.20)$$

$$\lambda_n = \frac{1}{n!} \left.\frac{\partial^n \Lambda(q)}{\partial q^n}\right|_{q=0} = \mathcal{D}_n\left[\Lambda(q)\right] \quad (14.21)$$

分别为 $\phi(x,y;q)$ 和 $\Lambda(q)$ 所谓的同伦导数，$\mathcal{D}_n$ 为 n 阶同伦导数算子. 需要强调的是，在同伦分析方法的框架下，我们有很大的自由来选择辅助线性算子 $\mathcal{L}$，初始猜测解 $w_0(x,y)$，尤其是收敛控制参数 c_0：它们都会影响级数 (14.18) 和 (14.19) 的收敛性. 假设所有这些都选择恰当，则同伦–Maclaurin 级数 (14.18) 和 (14.19) 在 $q=1$ 时绝对收敛. 然后，由 (14.17)，我们有同伦级数解

$$w(x,y) = w_0(x,y) + \sum_{n=1}^{+\infty} w_n(x,y), \quad (14.22)$$

$$\lambda = \lambda_0 + \sum_{n=1}^{+\infty} \lambda_n. \quad (14.23)$$

未知的 $w_n(x,y)$ 和 λ_{n-1} ($n \geqslant 1$) 的微分方程可以直接从零阶形变方程 (14.11) 至 (14.14) 推导得出：在零阶形变方程 (14.11) 至 (14.14) 的两边取 n 阶同伦导数，我们有所谓的 n 阶形变方程

$$\mathcal{L}\left[w_n(x,y) - \chi_n\, w_{n-1}(x,y)\right] = c_0\, \delta_{n-1}(x,y), \quad (14.24)$$

满足四个壁面处的边界条件

$$w_n(x,\pm 1) = \mu_n(x,\pm 1), \quad w_n(\pm 1,y) = \mu_n(\pm 1,y), \quad (14.25)$$

以及在原点处的额外限制条件

$$w_n(0,0) = (\chi_n + c_0)\, w_{n-1}(0,0), \quad (14.26)$$

其中

$$\chi_n = \begin{cases} 1, & n > 1, \\ 0, & n = 1 \end{cases} \quad (14.27)$$

和

$$\delta_k(x,y) = \mathcal{D}_k\left\{\mathcal{N}\left[\phi(x,y;q)\right]\right\}$$

$$= \Delta w_k(x,y) + e^A \sum_{j=0}^{k} \lambda_{k-j} \, \mathcal{D}_j \left[e^{\phi(x,y;q)} \right], \tag{14.28}$$

$$\mu_n(x,y) = (\chi_n + c_0) \, w_{n-1}(x,y) + c_0 \, (1-\chi_n) \, A, \tag{14.29}$$

由定理 4.1 获得. 根据定理 4.7, 我们有递推公式

$$\mathcal{D}_0 \left[e^{\phi(x,y;q)} \right] = e^{w_0(x,y)}, \quad \mathcal{D}_k \left[e^{\phi(x,y;q)} \right] = \sum_{j=0}^{k-1} \left(1 - \frac{j}{k} \right) w_{k-j} \, \mathcal{D}_j \left[e^{\phi(x,y;q)} \right]. \tag{14.30}$$

因此, 通过像 Mathematica 这样的科学计算软件, 很容易计算出 (14.24) 的 $\delta_{n-1}(x,y)$. 注意, (14.24) 可以直接通过定理 4.15 得到. 详情请参考第四章.

如前所述, 在同伦分析方法的框架下, 我们有极大的自由来选择辅助线性算子 $\mathcal{L}$ 和初始猜测解 $w_0(x,y)$. 注意, Gelfand 方程包含一个线性算子, 即拉普拉斯算子 Δ. 然而, 即使是二维线性偏微分方程

$$\Delta u(x,y) = 0, \quad -1 \leqslant x \leqslant 1, \quad -1 \leqslant y \leqslant 1,$$

都具有一个复杂的通解

$$u(x,y) = \sum_{k=0}^{+\infty} \left(B_{k,1} \, e^{-\alpha k x} + B_{k,2} \, e^{\alpha k x} \right) [B_{k,3} \, \sin(\alpha k y) + B_{k,4} \, \cos(\alpha k y)]$$

$$+ \sum_{k=1}^{+\infty} \left(B_{k,5} \, e^{-\beta k y} + B_{k,6} \, e^{\beta k y} \right) [B_{k,7} \, \sin(\beta k x) + B_{k,8} \, \cos(\beta k x)],$$

其中系数 α, β 和 $B_{k,i}$ 由边界条件确定. 例如, 在边界 $x=1$ 上, 上述表达式为

$$u(1,y) = \sum_{k=0}^{+\infty} \left(B_{k,1} \, e^{-\alpha k} + B_{k,2} \, e^{\alpha k} \right) [B_{k,3} \, \sin(\alpha k y) + B_{k,4} \, \cos(\alpha k y)]$$

$$+ \sum_{k=1}^{+\infty} \left(B_{k,5} \, e^{-\beta k y} + B_{k,6} \, e^{\beta k y} \right) [B_{k,7} \, \sin(\beta k) + B_{k,8} \, \cos(\beta k)].$$

所以, 通过上述表达式来满足 $x=1$ 上的边界条件并不容易. 因此, 如果我们选择拉普拉斯算子 Δ 作为辅助线性算子 $\mathcal{L}$, 则很难求解高阶形变方程 (14.24) 至 (14.26). 因此, 我们应该选择一个比拉普拉斯算子 Δ 更好的辅助线性算子 $\mathcal{L}$. 幸运的是, 在同伦分析方法的框架下, 我们有极大的自由来选择辅助线性算子 $\mathcal{L}$. 基于这种自由, 我们确实可以选择这样一个辅助线性算子 $\mathcal{L}$ 使得高阶形变方程很容易求解, 如下所述.

需要注意的是, 边界条件 (14.5) 和边界

$$\partial\Omega : (x,y) \in [-1,1] \times [-1,1]$$

关于 x 轴和 y 轴对称. 此外, 很容易证明, 如果 $w(x,y)$ 是二维 Gelfand 方程的解, 那么 $w(\pm x, \pm y)$ 也是该方程的解. 所以, $w(x,y)$ 关于 x 轴和 y 轴对称, 因此 $w(x,y)$ 可以用基函数

$$\{x^{2m} y^{2n} \mid m = 1, 2, 3, \cdots, \; n = 1, 2, 3, \cdots\} \tag{14.31}$$

表示为

$$w(x,y) = \sum_{m=1}^{+\infty} \sum_{n=1}^{+\infty} b_{m,n} \, x^{2m} \, y^{2n}, \tag{14.32}$$

其中 $b_{m,n}$ 为待定系数. 它为我们提供了 $w(x,y)$ 所谓的解表达. 我们的目标是在给定的 A 下找到具有 (14.32) 形式的特征函数 $w(x,y)$ 的收敛级数解, 以及对应的特征值 λ 的收敛级数.

为了满足附加限制条件 (14.10) 和解表达 (14.32), 我们选择最简单的初始猜测解

$$w_0(x,y) = 0. \tag{14.33}$$

注意, 此初始猜测解满足限制条件 (14.10), 但不满足四个壁面处的边界条件 (14.9).

接下来, 我们应该选择辅助线性算子 $\mathcal{L}$. 为了符合 $w(x,y)$ 的解表达 (14.32), 其对于任意非零常数 C_1 应该满足

$$\mathcal{L}(C_1) = 0. \tag{14.34}$$

此外, 由于 $w_0(x,y) = 0$, 有 $\delta_0(x,y) = \lambda_0 \, e^A$, 因此, $\delta_{n-1}(x,y)$ 可能含有一个非零常数. 因此, 为了符合解表达 (14.32), 辅助线性算子 $\mathcal{L}$ 的逆算子 $\mathcal{L}^{-1}$ 应该具有如下性质

$$\mathcal{L}^{-1}(1) = C_2 \, x^2 \, y^2, \tag{14.35}$$

其中 C_2 是非零常数. 特别地, 应选择适当的辅助线性算子 $\mathcal{L}$, 以便在满足四个壁面处的边界条件 (14.25) 时较容易地求解高阶形变方程 (14.24). 令 $w_m^*(x,y)$ 表示 (14.24) 的特解. 显然

$$w_n^*(x,y) - w_n^*(x, \pm 1) - w_n^*(\pm 1, y) + w_n^*(\pm 1, \pm 1)$$

在四个壁面处为零, 而且

$$\mu_n(x, \pm 1) + \mu_n(\pm 1, y) - \mu_n(\pm 1, \pm 1)$$

满足四个壁面处的边界条件 (14.25), 其中 $\mu_n(x, y)$ 由 (14.29) 定义. 所以

$$w_n(x, y) = w_n^*(x, y) - w_n^*(x, \pm 1) - w_n^*(\pm 1, y) + w_n^*(\pm 1, \pm 1)$$
$$+ \mu_n(x, \pm 1) + \mu_n(\pm 1, y) - \mu_n(\pm 1, \pm 1) \tag{14.36}$$

满足 n 阶形变方程 (14.24) 和边界条件 (14.25), 只要辅助线性算子 $\mathcal{L}$ 对任意光滑函数 $f(x)$ 和 $g(y)$ 具有如下性质

$$\mathcal{L}[f(x)] = \mathcal{L}[g(y)] = 0. \tag{14.37}$$

存在无穷多个满足上述性质 (14.34), (14.35) 和 (14.37) 的线性算子, 例如二阶线性算子

$$\mathcal{L}(u) = c_2 \frac{1}{xy} \frac{\partial^2 u}{\partial x \partial y} \tag{14.38}$$

和四阶线性算子

$$\mathcal{L}(u) = c_4 \frac{\partial^4 u}{\partial x^2 \partial y^2}, \tag{14.39}$$

其中 c_2 和 c_4 为常数. 这两个线性算子是更一般的线性算子

$$\mathcal{L}(u) = \frac{c_2}{xy} \frac{\partial^2 u}{\partial x \partial y} + c_4 \frac{\partial^4 u}{\partial x^2 \partial y^2} \tag{14.40}$$

的特例, 其逆算子为

$$\mathcal{L}^{-1}(x^k y^n) = \frac{x^{k+2} y^{n+2}}{(k+2)(n+2)[c_2 + c_4(k+1)(n+1)]}. \tag{14.41}$$

使用上面的逆算子 $\mathcal{L}^{-1}$, 很容易得到 n 阶形变方程 (14.24) 的一个特解

$$w_n^*(x, y) = c_0 \mathcal{L}^{-1}[\delta_{n-1}(x, y)] + \chi_n w_{n-1}(x, y). \tag{14.42}$$

然后, 通过 (14.36) 得到高阶形变方程 (14.24) 至 (14.25) 的解 $w_n(x, y)$. 之后, λ_{n-1} 由线性代数方程 (14.26) 确定. 详情请参考廖世俊和谭越 [321].

需要注意的是, 上述方法只需要进行代数计算. 因此, 很容易获得特征函数 $w(x, y)$ 和特征值 λ 的高阶近似值, 尤其是借助如 Mathematica, Maple 等科学计算软件. 通过这种方式, 我们大大简化了二维 Gelfand 方程的求解过程, 如下所述.

求解二维 Gelfand 方程的 Mathematica 程序见附录 14.1.

14.2.2 同伦近似

在同伦分析方法的框架下, 收敛控制参数 c_0 为我们提供了一个便捷的方式来保证级数解的收敛性. 这里需要强调的是, 辅助线性算子 (14.40) 包含两个参数 c_2 和 c_4, 它们也可以像 c_0 一样被视为收敛控制参数. 因此, 我们现在有三个收敛控制参数 c_0, c_2 和 c_4 来保证特征函数 (14.22) 和特征值 (14.23) 的收敛性.

我们发现, c_2 和 c_4 取任意值时, 四个壁面处的特征函数 $u(x,y)$ 的 n 阶近似为

$$A(1+c_0)^n,$$

其仅在

$$|1+c_0| < 1 \tag{14.43}$$

且 $n \to +\infty$ 时为零. 上述表达式限制了收敛控制参数 c_0 的选择. 特别地, 当 $c_0 = -1$ 时, 四个壁面处的边界条件在每个近似阶上都是完全满足的. 为了简单起见, 我们选取 $c_0 = -1$. 然后, 还剩下两个未知的收敛控制参数 c_2 和 c_4.

定义二维 Gelfand 方程的平均平方残差

$$E_m = \frac{1}{100} \sum_{i=0}^{9} \sum_{j=0}^{9} \left[\Delta u(i\Delta x, j\Delta y) + \lambda\, e^{u(i\Delta x, j\Delta y)} \right]^2, \quad \Delta x = \Delta y = \frac{1}{10}, \tag{14.44}$$

其中 u 和 λ 分别是特征函数和特征值的 m 阶同伦近似. 显然, E_m 取决于两个收敛控制参数 c_2 和 c_4.

由于原 Gelfand 方程 (14.8) 是二阶的, 我们先使用二阶线性算子 (14.38) 作为辅助线性算子, 对应于线性算子 (14.40) 中的 $c_4 = 0$. 由于 $c_0 = -1$ 和 $c_4 = 0$, 现在只存在一个非零收敛控制参数 c_2. 不失一般性, 考虑 $A = 1$ 的情况. 我们发现控制方程 (14.8) 的平均平方残差 E_m 的最小值并不随着 m 的增加而变小, 如表 14.1 所示. 此外, 无法确定 c_2 的范围使得控制方程 (14.8) 的平均平方残差 E_m 随着 m 的增加而变小, 如图 14.1 所示. 例如, 在 $A = 1$ 和 $c_2 = -20$ 的情况下, 对应的同伦级数是发散的. 因此, 我们不能通过使用二阶线性算子 (14.38) 作为辅助线性算子来得到二维 Gelfand 方程 (14.8) 的收敛级数解.

表 14.1　当 $A = 1$ 时，选取 $c_0 = -1$ 和二阶辅助线性算子 (14.38)，二维 Gelfand 方程 (14.8) 的平均平方残差的最小值

近似阶数 m	平均平方残差的最小值	c_2 的最优值
10	1.23×10^{-2}	-5.3545
15	1.57×10^{-2}	-7.9349
20	1.89×10^{-2}	-10.5021

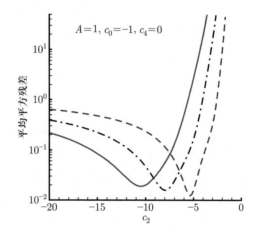

图 14.1　当 $A = 1$ 时，选取 $c_0 = -1$ 并使用二阶线性算子 (14.38) 作为辅助线性算子，(14.8) 相对于 c_2 的平均平方残差．虚线: 10 阶同伦近似; 点划线: 15 阶同伦近似; 实线: 20 阶同伦近似

然后，我们使用四阶线性算子 (14.39) 作为辅助线性算子，对应于线性算子 (14.40) 中的 $c_2 = 0$. 通过选取最优的收敛控制参数 $c_4^* = -0.2494$, (14.8) 的平均平方残差的最小值在 20 阶同伦近似时变小到 2.85×10^{-4}, 如表 14.2 中所示. 此外，任意收敛控制参数 $c_4 \leqslant -1/4$, (14.8) 的平均平方残差 E_m 随着近似阶数 m 的增加而变小，如图 14.2 所示. 这表明我们可以通过使用具有收敛控制参数 $c_4 \leqslant -1/4$ 的四阶线性算子 (14.39) 作为辅助线性算子来获得特征函数和特征值的收敛级数解. 这确实是正确的: 当 $A = 1$ 时，选取 $c_4 = -1/4$ 的四阶线性算子 (14.39), 二维 Gelfand 方程 (14.8) 的平均平方残差在 100 阶同伦近似时单调递减到 1.5×10^{-7}, 以及相应的特征值 λ 收敛到 1.62311584, 如表 14.3 所示. 注意，即使是特征函数 $u(x, y)$ 的 10 阶同伦近似也是足够精确的，如图 14.3 所示.

表 14.2 当 $A=1$ 时，选取 $c_0=-1$ 和四阶辅助线性算子 (14.39)，二维 Gelfand 方程 (14.8) 的平均平方残差的最小值

近似阶数 m	平均平方残差的最小值	c_4 的最优值
3	2.65×10^{-2}	-0.4443
5	9.07×10^{-3}	-0.3144
8	3.17×10^{-3}	-0.2866
10	1.83×10^{-3}	-0.2763
15	6.31×10^{-4}	-0.2555
20	2.85×10^{-4}	-0.2494

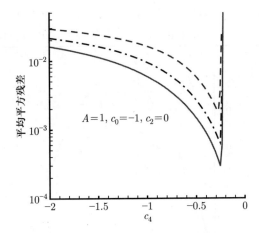

图 14.2 当 $A=1$ 时，选取 $c_0=-1$ 并使用四阶线性算子 (14.39) 作为辅助线性算子，(14.8) 相对于 c_4 的平均平方残差。虚线：10 阶同伦近似；点划线：15 阶同伦近似；实线：20 阶同伦近似

表 14.3 当 $A=1, c_0=-1$ 时，选取 $c_4=-1/4$ 时的四阶线性算子 (14.39) 作为辅助线性算子，(14.8) 的特征值 λ 和平均平方残差 E_m

近似阶数 m	平均平方残差	特征值 λ
3	0.574	1.5765
5	9.4×10^{-2}	1.6059
10	4.9×10^{-3}	1.6212

续表

近似阶数 m	平均平方残差	特征值 λ
15	6.7×10^{-4}	1.6237
20	2.8×10^{-4}	1.6233
25	1.4×10^{-4}	1.6231
30	7.6×10^{-5}	1.6231
40	2.5×10^{-5}	1.62311
50	8.9×10^{-6}	1.6231158
60	3.3×10^{-6}	1.6231158
70	1.3×10^{-6}	1.6231158
80	5.5×10^{-7}	1.62311584
90	2.6×10^{-7}	1.62311584
100	1.5×10^{-7}	1.62311584

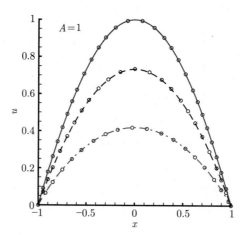

图 14.3 当 $A = 1$, $c_0 = -1$ 时, 选取 $c_4 = -1/4$ 时的四阶线性算子 (14.39) 作为辅助线性算子, 二维 Gelfand 方程 (14.8) 的 $u(x,y)$ 的剖面图. 线: 20 阶同伦近似; 空心圆: 10 阶同伦近似; 实线: $y = 0$; 虚线: $y = 1/2$; 点划线: $y = 1/4$.

我们发现, A 取其他值时, 选取适当的收敛控制参数 c_4 的四阶辅助线性算子 (14.39) 可以以类似的方式得到收敛的特征函数和特征值. 例如, 当 $A = 5$ 时, 我们选取最优收敛控制参数 $c_4 = -2/5$ 得到收敛特征函数和特征值: 如表 14.4 所示, (14.8) 的平均平方残差单调递减, 对应的特征值收敛到 $\lambda = 0.516$. 此外, 特征函数的 30 阶同伦近似与 50 阶同伦近似吻合良好, 如图 14.4 所示.

表 14.4 当 $A=5$, $c_0=-1$ 时,选取 $c_4=-2/5$ 时的四阶线性算子 (14.39) 作为辅助线性算子, (14.8) 的特征值 λ 和平均平方残差 E_m

近似阶数 m	平均平方残差	特征值 λ
5	21.1	0.471
10	3.31	0.490
15	1.11	0.501
20	0.56	0.507
30	0.18	0.512
40	7.3×10^{-2}	0.514
50	3.3×10^{-2}	0.515
60	1.6×10^{-2}	0.516
70	8.3×10^{-3}	0.516
80	4.6×10^{-3}	0.516

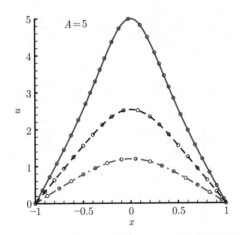

图 14.4 当 $A=5$, $c_0=-1$ 时,选取 $c_4=-2/5$ 时的四阶线性算子 (14.39) 作为辅助线性算子,二维 Gelfand 方程 (14.8) 的 $u(x,y)$ 的剖面图. 线: 50 阶同伦近似; 空心圆: 30 阶同伦近似; 实线: $y=0$; 虚线: $y=1/2$; 点划线: $y=1/4$

将 A 看作未知参数,选取 $c_0=-1$ 和 $c_4=-1$ 时的四阶辅助线性算子 (14.39), 我们得到了特征值的 30 阶同伦近似. 我们发现二维 Gelfand 方程的特征值在 $A=1.391$ 处具有最大值 1.702, 这与 Boyd 数值结果在 $A=1.39$ 处的 $\lambda_{\max}=1.702$ 非常吻合,如表 14.5 所示. 此外,特征值的 30 阶同伦近似的前 7 项给出的简化公式

$$\lambda \approx e^{-A} \big(3.39403927\,A + 0.85308129\,A^2 + 0.14514688\,A^3$$
$$+ 1.83020402 \times 10^{-2}\,A^4 + 1.65401606 \times 10^{-3}\,A^5$$
$$+ 1.03116169 \times 10^{-4}\,A^6 + 5.35313091 \times 10^{-6}\,A^7\big) \quad (14.45)$$

与 20 阶同伦近似和 Boyd [301] 在 $0 \leqslant A \leqslant 10$ 时的数值结果非常吻合, 如图 14.5 所示.

表 14.5　二维 Gelfand 方程的最大特征值 $\lambda_{\max}$ 与 Boyd 的解析和数值结果的比较. 同伦近似是通过 $c_0 = -1$ 和 $c_4 = -1$ 时的四阶辅助线性算子 (14.39) 得到的

	$\lambda_{\max}$	对应的 A 的值
5 阶同伦近似	1.701	1.383
10 阶同伦近似	1.702	1.389
15 阶同伦近似	1.702	1.391
20 阶同伦近似	1.702	1.391
25 阶同伦近似	1.702	1.391
30 阶同伦近似	1.702	1.391
Boyd 一点公式 (14.2)	1.84	1.56
Boyd 三点公式 (14.3)	1.735	1.465
Boyd 数值结果	1.702	1.39

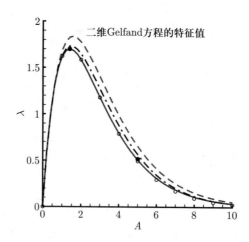

图 14.5　不同方法给出的二维 Gelfand 方程的特征值的比较. 实线: 30 阶同伦近似前 7 项给出的简化公式 (14.45); 空心圆: 20 阶同伦近似; 实心圆: Boyd [301] 给出的数值结果; 虚线: Boyd 一点近似 (14.2); 点划线: Boyd 三点近似 (14.3)

所有这些都验证了, 在同伦分析方法的框架下, 二阶非线性偏微分方程 (14.4) 可以转化为 (14.24) 由四阶辅助线性算子

$$\mathcal{L}(u) = c_4 \frac{\partial^4 u}{\partial x^2 \partial y^2}$$

控制的无穷多个四阶线性偏微分方程. 注意, 使用上述四阶辅助线性算子, 存在无穷多个光滑函数, 例如

$$\sin x, \ \exp(y), \ x\,f(y), \ y\,g(x)$$

等, 满足

$$\mathcal{L}[\sin x] = \mathcal{L}[\exp(y)] = \mathcal{L}[x\,f(y)] = \mathcal{L}[y\,g(x)] = 0,$$

其中 $f(y)$ 和 $g(x)$ 是除 y 和 x 的多项式外的任意光滑函数. 然而, 它们都不允许出现在高阶形变方程 (14.24) 的解中, 因为它们不符合解表达 (14.32). 换句话说, 如果 $w_n^*(x,y)$ 是 (14.24) 的特解, 那么

$$w_n^*(x,y) + B_1\,\sin x + B_2\,\exp(y) + B_3\,x\,f(y) + B_4\,y\,g(x) + \cdots$$

也满足 (14.24), 其中 B_1, B_2, B_3 和 B_4 为系数, $f(y)$ 和 $g(x)$ 不是 y 和 x 的多项式. 然而, 为了符合解表达 (14.32), 我们不得不在同伦分析方法的框架下强迫

$$B_1 = B_2 = B_3 = B_4 = \cdots = 0.$$

这说明解表达在同伦分析方法的框架中确实起着重要的作用, 但是如果解表达使用得当, 它可以大大简化一些非线性问题的求解过程.

这里需要强调的是, 特征函数 (14.22) 和特征值 (14.23) 的收敛性由两个收敛控制参数 c_0 和 c_4 保证. 事实上, 正是这两个收敛控制参数为同伦分析方法在选择辅助线性算子 $\mathcal{L}$ 时的极大自由度提供了强有力的支撑, 这是因为发散级数根本没有任何意义.

需要注意的是, 通过摄动方法 [300, 302, 303, 306, 324–327] 用最高导数乘以摄动量时, 可以将非线性微分方程转化为无穷多个低阶线性微分方程. 然而, 据作者所知, 二阶非线性偏微分方程从未通过任何其他解析和数值方法以这种方式转化为无穷多个四阶线性偏微分方程! 这表明我们可能有比传统上认为和相信的更大的自由来解决非线性问题! 这显示了同伦分析方法求解非线性问题的原创性和极大的灵活性.

事实上, 一种全新的方法总会带来一些新的和 (或) 不同的东西.

14.3 三维 Gelfand 方程的同伦近似

考虑三维 Gelfand 方程

$$\Delta u + \lambda \, e^u = 0, \quad -1 \leqslant x,y,z \leqslant 1 \tag{14.46}$$

满足边界条件

$$u(\pm 1,y,z) = 0, \quad u(x,\pm 1,z) = 0, \quad u(x,y,\pm 1) = 0. \tag{14.47}$$

上述二阶非线性偏微分方程很容易通过同伦分析方法以与上述类似的方式进行求解. 相应的 Mathematica 程序见附录 14.2.

写作

$$A = u(0,0,0) \tag{14.48}$$

和

$$u(x,y,z) = A + w(x,y,z). \tag{14.49}$$

原始的三维 Gelfand 方程变为

$$\Delta w + \lambda \, e^A \, e^w = 0, \quad -1 < x,y,z < 1, \tag{14.50}$$

满足六个壁面处的边界条件

$$w(\pm 1,y,z) = w(x,\pm 1,z) = w(x,y,\pm 1) = -A, \tag{14.51}$$

以及限制条件

$$w(0,0,0) = 0. \tag{14.52}$$

令 $w_0(x,y,z)$ 和 λ_0 分别表示特征函数和特征值的初始猜测解, $q \in [0,1]$ 表示嵌入变量. 在同伦分析方法的框架下, 我们先构造两个连续变化 $\phi(x,y,z;q)$ 和 $\Lambda(q)$, 随着 $q \in [0,1]$ 从 0 增加到 1, $\phi(x,y,z;q)$ 从初始猜测解 $w_0(x,y,z)$ 连续变化到特征函数 $w(x,y,z)$, 与此同时, $\Lambda(q)$ 从初始猜测解 λ_0 连续变化到特征值 λ. 这两个连续变化 $\phi(x,y,z;q)$ 和 $\Lambda(q)$ 由零阶形变方程

$$(1-q)\,\mathcal{L}\,[\phi(x,y,z;q) - w_0(x,y,z)] = q\,c_0\,\mathcal{N}\,[\phi(x,y,z;q),\Lambda(q)] \tag{14.53}$$

控制，在立方区域 $-1 < x, y, z < +1$，满足六个壁面处的边界条件

$$(1-q)\mathcal{L}\left[\phi(\pm 1, y, z; q) - w_0(\pm 1, y, z)\right] = c_0\, q\, \left[\phi(\pm 1, y, z; q) + A\right], \quad (14.54)$$

$$(1-q)\mathcal{L}\left[\phi(x, \pm 1, z; q) - w_0(x, \pm 1, z)\right] = c_0\, q\, \left[\phi(x, \pm 1, z; q) + A\right], \quad (14.55)$$

$$(1-q)\mathcal{L}\left[\phi(x, y, \pm 1; q) - w_0(x, y, \pm 1)\right] = c_0\, q\, \left[\phi(x, y, \pm 1; q) + A\right], \quad (14.56)$$

以及在原点的额外限制条件

$$(1-q)\mathcal{L}\left[\phi(0, 0, 0; q) - w_0(0, 0, 0)\right] = c_0\, q\, \phi(0, 0, 0; q), \quad (14.57)$$

其中

$$\begin{aligned}
&\mathcal{N}[\phi(x,y,z;q), \Lambda(q)] \\
&= \frac{\partial^2 \phi(x,y,z;q)}{\partial x^2} + \frac{\partial^2 \phi(x,y,z;q)}{\partial y^2} + \frac{\partial^2 \phi(x,y,z;q)}{\partial z^2} \\
&\quad + e^A\, \Lambda(q)\, \exp[\phi(x,y,z;q)]
\end{aligned} \quad (14.58)$$

是 (14.50) 的非线性算子，$\mathcal{L}$ 为具有性质 $\mathcal{L}(0) = 0$ 的辅助线性算子，$c_0 \neq 0$ 为收敛控制参数. 这里需要强调的是，在同伦分析方法的框架下，我们有极大的自由度来选择辅助线性算子 $\mathcal{L}$ 和收敛控制参数 c_0.

类似地，我们有同伦级数解

$$w(x, y, z) = w_0(x, y, z) + \sum_{n=1}^{+\infty} w_n(x, y, z), \quad (14.59)$$

$$\lambda = \lambda_0 + \sum_{n=1}^{+\infty} \lambda_n. \quad (14.60)$$

通过定理 4.15，$w_n(x, y, z)$ 由 n 阶形变方程

$$\mathcal{L}\left[w_n(x, y, z) - \chi_n\, w_{n-1}(x, y, z)\right] = c_0\, \delta_{n-1}(x, y, z), \quad -1 < x, y, z < 1 \quad (14.61)$$

控制，满足六个壁面处的边界条件

$$w_n(\pm 1, y, z) = \mu_n(\pm 1, y, z),$$

$$w_n(x, \pm 1, z) = \mu_n(x, \pm 1, z),$$

$$w_n(x, y, \pm 1) = \mu_n(x, y, \pm 1),$$

其中

$$\mu_n(x,y,z) = (\chi_n + c_0)\, w_{n-1}(x,y,z) + c_0\,(1-\chi_n)\, A, \tag{14.62}$$

$$\delta_k(x,y,z) = \Delta w_k(x,y,z) + e^A \sum_{j=0}^{k} \lambda_{k-j}\, \mathcal{D}_j\left(e^\phi\right) \tag{14.63}$$

由定理 4.1 获得, $\mathcal{D}_j\left(e^\phi\right)$ 由递推公式 (14.30) 给出.

同理, 上述线性偏微分方程的解写作

$$\begin{aligned} w_n(x,y,z) = &\, w_n^*(x,y,z) - w_n^*(\pm 1, y, z) - w_n^*(x, \pm 1, z) - w_n^*(x, y, \pm 1) \\ &+ w_n^*(x, \pm 1, \pm 1) + w_n^*(\pm 1, y, \pm 1) + w_n^*(\pm 1, \pm 1, z) \\ &- w_n^*(\pm 1, \pm 1, \pm 1) \\ &+ \mu_n(\pm 1, y, z) + \mu_n(x, \pm 1, z) + \mu_n(x, y, \pm 1) \\ &- \mu_n(x, \pm 1, \pm 1) - \mu_n(\pm 1, y, \pm 1) - \mu_n(\pm 1, \pm 1, z) \\ &+ \mu_n(\pm 1, \pm 1, \pm 1), \end{aligned} \tag{14.64}$$

其中

$$w_n^*(x,y,z) = c_0\, \mathcal{L}^{-1}[\delta_{n-1}(x,y,z)] + \chi_n\, w_{n-1}(x,y,z) \tag{14.65}$$

是 (14.61) 的特解. 此外, 特征值 λ_{n-1} 由以下线性代数方程确定

$$w_n(0,0,0) = (\chi_n + c_0)\, w_{n-1}(0,0,0). \tag{14.66}$$

对于三维 Gelfand 方程, 我们也选择相同的初始猜测解

$$w_0(x,y,z) = 0.$$

类似地, 我们选择辅助线性算子 $\mathcal{L}$ 为

$$\mathcal{L}(w) = \frac{c_3}{xyz}\, \frac{\partial^3 w}{\partial x \partial y \partial z} + c_6\, \frac{\partial^6 w}{\partial x^2 \partial y^2 \partial z^2}, \tag{14.67}$$

其中 c_3 和 c_6 为常数. 对于任意正整数 m, n, k, 其逆算子为

$$\mathcal{L}^{-1}(x^m\, y^n\, z^k) = \frac{x^{m+2}\, y^{n+2}\, z^{k+2}}{(m+2)(n+2)(k+2)[c_3 + c_6(m+1)(n+1)(k+1)]}. \tag{14.68}$$

特别是当 $c_3 = 0$ 时, 我们有六阶辅助线性算子

$$\mathcal{L}(w) = c_6\, \frac{\partial^6 w}{\partial x^2 \partial y^2 \partial z^2}. \tag{14.69}$$

详情请参考廖世俊和谭越 [321].

需要注意的是，这里存在三个收敛控制参数 c_0, c_3 和 c_6. 虽然它们都没有物理意义，但提供了一种便捷的方式来保证同伦级数 (14.59) 和 (14.60) 的收敛性. 不失一般性，首先考虑 $A = 1$ 的特殊情况. 同理，发现六个壁面处的 $u(x, y, z)$ 的 n 阶近似为

$$A(1 + c_0)^n,$$

其在 $|1 + c_0| < 1$，即 $-2 < c_0 < 0$，且 $n \to +\infty$ 时会消失. 同样，我们也发现当 $c_3 \neq 0$ 且 $-2 < c_0 < 0$ 时，无法得到特征值和特征函数的收敛级数. 因此，我们选择 $c_0 = -1$ 和 $c_3 = 0$，以便满足六个壁面处的边界条件. 然后，我们将注意力集中在 (14.69) 中的收敛控制参数 c_6 对同伦级数 (14.59) 和 (14.60) 收敛性的影响上.

类似地，我们用与 (14.44) 类似的方式定义三维 Gelfand 方程的平均平方残差 E_m，当 $A = 1$, $c_0 = -1$ 时，选取六阶辅助线性算子 (14.69)，其只依赖于 c_6，如图 14.6 所示. 三维 Gelfand 方程的平均平方残差 E_m 的最小值和对应的收敛控制参数 c_6 的最优值在表 14.6 中列出. 我们发现，随着近似阶数的增加，区间 $0.1 \leqslant c_6 < +\infty$ 上任意收敛控制参数 c_6 的平均平方残差 E_m 不断减小，并且 c_6 的最优值接近 0.1. 确实如此：当 $A = 1$ 时，我们通过选取 $c_0 = -1$ 和 $c_6 = 1/8$ 的六阶辅助线性算子 (14.69) 可以得到收敛的特征函数及其对应的特征值 $\lambda = 2.2636$，如表 14.7 所示.

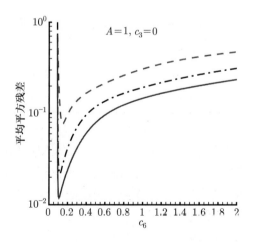

图 14.6 当 $A = 1$, $c_0 = -1$ 时，选取六阶辅助线性算子 (14.69)，三维 Gelfand 方程相对于 c_6 的平均平方残差. 虚线：5 阶同伦近似; 点划线：10 阶同伦近似; 实线：15 阶同伦近似

表 14.6 当 $A=1$, $c_0=-1$ 时,选取六阶辅助线性算子 (14.69),三维 Gelfand 方程的平均平方残差的最小值

近似阶数 m	平均平方残差的最小值	c_6 的最优值
3	0.127	0.1614
5	7.7×10^{-2}	0.1478
10	2.2×10^{-2}	0.1185
15	1.2×10^{-2}	0.1088
20	7.9×10^{-3}	0.0994
25	5.8×10^{-3}	0.0982

表 14.7 当 $A=1$, $c_0=-1$ 时,选取 $c_6=1/8$ 时的六阶辅助线性算子 (14.69),三维 Gelfand 方程的特征值 λ 和平均平方残差 E_m

近似阶数 m	平均平方残差	特征值 λ
1	0.77	1.6878
3	0.33	2.1757
5	9.6×10^{-2}	2.2935
10	2.2×10^{-2}	2.2668
15	1.3×10^{-2}	2.2635
20	9.5×10^{-3}	2.2636
25	7.3×10^{-3}	2.2636
30	5.9×10^{-3}	2.2636

将 A 看作未知参数,选取 $c_0=-1$ 和 $c_6=1$ 的六阶辅助线性算子 (14.69),使用附录 14.2 中的 Mathematica 程序我们可以得到特征值的 25 阶近似. 我们发现三维 Gelfand 方程的特征值在 $A \approx 1.610$ 处具有最大值 2.476,如表 14.8 所示. 特征值的 25 阶同伦近似在区间 $0 \leqslant A \leqslant 12$ 上有效,并且由其前八项给出的简化公式,即

$$\lambda \approx e^{-A} \left(4.48514605 A + 1.30348867 A^2 \right.$$
$$\left. + 0.31378876 A^3 + 0.056269253 A^4 \right.$$

$$+ 7.77343016 \times 10^{-3} A^5 + 8.58885688 \times 10^{-4} A^6$$
$$+ 6.90477890 \times 10^{-5} A^7 + 2.20710623 \times 10^{-6} A^8) \tag{14.70}$$

是 λ 的良好近似, 如图 14.7 所示.

表 14.8 当 $c_0 = -1$ 时, 选取 $c_6 = 1$ 时的六阶辅助线性算子 (14.69), 三维 Gelfand 方程的最大特征值 $\lambda_{\max}$

	$\lambda_{\max}$	对应的 A 的值
4 阶同伦近似	2.477	1.603
8 阶同伦近似	2.476	1.600
12 阶同伦近似	2.476	1.602
16 阶同伦近似	2.476	1.605
20 阶同伦近似	2.476	1.607
25 阶同伦近似	2.476	1.610

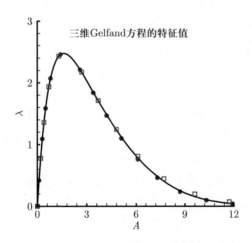

图 14.7 当 $c_0 = -1$ 时, 选取 $c_6 = 1$ 时的六阶辅助线性算子 (14.69), 三维 Gelfand 方程的特征值. 实线: 25 阶同伦近似; 空心圆: 20 阶同伦近似; 方块: 25 阶近似前 8 项给出的简化公式 (14.70)

所有这些都证明了, 在同伦分析方法的框架下, 二阶三维非线性偏微分方程 (14.46) 可以转化为无穷多个六阶线性偏微分方程, 其解很容易通过代数计算得到. 这样, 原始的三维非线性 Gelfand 方程就以相当简单的方式进行求解了. 这主要是因为同伦分析方法为我们选择辅助线性算子提供了极大的自由度, 同时也为我们提供了一种通过收敛控制参数来保证级数解收敛的便捷方式. 据我们所知, 这种变换从未被其他解析和数值方法报道过. 这表明了我们在解决非线性问题上的自由度可能比我们传统上认为的大得多!

14.4 本章小结

本章提出了一种简单有效的解析方法来求解具有强非线性的高维 Gelfand 方程. 在同伦分析方法的框架下, 二阶非线性偏微分方程被转化为无穷多个四阶二维或六阶三维线性偏微分方程, 使其在解表达的限制下非常容易求解.

基于同伦分析方法, 可以将描述非相似边界层流动的三阶非线性偏微分方程转化为无穷多个三阶线性常微分方程, 使得非相似边界层流动可以用与求解相似边界层流动类似的方式进行求解, 如第十一章所述. 此外, 描述非定常边界层流动的三阶非线性偏微分方程可以转化为无穷多个三阶线性常微分方程, 从而使得非定常边界层流动可以用与求解定常边界层流动类似的方式进行求解, 如第十二章所述. 这里, 我们进一步说明了在同伦分析方法的框架下, 二阶非线性偏微分方程 (Gelfand 方程) 可以转化为无穷多个四阶二维或六阶三维线性偏微分方程. 所有这些都验证了同伦分析方法确实为我们选择辅助线性算子 $\mathcal{L}$ 提供了极大的自由. 基于这种在辅助线性算子 $\mathcal{L}$ 选择上的极大自由, 一些非线性微分方程可以用更简单的方式求解, 如上所述.

需要强调的是, 收敛控制参数 c_0, c_4 和 c_6 对保证 Gelfand 方程特征函数和特征值级数解的收敛性起到了非常重要的作用. 二维 Gelfand 方程通过选取 $c_0 = -1$ 和包含负收敛控制参数 c_4 的四阶辅助线性算子 (14.39) 可以得到收敛的结果. 然而, 三维 Gelfand 方程则是通过选取 $c_0 = -1$ 和包含正收敛控制参数 c_6 的六阶辅助线性算子 (14.69) 来得到收敛的结果. 因此, 如果没有这些选择适当的收敛控制参数, 则很难获得收敛的结果. 注意, 发散级数解没有任何意义. 因此, 在同伦分析方法的框架下, 选取辅助线性算子 $\mathcal{L}$ 的自由本质上是基于保证同伦级数解的收敛性, 否则这种自由是没有意义的. 这再次表明收敛控制参数 [307] 的重要性: 收敛控制参数从本质上将同伦分析方法与其他所有解析方法区分开来.

据我们所知, 本章使用的变换从未被任何其他解析和数值方法报道过, 它揭示了同伦分析方法的原创性和极大的灵活性. 此外, 这还表明, 我们可能有比传统上认为和相信的更大的自由度来解决非线性问题. 事实上, 这个例子很好地让我们对非线性问题保持一个开放的心态: 我们的一些 "传统" 想法可能是我们思想的最大限制.

遗憾的是, 对于一般的非线性微分方程, 关于同伦分析方法的许多机理现在仍然不清楚. 例如, 如何在无穷多个可能的辅助线性算子中找到最好的或最优的辅助线性算子? 我们可以给出一些严格的数学证明吗? 选择辅助线性算子的自由可能会给应用数学和纯数学带来一些新的和有趣的问题, 如果这种自由真的存在的话, 本人希望最终能够给我们解决强非线性微分方程的 "真正" 自由.

参考文献

[297] Adomian, G.: Nonlinear stochastic differential equations. J. Math. Anal. Applic. **55**, 441–452 (1976).

[298] Adomian, G.: A review of the decomposition method and some recent results for nonlinear equations. Comput. Math. Appl. **21**, 101–127 (1991).

[299] Adomian, G.: Solving Frontier Problems of Physics: The Decomposition Method. Kluwer Academic Publishers, Boston (1994).

[300] Awrejcewicz, J., Andrianov, I.V., Manevitch, L.I.: Asymptotic Approaches in Nonlinear Dynamics. Springer-Verlag, Berlin (1998).

[301] Boyd, J.P.: An analytical and Numerical Study of the Two-dimensional Bratu Equation. Journal of Scientific Computing. **1**, 183–206 (1986).

[302] Cole, J.D.: Perturbation Methods in Applied Mathematics. Blaisdell Publishing Company,Waltham (1992).

[303] Hinch, E.J.: Perturbation Methods. In seres of Cambridge Texts in Applied Mathematics, Cambridge University Press, Cambridge (1991).

[304] Jacobsen, J., Schmitt, K.: The Liouville-Bratu-Gelfand problem for radial operators. Journal of Differential Equations. **184**, 283–298 (2002).

[305] Karmishin, A.V., Zhukov, A.T., Kolosov, V.G.: Methods of Dynamics Calculation and Testing for Thin-walled Structures (in Russian). Mashinostroyenie, Moscow (1990).

[306] Lagerstrom, P.A.: Matched Asymptotic Expansions: Ideas and Techniques. In series of Applied Mathematical Sciences, **76**, Springer-Verlag , New York (1988).

[307] Liang, S.X., Jeffrey, D.J.: Comparison of homotopy analysis method and homotopy perturbation method through an evelution equation. Commun. Nonlinear Sci. Numer. Simulat. **14**, 4057–4064 (2009).

[308] Liao, S.J.: The Proposed Homotopy Analysis Technique for the Solution of Nonlinear Problems. PhD dissertation, Shanghai Jiao Tong University (1992).

[309] Liao, S.J.: A kind of approximate solution technique which does not depend upon small parameters—(II) An application in fluid mechanics. Int. J. Nonlin. Mech. **32**, 815–822 (1997).

[310] Liao, S.J.: An explicit, totally analytic approximation of Blasius viscous flow problems. Int. J. Nonlin. Mech. **34**, 759–778 (1999a).

[311] Liao, S.J.: A uniformly valid analytic solution of 2D viscous flowpast a semi-infinite flat plate. J. Fluid Mech. **385**, 101–128 (1999b).

[312] Liao, S.J.: On the analytic solution of magnetohydrodynamic flows of non-Newtonian fluids over a stretching sheet. J. Fluid Mech. **488**, 189–212 (2003a).

[313] Liao, S.J.: Beyond Perturbation - Introduction to the Homotopy Analysis Method. Chapman & Hall/ CRC Press, Boca Raton (2003b).

[314] Liao, S.J.: On the homotopy analysis method for nonlinear problems. Appl. Math. Comput. **147**, 499–513 (2004).

[315] Liao, S.J.: A new branch of solutions of boundary-layer flows over an impermeable stretched plate. Int. J. Heat Mass Tran. **48**, 2529–2539 (2005).

[316] Liao, S.J.: Series solutions of unsteady boundary-layer flows over a stretching flat plate. Stud. Appl. Math. **117**, 2529–2539 (2006).

[317] Liao, S.J.: Notes on the homotopy analysis method - Some definitions and theorems. Commun. Nonlinear Sci. Numer. Simulat. **14**, 983–997 (2009).

[318] Liao, S.J.: On the relationship between the homotopy analysis method and Euler transform. Commun. Nonlinear Sci. Numer. Simulat. **15**, 1421–1431 (2010a). doi:10.1016/j.cnsns.2009.06.008.

[319] Liao, S.J.: An optimal homotopy-analysis approach for strongly nonlinear differential equations. Commun. Nonlinear Sci. Numer. Simulat. **15**, 2003–2016 (2010b).

[320] Liao, S.J., Campo, A.: Analytic solutions of the temperature distribution in Blasius viscous flow problems. J. Fluid Mech. **453**, 411–425 (2002).

[321] Liao, S.J., Tan, Y.: A general approach to obtain series solutions of nonlinear differential equations. Stud. Appl. Math. **119**, 297–355 (2007).

[322] Liouville, J.: Sur léquation aux dérivées partielles $(\partial \ln \lambda)/(\partial u \partial v) \pm 2\lambda a^2 = 0$. J. de Math. **18**, 71–72 (1853).

[323] McGough, J.S.: Numerical continuation and the Gelfand problem. Applied Mathematics and Computation. **89**, 225–239 (1998).

[324] Murdock, J.A.: Perturbations: - Theory and Methods. John Wiley & Sons, New York (1991).

[325] Nayfeh, A.H.: Perturbation Methods. John Wiley & Sons, New York (1973).

[326] Nayfeh, A.H.: Perturbation Methods. John Wiley & Sons, New York (2000).

[327] Von Dyke, M.: Perturbation Methods in Fluid Mechanics. The Parabolic Press, Stanford (1975).

[328] Xu, H., Lin, Z.L., Liao, S.J., Wu, J.Z., Majdalani, J.: Homotopy-based solutions of the Navier-Stokes equations for a porous channel with orthogonally moving walls. Physics of Fluids. **22**, 053601 (2010). doi:10.1063/1.3392770.

附录 14.1 二维 Gelfand 方程的 Mathematica 程序

基于同伦分析方法求解二维 Gelfand 方程

$$\Delta u + \lambda\, e^u = 0, \quad -1 < x < 1, \quad -1 < y < 1,$$

其满足四个壁面处的边界条件

$$u(x, \pm 1) = 0, \quad u(\pm 1, y) = 0.$$

二维 Gelfand 方程的 Mathematica 程序

廖世俊编写
上海交通大学
2010 年 8 月

```
(***************************************************************)
(*                                                             *)
(* For given A, we find such an eigenvalue lambda and          *)
(*    a normalized eigenfunction w(x,y) satisfying:            *)
(*         w_{xx} + w_{yy} + lambda * Exp[w(x)] = 0            *)
(*    subject to the boundary conditions:                      *)
(*         w(1,y)=w(-1,y)=w(x,1)=w(x,-1)=-A,w(0,0)=0,          *)
(*                                                             *)
(***************************************************************)
<<Calculus`Pade`;
<<Graphics`Graphics`;

(***************************************************************)
(* Define initial guess of w(x)                                *)
(***************************************************************)
w[0] = 0;
U[0] = A + w[0];
```

```
(***************************************************************)
(* Define the function chi[k]                                 *)
(***************************************************************)
chi[k_]:=If[ k <= 1, 0, 1];


(***************************************************************)
(* Define the the auxiliary linear operator L                 *)
(***************************************************************)
L[f_] := Module[{temp},
Expand[c2*D[f,{x,1},{y,1}]/x/y + c4*D[f,{x,2},{y,2}]]
];



(***************************************************************)
(* Define the inverse operator of L                           *)
(***************************************************************)
Linv[x^m_. * y^n_.] := x^(m+2)*y^(n+2)/(m+2)/(n+2)
                              /(c2+c4*(n+1)*(m+1));
Linv[x^m_.] := x^(m+2) * y^2/(m+2)/2/(c2+c4*(m+1));
Linv[y^n_.] := x^2 * y^(n+2)/(n+2)/2/(c2+c4*(n+1));
Linv[c_]    := c*x^2*y^2/4/(c2+c4)/;FreeQ[c,x] && FreeQ[c,y];


(***************************************************************)
(* The linear property of the inverse operator of L           *)
(*     Linv[f_+g_]  := Linv[f]+Linv[g];                       *)
(***************************************************************)
Linv[p_Plus]  := Map[Linv,p];
Linv[c_*f_]   := c*Linv[f]   /;   FreeQ[c,x] && FreeQ[c,y];


(***************************************************************)
(* Define Getdelta[k]                                         *)
(***************************************************************)
Getdelta[k_]:=Module[{temp,n},
temp[1] = D[w[k],{x,2}] + D[w[k],{y,2}];
temp[2] = Sum[lambda[n]*Dexp[k-n],{n,0,k}];
delta[k]    = Expand[ temp[1] + temp[2] ];
];
```

附录 14.1 二维 Gelfand 方程的 Mathematica 程序

```
(*************************************************************)
(* Define GetDexp[n]                                        *)
(*************************************************************)
GetDexp[n_]:=Module[{j},
If[n==0,Dexp[0]=Exp[w[0]]];
If[n >0,Dexp[n]=Sum[(1-j/n)*Dexp[j]*w[n-j],{j,0,n-1}]//Expand];
];


(*************************************************************)
(* Define Getlambda                                         *)
(*   This module gets lambda[k-1]                           *)
(*************************************************************)
Getlambda[k_]:=Module[{eq,temp},
eq    = w[k]-(c0+chi[k])*w[k-1] /. {x -> 0,y->0};
temp = Solve[eq == 0, lambda[k-1]];
lambda[k-1]   = temp[[1,1,2]]//Expand;
];


(*************************************************************)
(* Define GetwSpecial                                        *)
(* This module gets a special solution of (14.24) and (14.25)*)
(*************************************************************)
GetwSpecial[k_]:=Module[{temp},
temp[0]  = Expand[RHS[k]];
temp[1]  = Linv[temp[0]];
temp[2]  = temp[1] + chi[k]*w[k-1]//Simplify;
wSpecial = temp[2]//Expand;
];


(*************************************************************)
(* Define Getw                                              *)
(*   This module gets a solution of (14.24) and (14.25)    *)
(*************************************************************)
Getw[k_]:=Module[{temp,alpha},
alpha    = (c0+chi[k])*w[k-1]+c0*(1-chi[k])*A;
temp[1]  = wSpecial /.  y->1;
temp[2]  = wSpecial /.  x->1;
temp[3]  = wSpecial /.  {x->1, y->1};
temp[4]  = alpha /.  x->1;
```

```
temp[5]   =   alpha /.  y->1;
temp[6]   =   alpha /.  {x->1,y->1};
temp[7]   =   wSpecial - temp[1] - temp[2] + temp[3]
                      + temp[4] + temp[5] - temp[6];
w[k] = Simplify[temp[7]]//Expand;
];


(*************************************************************)
(* Define GetErr[k]                                          *)
(*************************************************************)
GetErr[k_]:=Module[{temp,sum,dx,dy,Num,i,j,X,Y},
err[k] = D[U[k],{x,2}] + D[U[k],{y,2}] + LAMBDA[k]*Exp[U[k]];
Nx  = 10;
Ny  = 10;
dx  = N[1/Nx,100];
dy  = N[1/Ny,100];
sum = 0;
Num = 0;
For[i = 0, i <= Nx-1, i++,
    X = i*dx;
    For[j = 0, j <= Ny - 1, j++,
        Y = j*dy;
        temp = err[k]^2 /. {x->X, y->Y};
        sum  = sum + temp;
        Num  = Num + 1;
        ];
    ];
Err[k] = sum/Num;
If[NumberQ[Err[k]], Print["Squared Residual of G.E. = ",
                          Err[k]//N]]
];


(*************************************************************)
(* Define Body1[A]                                           *)
(*      Boyd's one-point formula                             *)
(*************************************************************)
Boyd1[A_] := 3.2*A*Exp[-0.64*A];


(*************************************************************)
```

```
(* Define Body3[A]                                         *)
(*        Boyd's three-point formula                       *)
(***********************************************************)
Boyd3[A_] := Module[{temp,B,C,G},
G = 0.2763 + Exp[0.463*A] + 0.0483*Exp[-0.209*A];
B = A*( 0.829-0.566*Exp[0.463*A]-0.0787*Exp[-0.209*A])/G;
C = A*(-1.934+0.514*Exp[0.463*A]+1.9750*Exp[-0.209*A])/G;
temp[1] =  2.667*A + 4.830*B + 0.127*C;
temp[2] =  0.381*A + 0.254*B + 0.018*C;
temp[1]*Exp[-temp[2]]
];


(***********************************************************)
(*                     Main Code                           *)
(***********************************************************)
ham[m0_,m1_]:=Module[{temp,k,j},
For[k=Max[1,m0], k<=m1, k=k+1,
Print[" k  =  ",k];
GetDexp[k-1];
    Getdelta[k-1];
    RHS[k] = c0*delta[k-1]//Expand;
GetwSpecial[k];
Getw[k];
U[k] = U[k-1] + w[k]//Simplify;
Getlambda[k];
Lambda[k-1]  = Sum[lambda[j],{j,0,k-1}]//Expand;
     LAMBDA[k-1]   = Lambda[k-1]*Exp[-A];
Print[k-1,"th approximation of LAMBDA = ",N[LAMBDA[k-1],10] ];
   ];
Print["Successful !"];
];


(***********************************************************)
(* Define the parameters                                   *)
(***********************************************************)
c0    =   -1;
A     =    1;
c2    =    0;
c4    =   -1/4;
```

```
(************************************************************)
(*  Print the parameters                                    *)
(************************************************************)
Print["     A    = ",A];
Print["     c0   = ",c0];
Print["     c2   = ",c2];
Print["     c4   = ",c4];

(* Get the 20th-order approximation *)
ham[1,21];

(* Get squared residual of up to 20th-order approximation *)
For[k=5, k<=20, k=k+5, Print["k = ",k]; GetErr[k]]
```

附录 14.2 三维 Gelfand 方程的 Mathematica 程序

基于同伦分析方法求解三维 Gelfand 方程

$$\Delta u + \lambda\, e^u = 0, \quad -1 \leqslant x,y,z \leqslant 1,$$

其满足边界条件

$$u(\pm 1, y, z) = 0, \quad u(x, \pm 1, z) = 0, \quad u(x, y, \pm 1) = 0.$$

三维 Gelfand 方程的 Mathematica 程序

廖世俊编写

上海交通大学

2010 年 8 月

```
(************************************************************)
(*                                                          *)
(*    For given A, we find such an eigenvalue lambda and    *)
(*    a normalized eigenfunction w(x,y) that:               *)
(*          w_{xx}+w_{yy}+w_{zz}+lambda*Exp[w(x)] = 0       *)
(*    subject to the boundary conditions:                   *)
```

```
(*          w(1,y)=w(-1,y)=w(x,1)=w(x,-1)=-A,w(0,0)=0,        *)
(**************************************************************)
<<Calculus`Pade`;
<<Graphics`Graphics`;
<<tool2010.nb;

(**************************************************************)
(* Define initial guess                                        *)
(**************************************************************)
w[0] = 0;
U[0] = A + w[0];

(**************************************************************)
(* Define  chi[k]                                              *)
(**************************************************************)
chi[k_]:=If[k<=1,0,1];

(**************************************************************)
(* Define inverse operator of auxiliary linear operator L      *)
(**************************************************************)
Linv[x^m_ * y^n_ * z^k_] := x^(m+2)*y^(n+2)*z^(k+2)/(m+2)
                 /(n+2)/(k+2)/(c3+c6*(n+1)*(m+1)*(k+1));
Linv[y^n_ * z^k_] := x^2*y^(n+2)*z^(k+2)/2/(n+2)/(k+2)
                 /(c3+c6*(n+1)*(k+1));
Linv[x^m_ * z^k_] := x^(m+2)*y^2*z^(k+2)/(m+2)/2/(k+2)
                 /(c3+c6*(m+1)*(k+1));
Linv[x^m_ * y^n_] := x^(m+2)*y^(n+2)*z^2/(m+2)/(n+2)/2
                 /(c3+c6*(n+1)*(m+1));
Linv[x^m_] := x^(m+2)*y^2*z^2/(m+2)/2/2/(c3+c6*(m+1));
Linv[y^n_] := x^2*y^(n+2)*z^2/2/(n+2)/2/(c3+c6*(n+1));
Linv[z^k_] := x^2*y^2*z^(k+2)/2/2/(k+2)/(c3+c6*(k+1));
Linv[c_]   := c*x^2*y^2*z^2/8/(c3+c6) /; FreeQ[c,x]
              && FreeQ[c,y] && FreeQ[c,z];

(**************************************************************)
(* The linear property of the inverse operator of L            *)
(*   Linv[f_+g_]  := Linv[f]+Linv[g];                          *)
(**************************************************************)
Linv[p_Plus] := Map[Linv,p];
```

```
Linv[c_*f_]   := c*Linv[f]/;FreeQ[c,x]&&FreeQ[c,y]&&FreeQ[c,z];


(***************************************************************)
(* Define GetR[k]                                              *)
(***************************************************************)
Getdelta[k_]:=Module[{temp,n},
temp[1]   = D[w[k],{x,2}] + D[w[k],{y,2}] + D[w[k],{z,2}];
temp[2]   = Sum[lambda[n]*Dexp[k-n],{n,0,k}];
delta[k]  = Expand[ temp[1] + temp[2] ];
];


(***************************************************************)
(* Define GetDexp[n]                                           *)
(***************************************************************)
GetDexp[n_]:=Module[{},
If[n == 0, Dexp[0] = 1 ];
If[n >0,Dexp[n]=Sum[(1-j/n)*Dexp[j]*w[n-j],{j,0,n-1}]//Expand];
];


(***************************************************************)
(* Define Getlambda                                            *)
(*    This module gets lambda[k-1]                             *)
(***************************************************************)
Getlambda[k_]:=Module[{eq,temp},
eq    = w[k]-(c0+chi[k])*w[k-1] /. {x->0, y->0, z->0};
temp = Solve[eq == 0, lambda[k-1]];
lambda[k-1]  = temp[[1,1,2]]//Expand;
];


(***************************************************************)
(* Define GetwSpecial                                          *)
(*    This module gets a special solution of (14.61)           *)
(***************************************************************)
GetwSpecial[k_]:=Module[{temp},
temp[0]   = Expand[RHS[k]];
temp[1]   = Linv[temp[0]];
temp[2]   = temp[1] + chi[k]*w[k-1];
wSpecial = temp[2]//Expand;
];
```

```
(***********************************************************)
(* Define Getw                                            *)
(*     This module gets a solution of the high-order EQ   *)
(***********************************************************)
Getw[k_]:=Module[{temp,mu},
mu       =  (c0+chi[k])*w[k-1] + c0*(1-chi[k])*A;
temp[1]  =  wSpecial /. x->1;
temp[2]  =  wSpecial /. y->1;
temp[3]  =  wSpecial /. z->1;
temp[4]  =  wSpecial /. {x->1, y->1};
temp[5]  =  wSpecial /. {x->1, z->1};
temp[6]  =  wSpecial /. {y->1, z->1};
temp[7]  =  wSpecial /. {x->1, y->1, z->1};
temp[8]  =  wSpecial-temp[1]-temp[2]-temp[3]+temp[4]
                 +temp[5]+temp[6]-temp[7];
temp[1]  =  mu /. x->1;
temp[2]  =  mu /. y->1;
temp[3]  =  mu /. z->1;
temp[4]  =  mu /. {x->1, y->1};
temp[5]  =  mu /. {x->1, z->1};
temp[6]  =  mu /. {y->1, z->1};
temp[7]  =  mu /. {x->1, y->1, z->1};
temp[9]  =  temp[8]+temp[1]+temp[2]+temp[3]-temp[4]
                 -temp[5]-temp[6]+temp[7];
w[k]     =  Expand[temp[9]];
];

(***********************************************************)
(* Define GetErr[k]                                       *)
(***********************************************************)
GetErr[k_]:=Module[
{temp,sum,dx,dy,dz,Num,i,j,m,X,Y,Z,Nx,Ny,Nz},
err[k] = D[U[k],{x,2}] + D[U[k],{y,2}] + D[U[k],{z,2}]
         + LAMBDA[k]*Exp[U[k]];
Nx  = 10;
Ny  = 10;
Nz  = 10;
dx  = N[1/Nx,100];
```

```
dy   = N[1/Ny,100];
dz   = N[1/Nz,100];
sum  = 0;
Num  = 0;
For[i = 0, i <= Nx-1, i++,
    X = i*dx;
    For[j = 0, j <= Ny - 1, j++,
        Y = j*dy;
        For[m = 1, m <= Nz-1, m++,
            Z = m*dz;
            temp = err[k]^2 /. {x->X, y->Y,z->Z};
            sum  = sum + temp;
            Num  = Num + 1;
            ];
        ];
    ];
Err[k] = sum/Num;
If[NumberQ[Err[k]], Print["Squared Residual of G.E. = ",
                          Err[k]//N]]
];


(*************************************************************)
(* Define HP[F,m,n]                                          *)
(*    This module gives [m,n] homotopy-Pade approximation    *)
(*    of the series : F[k]   =   sum[f[i],{i,0,k}]           *)
(*************************************************************)
hp[F_,m_,n_]:=Block[{i,k,dF,temp,q},
dF[0] = F[0];
For[k = 1, k <= m+n, k = k+1, dF[k] = Expand[F[k]-F[k-1]]];
temp = dF[0]+Sum[dF[i]*q^i,{i,1,m+n}];
Pade[temp,{q,0,m,n}]/.q->1
];


(*************************************************************)
(*                        Main Code                          *)
(*************************************************************)
ham[m0_,m1_]:=Module[{temp,k,j},
For[k=Max[1,m0], k<=m1, k=k+1,
    Print[" k =  ",k];
```

```
      GetDexp[k-1];
      Getdelta[k-1];
      RHS[k] = c0 * delta[k-1]//Expand;
      GetwSpecial[k];
      Getw[k];
      U[k] = U[k-1] + w[k]//Expand;
      Getlambda[k];
      Lambda[k-1]  = Expand[Sum[lambda[j],{j,0,k-1}]];
      LAMBDA[k-1]  = Lambda[k-1]*Exp[-A];
      Print[k-1,"th approximation of lambda = ",
                N[LAMBDA[k-1],10] ];
   ];
Print["Successful !"];
];


(****************************************************************)
(* Physical and control parameters                             *)
(****************************************************************)
A     =    1;
c0    =    -1;
c3    =    0;
c6    =    1/8;


(****************************************************************)
(* Print physical and control parameters                       *)
(****************************************************************)
Print["  A  = ",A];
Print["  c0 = ",c0];
Print["  c3 = ",c3];
Print["  c6 = ",c6];

(* Get the 10th-order approximation of LAMBDA *)
ham[1,11];

(* Get squared residual of up to10th-order approximation *)
For[k=2, k<=10, k=k+2, Print["k = ",k]; GetErr[k]]
```

第十五章　非线性水波与非均匀来流的相互作用

本章说明了同伦分析方法对周期性行进波与具有指数性质分布涡量的非均匀来流的非线性相互作用的复杂非线性偏微分方程的有效性. 在同伦分析方法的框架下, 原始的变系数强非线性偏微分方程被转化为无穷多个简单的线性偏微分方程, 这些线性偏微分方程很容易求解. 从物理方面, 我们发现 Stokes 波浪破碎准则对于非均匀来流上的行进波仍然是正确的. 这证明了同伦分析方法可以用于求解一些复杂的非线性偏微分方程来加深和丰富我们对一些有趣的非线性现象的物理理解.

15.1　绪论

科学和工程领域中, 存在许多复杂的非线性偏微分方程. 本章将基于同伦分析方法 [339, 341–355, 368] 来求解此类非线性问题.

这里我们考虑在定常的, 非均匀的, 具有垂直分布涡量的剪切流上传播的一列稳定深水波. 假设水是无黏不可压缩的二维流动, 因而存在流函数 ψ. 此外, 涡量的大小随流函数 ψ 的大小呈指数变化, 但沿同一流线保持不变. 这主要是因为在实际中, 来流的分布在深度上大多呈指数分布.

从数学方面, 该问题是由一个在未知波面上且具有非线性边界条件的非线性偏微分方程来描述的, 如下所述. 由于其在数学上求解的困难性, 早期的工作要么假设来流是均匀的, 因此忽略涡量分布的影响; 要么假设波幅很小, 来流很弱, 因此简单地忽略它们之间的非线性相互作用. 在前一种情况下, 波被视为在参考系中与来流以相同的速度移动的静水中传播, 因此均匀来流可以线性地叠加到纯波中. Fenton [337] 研究过这个问题, 并且给出了 Stokes 深水波的五阶近似解. 对于具有均匀来流的浅水波, Clamond [330,331], Clamond 和 Fructus [332] 利用重整化方法获得了其高精度的近似值. 遗憾的是, 在实践中, 来流大多是不均匀的, 因此会产生涡量. 在这种情况下, 如果是小幅波, 或者是弱来流, 通常会使用线性叠加理论. 其中, Thompson [364], Kishida 和 Sobey [338] 研究了线性剪切流, Eastwood 和 Watson [336] 研究了双线性来流. 详情请参考 Peregrine [358], Toumazis [365] 和 Wang [366,367]. 但是, 如果波幅不是很小, 而且来流也不是很弱, 上述的方法就变得不可靠了: 在波流同向的时候, 上述方法往往会高估流体速度; 在波流反向的时候, 则会低估.

应用非线性方法处理波流相互作用问题起始于 Dalrymple [333]. Dalrymple 运用数值方法研究了波浪行进于一近似于 1/7 次幂来流之上的数学模型. 之后 Thomas [363] 也应用了数值方法计算了波浪与具有指数性质分布涡量的来流的相互作用. Wang 等人 [366] 采用摄动方法对波高和来流涡量参数分别进行泰勒展开, 得到了五阶级数解. 我们知道, 摄动结果通常只适用于弱非线性情况.

2003 年, 廖世俊和 Cheung [354] 应用同伦分析方法 [339,341-355,368] 得到了二维非线性行进深水波的高精度解析近似. 2009 年, 成钧、仓杰和廖世俊 [329] 进一步应用同伦分析方法研究了二维非线性深水波与指数剪切流之间的相互作用.

15.2 数学模型

15.2.1 原始边值方程

令 (x,y) 表示笛卡儿坐标, 水平 x 轴指向波传播方向, y 轴垂直向上, 原点位于平均水位. 使用移动参考系 $(x'-ct, y') \mapsto (x,y)$, 其中 c 为波速, (x',y') 为静止参考系中的坐标, 因此该问题是定常的. 令 $y = \eta(x)$ 表示自由表面, H 表示波高, λ 表示波长, $\mathbf{u} = \{u(x,y), v(x,y)\}$ 表示波流共同作用时的波流场.

由于流体是不可压缩的并且流动是二维的, 引入流函数 $\psi(x,y)$, 定义如下

$$\psi_y(x,y) = u(x,y) - c, \quad \psi_x(x,y) = -v(x,y). \tag{15.1}$$

令 $\Omega(\psi)$ 表示给定来流的涡量分布. 由涡量的定义, 我们有控制方程

$$\psi_{xx} + \psi_{yy} = -\Omega(\psi), \tag{15.2}$$

其中下标表示偏导.

这里, 我们研究由 Phillips [359] 指出的一个特殊例子

$$\Omega(\psi) = \varepsilon \exp(-\psi), \tag{15.3}$$

即涡量具有沿水深方向呈指数形式分布的情况, 其中 ε 为确定涡量强度的物理参数. 当 $\varepsilon < 0$ 时, 来流方向与波传播方向相同, 称为助流; 当 $\varepsilon > 0$ 时, 来流方向与波传播方向相反, 称为逆流. 注意, 所有均匀来流的涡量都为零. 因此, 给定的涡量分布对应于无穷多个来流. 为了避免这种情况, 我们定义 $\varepsilon = 0$ 对应于静水.

在自由表面上, 我们有运动学条件

$$\psi = 0, \quad y = \eta(x) \tag{15.4}$$

和动力学条件

$$\frac{1}{2}(\psi_x^2 + \psi_y^2) + g y = Q, \quad y = \eta(x), \tag{15.5}$$

其中 Q 是所谓的伯努利常数, 为稍后确定的未知数. 此外, 在无穷深处, 我们有边界条件

$$\psi_x \to 0, \quad y \to -\infty. \tag{15.6}$$

给定涡量分布 $\Omega(\psi) = \varepsilon \exp(-\psi)$, 则上述非线性边值问题包含四个未知数: 流函数 $\psi(x,y)$, 自由表面 $\eta(x)$, 波速 c 和伯努利常数 Q. 注意偏微分方程 (15.2) 包含指数项 $\Omega(\psi) = \varepsilon \exp(-\psi)$, 其对于未知流函数 ψ 具有很强的非线性. 此外, 在未知的自由表面 $y = \eta(x)$ 上满足非线性动力学条件 (15.5). 从数学上来说, 控制方程 (15.2) 和边界条件 (15.5) 的强非线性, 尤其是未知的自由表面 $y = \eta(x)$ 处的非线性, 给我们求解该非线性边值问题带来了很大的困难.

15.2.2 Dubreil-Jacotin 变换

为了处理未知的自由表面, 可采用 Dubreil-Jacotin 变换 [334, 335, 363] 在已知域中重新表述问题. 将流函数 ψ 视为自变量 (如 x), 将 y 视为同时依赖

于 x 和 ψ 的未知函数, 用 $y(x,\psi)$ 表示. 然后, 根据微积分基本定理, 可从 (15.1) 得到局部速度分量

$$u - c = \frac{1}{y_\psi}, \quad v = \frac{y_x}{y_\psi}. \tag{15.7}$$

为了简单起见, 引入如下无量纲变量

$$x = k\, x^*, \quad y = k\, y^*, \quad \psi = \frac{k}{c}\, \psi^*, \quad \Omega(\psi) = \frac{1}{k\, c}\, \Omega^*(\psi), \tag{15.8}$$

其中带星号的变量是有量纲的, $k = 2\pi/\lambda$ 为波数. 因此, 未知域

$$-\infty < y \leqslant \eta(x), \quad -\infty < x < +\infty$$

中的原始偏微分方程 (15.2) 至 (15.5) 控制的问题可在已知域

$$0 \leqslant \psi < +\infty, \quad -\infty < x < +\infty$$

中通过复杂的非线性偏微分方程

$$y_{xx}\, y_\psi^2 - 2\, y_x\, y_\psi\, y_{x\psi} + (1 + y_x^2)\, y_{\psi\psi} = y_\psi^3\, \Omega(\psi) \tag{15.9}$$

重新表示, 且满足下列边界条件

$$\mu\, (1 + y_x^2) + 2\, (y - \gamma)\, y_\psi^2 = 0, \quad \psi = 0, \tag{15.10}$$

$$y_x \to 0, \quad \psi \to +\infty, \tag{15.11}$$

其中 $\mu = k\, c^2/g$ 和 $\gamma = k\, Q/g$ 是未知常数.

注意, 通过 Dubreil-Jacotin 变换 [334,335,363], 包含指数非线性项 $\Omega(\psi) = \varepsilon \exp(-\psi)$ 的原始非线性偏微分方程 (15.2) 变为非线性偏微分方程 (15.9). 虽然 (15.9) 看上去比 (15.2) 要复杂得多, 但它的非线性实际上更弱, 因为 (15.9) 中的指数项 $\Omega(\psi) = \varepsilon \exp(-\psi)$ 通过变换可视为一个已知的可变系数. 更重要的是, 此时在已知边界 $\psi = 0$ 上 (15.9) 满足自由表面边界条件. 此外, 上述偏微分方程仅包含三个未知数: 一个未知函数 $y(x,\psi)$ 和两个未知常数 μ 和 γ. 因此, 与原始方程相比, 未知数减少了一个, 即从四个减少到三个. 综上所述, 采用 Dubreil-Jacotin 变换 [334,335,363], 虽然 (15.9) 看上去比原始偏微分方程 (15.2) 更复杂, 但实质上该变换简化了原问题.

自由表面 $\eta(x)$ 对应于流线 $\psi = 0$. 由于原点定义在平均水位上, 因此自由表面为

$$\eta(x) = y(x,0) - \frac{1}{\pi} \int_0^\pi y(x,0)\, \mathrm{d}x. \tag{15.12}$$

无量纲波高 H (按 k^{-1} 缩放) 定义为

$$H = y(0,0) - y(\pi,0). \tag{15.13}$$

因此, 只要获得 $y(x,\psi)$, 就可以直接获得自由表面 $y = \eta(x)$ 和波高.

15.3 简明数学公式

本小节采用同伦分析方法来获得包含变系数 $\Omega(\psi) = \varepsilon\exp(-\psi)$, 满足边界条件 (15.10) 和 (15.11) 的复杂非线性偏微分方程 (15.9) 的高精度解析近似.

15.3.1 解表达

如前几章所述, 所谓的解表达在同伦分析方法的框架中起着重要作用. 从数学上来讲, 对于强非线性偏微分方程 (15.9), 很难猜测 $y(x,\psi)$ 的表达式形式. 幸运的是, 从问题的物理背景来看, 很容易找到它的合适解表达. 从物理上来讲, 在随波移动的坐标系中, 波流共同场是由一列静水深水波, 由坐标轴随波运动而产生的一均匀流, 以及具有指数分布涡量的任意来流构成. 对于静水深水波 (即 $\Omega = 0$), 波高是周期性的, 其周期解可以表达为

$$y = \sum_{m=0}^{+\infty} a_m \cos(mx),$$

其中 a_m 为常系数. 即使存在非均匀的来流, 波高在 x 方向仍然是周期性的, 因此应始终由上述表达式表示. 此外, 由于 (15.9) 中包含 $\exp(-\psi)$, 所以解 $y(x,\psi)$ 中也应该包含 $\exp(-n\psi)$, 其中 $n \geqslant 1$ 是整数. $\exp(-n\psi)$ ($n \geqslant 1$) 这种类型的项对于 $\psi \to +\infty$ 的边界条件 (15.11) 也是必要的. 此外, 众所周知, 由坐标轴移动而产生的均匀来流是不会对波流相互作用产生影响的. 因此, 从物理上来看, 即使不求解非线性偏微分方程 (15.9) 至 (15.13), $y(x,\psi)$ 也可以表达为

$$y = -\psi + \sum_{n=1}^{+\infty}\sum_{m=0}^{+\infty} \alpha_{n,m} \exp(-n\psi) \cos(mx), \tag{15.14}$$

其中第一部分是由坐标轴移动导致的均匀来流所产生的, 第二部分是由非线性波和非均匀来流的共同作用产生的. 上述公式为我们提供了 $y(x,\psi)$ 的所谓解表达, 它在同伦分析方法的框架中起着重要的作用, 如下所述. 注意, (15.14) 自动满足边界条件 (15.11).

15.3.2 零阶形变方程

令 $q \in [0,1]$ 表示嵌入变量, $y_0(x,\psi)$, μ_0 和 γ_0 分别表示 $y(x,\psi)$, μ 和 γ 的初始猜测解. 在同伦分析方法的框架下, 我们首先构造三个连续变化 $Y(x,\psi;q)$, $\Delta(q)$ 和 $\Gamma(q)$, 随着嵌入变量 q 从 0 增加到 1, $Y(x,\psi;q)$ 从初始猜测解 $y_0(x,\psi)$ 连续变化到波浪超高 $y(x;\psi)$, 与此同时, $\Delta(q)$ 从初始猜测解 μ_0 连续变化到 μ, $\Gamma(q)$ 从初始猜测解 γ_0 连续变化到 γ. 这种连续变化 (或形变) 由零阶形变方程

$$(1-q)\,\mathcal{L}\,[\,Y(x,\psi;q) - y_0(x,\psi)\,] = c_0\,q\,\mathcal{N}[Y(x,\psi;q)] \tag{15.15}$$

定义, 满足边界条件

$$(1-q)\,\mathcal{L}_b\,[\,Y(x,\psi;q) - y_0(x,\psi)\,]$$
$$= c_0\,q\,\mathcal{N}_b[Y(x,\psi;q),\,\Delta(q),\,\Gamma(q)], \quad \psi = 0, \tag{15.16}$$

$$\frac{\partial Y(x,\psi;q)}{\partial x} = 0, \quad \psi \to +\infty, \tag{15.17}$$

以及波高条件

$$Y(0,0;q) - Y(\pi,0;q) = H, \tag{15.18}$$

其中 $c_0 \neq 0$ 为收敛控制参数, $\mathcal{L}$ 和 $\mathcal{L}_b$ 为两个辅助线性算子, $\mathcal{N}$ 和 $\mathcal{N}_b$ 为两个非线性算子, 定义如下

$$\mathcal{N}\,[Y(x,\psi;q)\,] = Y_{xx}\,Y_\psi^2 - 2\,Y_x\,Y_{x\psi}Y_\psi + (1+Y_x^2)\,Y_{\psi\psi} - \Omega\,Y_\psi^3, \tag{15.19}$$

$$\mathcal{N}_b\,[Y(x,\psi;q),\,\Delta(q),\,\Gamma(q)] = \Delta\,\left(1+Y_x^2\right) + 2\,(Y-\Gamma)\,Y_\psi^2, \tag{15.20}$$

分别对应于 (15.9) 和 (15.10). 辅助线性算子 $\mathcal{L}$ 和 $\mathcal{L}_b$ 具有性质 $\mathcal{L}(0) = 0$ 和 $\mathcal{L}_b(0) = 0$, 将在稍后确定.

当 $q = 0$ 时, 因为 $\mathcal{L}(0) = 0$ 和 $\mathcal{L}_b(0) = 0$, 零阶形变方程 (15.15) 至 (15.18) 有解

$$Y(x,\psi;0) = y_0(x,\psi), \quad \Delta(0) = \mu_0, \quad \Gamma(0) = \gamma_0. \tag{15.21}$$

当 $q = 1$ 时, 因为 $c_0 \neq 0$, 零阶形变方程 (15.15) 至 (15.18) 等价于原始偏微分方程 (15.9) 至 (15.13), 有

$$Y(x,\psi;1) = y(x,\psi), \quad \Delta(1) = \mu, \quad \Gamma(1) = \gamma. \tag{15.22}$$

因此，随着嵌入变量 $q \in [0,1]$ 从 0 增加到 1，$Y(x,\psi;q)$，$\Delta(q)$ 和 $\Gamma(q)$ 确实分别从初始猜测解 $y_0(x,\psi)$，μ_0 和 γ_0 连续变化到精确解 $y(x,\psi)$，μ 和 γ. 所以，从数学上讲，零阶形变方程 (15.15) 至 (15.18) 构造了三个同伦

$$Y(x,\psi;q) : y_0(x,\psi) \sim y(x,\psi), \quad \Delta(q) : \mu_0 \sim \mu, \quad \Gamma(q) : \gamma_0 \sim \gamma. \quad (15.23)$$

根据 (15.21) 并将 $Y(x,y;q)$，$\Delta(q)$ 和 $\Gamma(q)$ 展开成关于嵌入变量 $q \in [0,1]$ 的 Maclaurin 级数，我们有如下同伦 – Maclaurin 级数

$$Y(x,\psi;q) = y_0(x,\psi) + \sum_{n=1}^{+\infty} y_n(x,\psi)\, q^n, \quad (15.24)$$

$$\Delta(q) = \mu_0 + \sum_{n=1}^{+\infty} \mu_n\, q^n, \quad (15.25)$$

$$\Gamma(q) = \gamma_0 + \sum_{n=1}^{+\infty} \gamma_n\, q^n, \quad (15.26)$$

其中

$$y_n(x,\psi) = \mathcal{D}_n\left[Y(x,\psi;q)\right] = \frac{1}{n!} \left.\frac{\partial^n Y(x,\psi;q)}{\partial q^n}\right|_{q=0}$$

和

$$\mu_n = \mathcal{D}_n\left[\Delta(q)\right], \quad \gamma_n = \mathcal{D}_n\left[\Gamma(q)\right]$$

分别为 $Y(x,\psi;q)$，$\Delta(q)$ 和 $\Gamma(q)$ 所谓的同伦导数，$\mathcal{D}_n$ 为 n 阶同伦导数算子. 众所周知，幂级数通常具有有限的收敛半径. 幸运的是，零阶形变方程 (15.15) 和 (15.16) 包含收敛控制参数 c_0，可用于调整和控制级数解的收敛半径. 此外，我们有很大的自由来选择辅助线性算子 $\mathcal{L}$ 和 $\mathcal{L}_b$. 假设所有这些都选择恰当，则上述同伦 – Maclaurin 级数在 $q=1$ 时绝对收敛. 然后，由于 (15.22)，我们有同伦级数解

$$y(x,\psi) = y_0(x,\psi) + \sum_{n=1}^{+\infty} y_n(x,\psi), \quad (15.27)$$

$$\mu = \mu_0 + \sum_{n=1}^{+\infty} \mu_n, \quad (15.28)$$

$$\gamma = \gamma_0 + \sum_{n=1}^{+\infty} \gamma_n. \quad (15.29)$$

15.3.3 高阶形变方程

未知数 $y_n(x,\psi)$ 的控制方程和边界条件可以直接从零阶形变方程 (15.15) 至 (15.18) 推导得出. 将级数 (15.24) 至 (15.26) 代入零阶形变方程 (15.15) 至 (15.18), 然后将 q 的相同次幂的系数取等式, 可以得到 n 阶形变方程

$$\mathcal{L}[\,y_n(x,\psi) - \chi_n\, y_{n-1}(x,\psi)\,] = c_0\, \delta_{n-1}(x,\psi), \tag{15.30}$$

满足边界条件

$$\mathcal{L}_b(\,y_n - \chi_n\, y_{n-1}\,) = c_0\, \delta_{n-1}^b(x), \quad \psi = 0, \tag{15.31}$$

$$\frac{\partial y_n}{\partial x} = 0, \quad \psi \to +\infty \tag{15.32}$$

和波高条件

$$y_n(0,0) - y_n(\pi,0) = 0, \tag{15.33}$$

其中

$$\chi_k = \begin{cases} 0, & k \leqslant 1, \\ 1, & k > 1, \end{cases} \tag{15.34}$$

以及

$$\begin{aligned}\delta_k = {}& \sum_{i=0}^{k} \frac{\partial^2 y_{k-i}}{\partial x^2} \sum_{j=0}^{i} \frac{\partial y_{i-j}}{\partial \psi} \frac{\partial y_j}{\partial \psi} - 2 \sum_{i=0}^{k} \frac{\partial y_{k-i}}{\partial x} \sum_{j=0}^{i} \frac{\partial^2 y_{i-j}}{\partial x \partial \psi} \frac{\partial y_j}{\partial \psi} + \frac{\partial^2 y_k}{\partial \psi^2} \\ & + \sum_{i=0}^{k} \frac{\partial^2 y_{k-i}}{\partial \psi^2} \sum_{j=0}^{i} \frac{\partial y_{j-i}}{\partial x} \frac{\partial y_i}{\partial x} - \Omega(\psi) \sum_{i=0}^{k} \frac{\partial y_{k-i}}{\partial \psi} \sum_{j=0}^{i} \frac{\partial y_{j-i}}{\partial \psi} \frac{\partial y_i}{\partial \psi},\end{aligned} \tag{15.35}$$

$$\delta_k^b = \mu_k + \sum_{i=0}^{k} \mu_{k-i} \sum_{j=0}^{i} \frac{\partial y_{i-j}}{\partial x} \frac{\partial y_j}{\partial x} + 2 \sum_{i=0}^{k} (y_{k-i} - \gamma_{k-i}) \sum_{j=0}^{i} \frac{\partial y_{i-j}}{\partial \psi} \frac{\partial y_j}{\partial \psi}. \tag{15.36}$$

由定理 4.1 获得. 方程 (15.30) 和 (15.31) 可以通过定理 4.15 直接得到. 详情请参考第四章和成钧等人 [329].

需要注意的是, 高阶形变方程包含两个辅助线性算子 $\mathcal{L}$ 和 $\mathcal{L}_b$. 幸运的是, 在同伦分析方法的框架下, 我们有极大的自由来选择辅助线性算子, 如第

十四章所述. 注意, 偏微分方程 (15.9) 只包含一个线性项 $y_{\psi\psi}$. 然而, 如果我们选择

$$\mathcal{L}[y(x,\psi)] = \frac{\partial^2 y(x,\psi)}{\partial \psi^2}$$

为辅助线性算子, 将得到一个以 ψ 的幂级数表示的解 $y(x,\psi)$, 这显然不符合解表达 (15.14). 由于幂级数通常具有有限的收敛半径, 因此此类解很难满足 $\psi \to +\infty$ 的边界条件 (15.11). 因此, 如果遵循摄动方法的传统思想, 即高度重视非线性控制方程的线性项, 那么原始偏微分方程 (15.9) 的线性项可能会极大地误导我们. 幸运的是, 同伦分析方法为我们提供了极大的自由来选择辅助线性算子, 因此我们可以完全忽略 (15.9) 中的线性项 $y_{\psi\psi}$ 并主要基于从物理角度考虑得到的解表达 (15.14) 来选择适当的辅助线性算子 $\mathcal{L}$. 注意, $u = \exp(-\psi)\cos(x)$ 满足

$$\frac{\partial^2 u}{\partial \psi^2} + \frac{\partial^2 u}{\partial x^2} = 0.$$

因此, 为了符合解表达 (15.14), 我们选择辅助线性算子

$$\mathcal{L}(u) = \frac{\partial^2 u}{\partial \psi^2} + \frac{\partial^2 u}{\partial x^2}. \tag{15.37}$$

注意, 边界条件 (15.10) 不包含任何线性项. 类似地, 为了符合解表达 (15.14), 对于非线性边界条件 (15.10) 我们选择辅助线性算子

$$\mathcal{L}_b(u) = u + \frac{\partial u}{\partial \psi}. \tag{15.38}$$

需要强调的是, 上述两个辅助线性算子 (15.37) 和 (15.38) 分别与控制方程 (15.9) 和边界条件 (15.10) 几乎没有任何联系! 这主要是因为同伦分析方法为我们提供了极大的自由来选择辅助线性算子 $\mathcal{L}$ 和 $\mathcal{L}_b$ 来满足解表达 (15.14). 正如前几章提到的, 这种自由是同伦分析方法相对于其他解析方法的一个明显优势, 可以极大地简化某些非线性问题的求解.

此外, 同伦分析方法还为我们提供了选择初始猜测解的极大自由. 初始猜测解 $y_0(x,\psi)$ 也应满足解表达 (15.14) 以及波高的定义 (15.13). 因此, 我们不妨选择初始猜测解

$$y_0(x,\psi) = -\psi + \frac{H}{2}\exp(-\psi)\,\cos x. \tag{15.39}$$

注意, 到目前为止, 初始解 μ_0 和 γ_0 是未知的.

在上面的 n 阶形变方程 (15.30) 至 (15.33) 中, 右端项 $\delta_{n-1}(x,\psi)$ 和 $\delta_{n-1}^b(x)$ 仅依赖于 $y_k(x,\psi), \mu_k$ 和 γ_k, 其中 $0 \leqslant k \leqslant n-1$, 所以它们都被认为是已知项.

因此,根据两个辅助线性算子 $\mathcal{L}$ 和 $\mathcal{L}_b$ 的定义 (15.37) 和 (15.38),得到高阶形变方程 (15.30) 是线性的,且在已知边界 $\psi=0$ 和 $\psi\to+\infty$ 上分别满足线性边界条件 (15.31) 和 (15.32).这种在固定区域内的线性边值问题比原始的具有变系数 $\exp(-\psi)$ 的强非线性偏微分方程 (15.9) 简单得多,因此更容易求解.

15.3.4 连续求解过程

需要注意的是,n 阶形变方程 (15.30) 至 (15.33) 包含三个未知数,$y_n(x,\psi)$,μ_{n-1} 和 γ_{n-1}.由于辅助线性算子 $\mathcal{L}$ 的定义 (15.37) 对于任何正整数 M 有

$$\mathcal{L}\left[\sum_{m=1}^{M} C_{n,m}\,\exp(-m\psi)\cos(mx)\right]=0,$$

其中 $C_{n,m}$ 为常数.因此,高阶形变方程 (15.30) 的通解为

$$y_n(x,\psi)=\chi_n\, y_{n-1}(x,\psi)+y_n^*(x,\psi)+\sum_{m=1}^{M} C_{n,m}\,\exp(-m\psi)\cos(mx),\quad(15.40)$$

其中 M 为待定的正整数,

$$y_n^*(x,\psi)=c_0\,\mathcal{L}^{-1}\left[\delta_{n-1}(x,\psi)\right] \quad (15.41)$$

为 (15.30) 的特解.这里,$\mathcal{L}^{-1}$ 是 $\mathcal{L}$ 的逆算子,定义如下

$$\mathcal{L}^{-1}\left[\exp(-m\psi)\cos(nx)\right]=\frac{\exp(-m\psi)\cos(nx)}{(m^2-n^2)},\quad m\neq n \quad(15.42)$$

$$\mathcal{L}^{-1}\left[\exp(-m\psi)\sin(nx)\right]=\frac{\exp(-m\psi)\sin(nx)}{(m^2-n^2)},\quad m\neq n. \quad(15.43)$$

使用上面的公式,很容易得到 (15.30) 的特解 $y_n^*(x,\psi)$,特别是借助科学计算软件,如 Mathematica, Maple 等.

为了确定未知的 μ_{n-1},γ_{n-1} 和 $C_{n,m}$,我们将表达式 (15.40) 代入边界条件 (15.31),即

$$\mathcal{L}_b\left[y_n^*+\sum_{m=1}^{M} C_{n,m}\,\exp(-m\psi)\cos(mx)\right]=c_0\,\delta_{n-1}^b(x),\quad \psi=0,\quad(15.44)$$

其中 M 待定.上述方程可以改写为

$$\sum_{m=2}^{M}(1-m)\,C_{n,m}\cos(mx)=\sum_{m=0}^{2n+1} B_{n,m}(\gamma_{n-1},\mu_{n-1})\cos(mx),\quad(15.45)$$

其中系数 $B_{n,m}(\gamma_{n-1}, \mu_{n-1})$ 由 (15.44) 的右端项 $c_0\, \delta_{n-1}^b(x)$ 确定. 平衡上述等式的两边, 我们有

$$M = 2n+1, \quad C_{n,m} = \frac{B_{n,m}(\gamma_{n-1}, \mu_{n-1})}{1-m}, \quad 1 < m \leqslant 2n+1,$$

以及

$$B_{n,0}(\gamma_{n-1}, \mu_{n-1}) = 0, \quad B_{n,1}(\gamma_{n-1}, \mu_{n-1}) = 0, \tag{15.46}$$

这恰好为我们提供了两个代数方程来确定未知的 γ_{n-1} 和 μ_{n-1}.

至此, 只有系数 $C_{n,1}$ 是未知的. 将 (15.40) 代入 (15.33) 得到关于 $C_{n,1}$ 的代数方程

$$y_n^*(0,0) - y_n^*(\pi,0) + \sum_{m=1}^{2n+1} C_{n,m} - \sum_{m=1}^{2n+1} C_{n,m}\cos(m\pi) = 0, \tag{15.47}$$

其解为

$$C_{n,1} = -\frac{1}{2}\left\{ y_n^*(0,0) - y_n^*(\pi,0) + \sum_{m=2}^{2n+1}[1-(-1)^m]C_{n,m} \right\}. \tag{15.48}$$

因此, 所有的未知量都已确定, 从而问题得到解决.

这样, 我们仅通过代数计算依次得到 $n = 1, 2, 3, \cdots$ 的 $y_n(x, \psi)$, μ_{n-1} 以及 γ_{n-1}. 例如, 求解一阶形变方程, 我们有

$$\gamma_0 = \frac{1}{320}\left[160 + 25\, H^2 + \varepsilon\left(80 + 3\, H^2\right)\right], \tag{15.49}$$

$$\mu_0 = \frac{1}{8+H^2}\left[8 - \frac{3}{4}H^2 + \frac{5}{32}H^4 + \varepsilon\left(4 - \frac{1}{60}H^2 + \frac{3}{160}H^4\right)\right] \tag{15.50}$$

和

$$y_1(x, \psi) = c_0\left[A_{1,0}(\psi) + A_{1,1}(\psi)\cos x + A_{1,2}(\psi)\cos 2x + A_{1,3}(\psi)\cos 3x\right], \tag{15.51}$$

其中

$$A_{1,0}(\psi) = -\frac{H^2}{8}\exp(-2\psi) + \varepsilon\exp(-\psi) + \frac{\varepsilon H^2}{24}\exp(-3\psi),$$

$$A_{1,1}(\psi) = \frac{3H^3}{64}\exp(-\psi) - \frac{H^3}{64}\exp(-3\psi) - \frac{\varepsilon H}{2}\exp(-\psi)$$

$$- \frac{9\varepsilon H^3}{2240}\exp(-\psi) + \frac{\varepsilon H}{2}\exp(-2\psi) + \frac{\varepsilon H^3}{160}\exp(-4\psi),$$

$$A_{1,2}(\psi) = -\frac{H^2(256+40H^2-5H^4)}{128(8+H^2)}\exp(-2\psi)$$
$$-\frac{\varepsilon H^2(384+136H^2-9H^4)}{1920(8+H^2)}\exp(-2\psi)+\frac{3\varepsilon H^2}{40}\exp(-3\psi),$$
$$A_{1,3}(\psi) = -\frac{H^3}{32}\exp(-3\psi)-\frac{3\varepsilon H^3}{448}\exp(-3\psi)+\frac{\varepsilon H^3}{224}\exp(-4\psi).$$

同理，我们可以得到 $\gamma_1, \mu_1, y_2(x,\psi)$ 等．我们发现，所有的解 $y_n(x,\psi)$ 都包含 $\exp(-m\psi)$，其中 $m \geqslant 1$，所以自动满足边界条件 (15.32)．只要得到 $y(x,\psi)$ 收敛的级数解，就可以分别由 (15.7) 和 (15.12) 求得速度场和波高．注意，无量纲波速 $k\,c^2/g$ 由 μ 给出，自由表面的能量由 γ 给出．

我们发现，$y(x,\psi)$ 可以表示为

$$y(x,\psi) = y_c(\psi) + y_w(x,\psi) + y_i(x,\psi),$$

其中

$$y_c(\psi) = y(x,\psi)|_{H=0}$$

为无波情况下纯流解，

$$y_w(x,\psi) = y(x,\psi)|_{\varepsilon=0}$$

为无流情况下纯波解，$y_i(x,\psi)$ 是与波流相互作用相关的解．

更多详情请参考成钧、仓杰和廖世俊 [329]．

15.4 同伦近似

根据上述公式，我们开发了相应的 Mathematica 程序，见附录 15.1．注意，$y(x,\psi)$ 和 $\mu = kc^2/g$，$\gamma = k\,Q/g$ 的同伦级数包含所谓的收敛控制参数 c_0，它为我们提供了一种便捷的方式来保证同伦级数解的收敛性，如下所述．

通过控制方程的平均平方残差

$$E_m = \frac{1}{(1+N_x)(1+N_\psi)}\sum_{m=0}^{N_x}\sum_{j=0}^{N_\psi}\left\{\mathcal{N}\left[\sum_{n=0}^{m}y_n(x_i,\psi_j)\right]\right\}^2 \tag{15.52}$$

和非线性边界条件的平均平方残差

$$E_m^b = \frac{1}{1+N_x}\sum_{i=0}^{N_x}\left\{\mathcal{N}_b\left[\sum_{n=0}^{m}y_n(x_i,0),\sum_{n=0}^{m}\mu_n,\sum_{n=0}^{m}\gamma_n\right]\right\}^2 \tag{15.53}$$

来表示 m 阶近似的精度, 其中

$$x_i = i\left(\frac{\pi}{N_x}\right), \quad \psi_j = j\left(\frac{2\pi}{N_\psi}\right),$$

$N_x = N_\psi = 10$. 从物理的角度来看, 波速场随水深的增加呈指数衰减. 所以, 在一个波长 2π 以下的水深中控制方程的残差很小, 可以忽略不计.

不失一般性, 让我们首先考虑 $H = 3/10$ 和 $\varepsilon = 1/5$ 的情况. 控制方程和非线性边界条件相对于收敛控制参数 c_0 的平均平方残差 E_m 和 E_m^b, 分别如图 15.1 和图 15.2 所示. 注意, 随着 m 的增加, E_m 和 E_m^b 均在区间 $-0.6 \leqslant c_0 < 0$

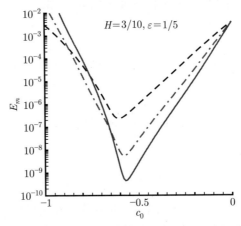

图 15.1 当 $H = 3/10$ 和 $\varepsilon = 1/5$ 时, 控制方程 (15.9) 相对于 c_0 的平方残差. 虚线: 5 阶同伦近似; 点划线: 8 阶同伦近似; 实线: 10 阶同伦近似

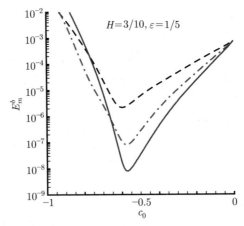

图 15.2 当 $H = 3/10$ 和 $\varepsilon = 1/5$ 时, 边界条件 (15.10) 相对于 c_0 的平方残差. 虚线: 5 阶同伦近似; 点划线: 8 阶同伦近似; 实线: 10 阶同伦近似

15.4 同伦近似

表 15.1 当 $H = 3/10$ 和 $\varepsilon = 1/5$ 时,控制方程 (15.9) 平方残差的最小值和相应的最优收敛控制参数 c_0

近似阶数 m	E_m 平方残差的最小值	c_0 的最优值
5	2.3×10^{-7}	-0.6049
8	5.5×10^{-9}	-0.5775
10	4.4×10^{-10}	-0.5686
15	1.3×10^{-10}	-0.5569

表 15.2 当 $H = 3/10$ 和 $\varepsilon = 1/5$ 时,边界条件 (15.10) 平方残差的最小值和相应的最优收敛控制参数 c_0

近似阶数 m	E_m^b 平方残差的最小值	c_0 的最优值
5	2.2×10^{-6}	-0.5976
8	8.1×10^{-8}	-0.5749
10	7.9×10^{-9}	-0.5723
15	5.7×10^{-11}	-0.5707

表 15.3 当 $H = 3/10$ 和 $\varepsilon = 1/5$ 时,选取 $c_0 = -11/20$,μ,γ 的 m 阶同伦近似和对应的平方残差 E_m,E_m^b

m	E_m	E_m^b	$\mu = k\,c^2/g$	$\gamma = k\,Q/g$
3	7.2×10^{-6}	3.6×10^{-5}	0.81670000	0.40780000
5	4.4×10^{-7}	3.2×10^{-6}	0.81310000	0.40420000
10	6.0×10^{-10}	1.0×10^{-8}	0.81420000	0.40530000
15	1.4×10^{-12}	8.3×10^{-11}	0.81440000	0.40540000
20	7.2×10^{-15}	7.4×10^{-13}	0.81442000	0.40556000
25	9.4×10^{-17}	5.6×10^{-15}	0.81442100	0.40556500
30	2.3×10^{-18}	3.5×10^{-17}	0.81442133	0.40556500
35	5.3×10^{-20}	1.6×10^{-19}	0.81442133	0.40556503
40	1.1×10^{-21}	5.2×10^{-22}	0.81442133	0.40556503

内减小. 此外, 由 E_m 和 E_m^b 的最小值给出 c_0 的最优值约为 -0.55, 如表 15.1 和表 15.2 所示. 这确实是正确的: 当 $H = 3/10$ 和 $\varepsilon = 1/5$ 时, 选取 $c_0 = -11/20$ 时的同伦近似收敛得很快, 如表 15.3 所示. 注意, 控制方程和非线性边界条件的平均平方残差分别在 40 阶近似值处单调减小到 1.1×10^{-21} 和 5.2×10^{-22}. 此外, 我们获得了收敛值 $\mu = 0.81442133$ 和 $\gamma = 0.40556503$. 通过同伦-帕德加速技术, 我们得到了更精确的收敛值 $\mu = 0.814421334285$ 和 $\gamma = 0.405565029876$, 如表 15.4 所示. 更进一步, 我们发现即使是波浪超高的 3 阶同伦近似也足够精确, 如图 15.3 所示. 所有这些都证明了我们确实可以通过上述同伦分析方法来得到满足非线性边界条件 (15.10) 的复杂非线性偏微分方程 (15.9) 的高精度解

表 15.4 当 $H = 3/10$ 和 $\varepsilon = 1/5$ 时, μ 和 γ 的 $[m, m]$ 阶同伦-帕德近似

m	$\mu = k\,c^2/g$	$\gamma = k\,Q/g$
2	0.802900000000	0.403500000000
4	0.815400000000	0.405600000000
6	0.814421000000	0.405560000000
8	0.814421000000	0.405565000000
10	0.814421330000	0.405565020000
12	0.814421334000	0.405565029000
14	0.814421334280	0.405565029870
16	0.814421334285	0.405565029876
18	0.814421334285	0.405565029876
20	0.814421334285	0.405565029876

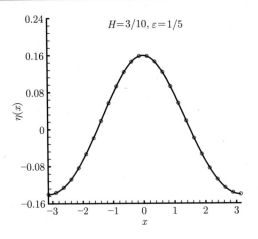

图 15.3 当 $H = 3/10$ 和 $\varepsilon = 1/5$ 时, 选取 $c_0 = -11/20$ 时的波浪超高. 实线: 15 阶同伦近似; 空心圆: 3 阶同伦近似

析近似. 因此, 给定 H 和 ε 在物理上的合理值, 我们可以通过附录 15.1 中给出的 Mathematica 程序以类似的方式获得收敛的同伦近似.

在讨论波流相互作用之前, 首先考虑两个极限情况: 纯水波, 对应于零涡量 $\Omega = 0$; 纯来流, 对应于零波高 $H = 0$. 当 $\Omega = 0$ 时, 为 Stokes 深水波模型 [361]. 根据 Stokes 理论 [357], 当波高达到最大值 $H = 0.886$ 时, 一列传播的深水波破裂, 对应的最大波峰角 $\alpha = 120$ 度. 我们发现, 即使对于接近极限波的波幅较大的波, 我们的 $[m,m]$ 阶同伦–帕德近似值 $\mu = k\,c^2/g$ 与波浪超高 $\eta_c = \eta(0)$ 都是收敛的, 如表 15.5 和表 15.6 所示. 在表 15.6 的最后一列给出了 η_c 与线性估计波峰高程 $\eta_c \approx H/2$ 的相对误差. 我们发现, 对于波幅较大的波的相对误差在 25% 左右. 此外, 波速的同伦–帕德近似与 Schwartz [360] 和 Longuet-Higgins [356] 给出的解析结果非常吻合, 如图 15.4 所示. 注意, 如表 15.5 和表 15.6 所示, 基于同伦分析方法给出的最大波高约为 $H = 0.8825$, 略小于由 Stokes 理论 [357] 给出的 $H = 0.886$.

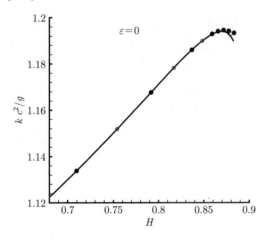

图 15.4 纯深水波的色散关系比较. 实线: [23,23] 阶同伦–帕德近似; 实心圆: Longuet-Higgins [356] 的结果; 空心圆: Schwartz [360] 的结果

表 15.5 无流情况下纯深水波 $\mu = k\,c^2/g$ 的 $[m,m]$ 阶同伦–帕德近似, 对应于 $\Omega = 0$

H	$m=15$	$m=17$	$m=19$	$m=20$	$m=21$
0.6500	1.1113	1.1113	1.1113	1.1113	1.1113
0.7000	1.1300	1.1300	1.1300	1.1300	1.1300
0.7500	1.1502	1.1502	1.1502	1.1502	1.1502
0.8000	1.1712	1.1712	1.1712	1.1712	1.1712

续表

H	$m=15$	$m=17$	$m=19$	$m=20$	$m=21$
0.8500	1.1905	1.1904	1.1905	1.1904	1.1904
0.8600	1.1926	1.1929	1.1930	1.1930	1.1930
0.8700	1.1939	1.1940	1.1941	1.1941	1.1941
0.8750	1.1933	1.1932	1.1936	1.1936	1.1936
0.8800	1.1915	1.1909	1.1917	1.1916	1.1916
0.8820	1.1902	1.1892	1.1902	1.1902	1.1902
0.8825	1.1898	1.1887	1.1898	1.1898	1.1897

表 15.6 与 $H/2$ 相比,无流情况下纯深水波波浪超高 η_c 的 $[m,m]$ 阶同伦-帕德近似,对应于 $\Omega=0$

H	$m=15$	$m=17$	$m=19$	$m=20$	$1-H/2\eta_c$
0.65	0.3875	0.3875	0.3875	0.3875	16.13%
0.7	0.4250	0.4250	0.4250	0.4250	17.65%
0.75	0.4649	0.4649	0.4649	0.4649	19.34%
0.8	0.5080	0.5079	0.5079	0.5079	21.24%
0.85	0.5566	0.5566	0.5565	0.5565	23.63%
0.86	0.5675	0.5672	0.5672	0.5671	24.18%
0.87	0.5793	0.5805	0.5798	0.5800	25.00%
0.875	0.5855	0.5858	0.5856	0.5858	25.32%
0.88	0.5923	0.5922	0.5922	0.5922	25.70%
0.882	0.5952	0.5953	0.5949	0.5949	25.87%
0.8825	0.5956	0.5956	0.5956	0.5956	25.93%

需要注意的是, $H=0$ 对应于无波的纯来流. 当 $c_0=-1$ 时, 纯来流的五阶同伦近似为

$$y(\psi) = -\psi - \varepsilon \exp(-\psi) + \frac{3}{4}\varepsilon^2 \exp(-2\psi) - \frac{5}{6}\varepsilon^3 \exp(-3\psi)$$
$$+ \frac{35}{32}\varepsilon^4 \exp(-4\psi) - \frac{63}{40}\varepsilon^5 \exp(-5\psi) + \cdots \qquad (15.54)$$

根据 (15.2), 我们得到了静止坐标系中纯来流的速度分布

$$u_c(\psi) = -\varepsilon \exp(-\psi) + \frac{1}{2}\varepsilon^2 \exp(-2\psi) - \frac{1}{2}\varepsilon^3 \exp(-3\psi) + \frac{5}{8}\varepsilon^4 \exp(-4\psi)$$
$$-\frac{7}{8}\varepsilon^5 \exp(-5\psi) + \frac{21}{16}\varepsilon^6 \exp(-6\psi), \tag{15.55}$$

这与摄动解 [366, 367] 完全相同. 以上两个极限情况表明了同伦分析方法对这种复杂的非线性偏微分方程的有效性.

对于 $\varepsilon \neq 0$ 和 $H \neq 0$ 的一般情况是最令人感兴趣的. 波速的 $[2,2]$ 阶同伦-帕德近似为

$$c^2/c_0^2 = \frac{\sum_{i=0}^{8} \alpha_i(H)\varepsilon^i}{1 + \sum_{j=0}^{8} \beta_j(H)\varepsilon^j}, \tag{15.56}$$

其中

$$\alpha_0 = 1 - 18.3545H^2 - 19.0783H^4 - 9.5071H^6 - 1.6706H^8$$
$$- 0.1686H^{10} - 0.3944H^{12} + 0.1497H^{14},$$

$$\alpha_1 = -19.8095H^{-2} + 94.9902 + 79.8629H^2 + 49.2679H^4 + 16.4719H^6$$
$$+ 1.3127H^8 + 0.7249H^{10} - 0.5257H^{12},$$

$$\alpha_2 = -0.8127H^{-4} - 153.8798H^{-2} - 17.4780 + 8.4334H^2 - 5.1748H^4$$
$$- 4.1199H^6 - 2.9438H^{10} + 0.7047H^{10},$$

$$\alpha_3 = 6.6032H^{-4} - 161.7258H^{-2} - 95.2033 - 20.2378H^2 + 3.0971H^4$$
$$+ 2.7794H^6 - 1.2655H^8 + 0.8404H^{10},$$

$$\alpha_4 = 5.2780H^{-4} - 79.0493H^{-2} - 48.4005 - 17.5988H^2 - 5.1302H^4$$
$$+ 0.4472H^6 + 0.4421H^8 + 0.0306H^{10},$$

$$\alpha_5 = 5.9728H^{-4} - 28.6492H^{-2} - 22.03866 - 3.7271H^2 + 3.03746H^4$$
$$+ 2.7544H^6 + 0.9785H^8 - 0.1480H^{10},$$

$$\alpha_6 = 0.4561H^{-4} - 4.9730H^{-2} - 6.5921 - 4.9442H^2 - 2.6846H^4$$
$$- 0.9577H^6 - 0.2226H^8 + 0.0393H^{10},$$

$$\alpha_7 = -1.0606H^{-4} - 2.26127H^{-2} - 2.5945 - 1.6613H^2 - 0.6098H^4$$
$$- 0.1298H^6 - 0.01597H^8 + 0.0014H^{10},$$

$$\alpha_8 = 0.4849H^{-4} + 1.2882H^{-2} + 1.1771 + 0.5258H^2 + 0.1341H^4$$
$$+ 0.0232H^6 + 0.0034H^8 + 0.0004H^{10}$$

和

$$\beta_0 = -18.6152H^2 - 14.3027H^4 - 4.7494H^6 + 0.3693H^8$$
$$- 0.1756H^{10} - 0.6505H^{12} - 0.1882H^{14},$$

$$\beta_1 = -19.8095H^{-2} + 101.3902 + 30.9370H^2 + 5.9367H^4 - 1.9934H^6$$
$$+ 0.5200H^8 + 3.2108H^{10} + 1.2336H^{12},$$

$$\beta_2 = -0.8127H^{-4} - 180.3940H^{-2} + 152.2704 + 118.3418H^2 + 26.6539H^4$$
$$- 5.0904H^6 - 9.6998H^8 - 3.3617H^{10},$$

$$\beta_3 = 5.5196H^{-4} - 364.0214H^{-2} - 98.3501 + 54.1467H^2 + 29.9964H^4$$
$$+ 5.5684H^6 - 1.2260H^8 - 0.4686H^{10},$$

$$\beta_4 = 14.4435H^{-4} - 241.5237H^{-2} - 201.5196 - 46.7589H^2 + 12.6025H^4$$
$$+ 11.9062H^6 + 2.9939H^8 + 0.1049H^{10},$$

$$\beta_5 = 10.2959H^{-4} - 54.8513H^{-2} - 80.1939 - 46.4665H^2 - 13.8617H^4$$
$$- 1.5235H^6 + 0.18215H^8 + 0.0065H^{10},$$

$$\beta_6 = 6.7965H^{-4} + 6.2383H^{-2} - 5.5078 - 9.7100H^2 - 5.4381H^4$$
$$- 1.4762H^6 - 0.17256H^8 - 0.0018H^{10},$$

$$\beta_7 = 1.28H^{-4} + 6.4458H^{-2} + 7.2395 + 3.6241H^2 + 0.9258H^4$$
$$+ 0.1129H^6 + 0.0041H^8 - 0.0002H^{10},$$

$$\beta_8 = -1.2567H^{-4} - 1.6932H^{-2} - 0.9140 - 0.2487H^2 - 0.0366H^4$$
$$- 0.0037H^6 - 0.0006H^8.$$

如图 15.5 至图 15.7 所示, 上述 [2,2] 阶同伦-帕德近似即使对于大波高 H 也足够精确, 这不仅表明了一般表达式 (15.56) 的精度, 也表明了对于大幅值波的同伦级数解的收敛性. 注意, 五阶摄动结果 [366] 只对小幅值的波有效 (如 $H = 0.1$), 并且随着波高 H 的增加变得越来越不精确, 如图 15.5 至图 15.7 所示. 因此, 对于大波高 ($H \geqslant 0.3$), 摄动近似通常会高估助流和逆流的波速.

对于给定的 ε, 我们可以通过同伦-帕德方法以类似的方式获得色散关系. 当 $\varepsilon = -0.25, -0.15, 0, 0.25, 0.5$ 时, $\mu = k\,c^2/g$ 和波高 H 的曲线关系, 如

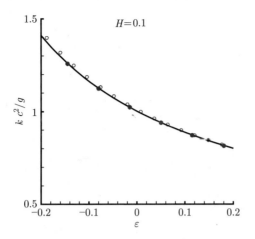

图 15.5 当 $H = 0.1$ 时, 涡量参数 ε 对深水波速 $k\,c^2/g$ 的影响. 实线: $[2,2]$ 阶同伦–帕德近似; 实心圆: $[8,8]$ 阶同伦–帕德近似; 空心圆: 5 阶摄动解 [366]

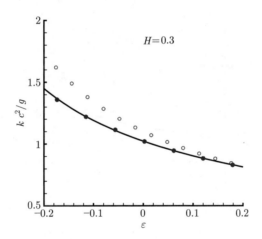

图 15.6 当 $H = 0.3$ 时, 涡量参数 ε 对深水波速 $k\,c^2/g$ 的影响. 实线: $[2,2]$ 阶同伦–帕德近似; 实心圆: $[8,8]$ 阶同伦–帕德近似; 空心圆: 5 阶摄动解 [366]

图 15.8 所示. 注意, 除非常接近波高最大值的几个点外, $[8,8]$ 和 $[12,12]$ 阶同伦–帕德近似吻合良好. 与纯波 ($\varepsilon = 0$) 相比, 同向的指数剪切流 ($\varepsilon < 0$) 倾向于增大给定波高 H 的波速, 但反向的指数剪切流 ($\varepsilon > 0$) 倾向于将其减小. 特别地, 如表 15.7 和图 15.9 所示, 最大波高强烈依赖于来流的涡量: 同向的指数剪切流 ($\varepsilon < 0$) 倾向于减小最大波高, 但反向的指数剪切流 ($\varepsilon > 0$) 倾向于将其增大. 注意, 无流情况下的纯行进波的最大波高约为 $H = 0.8825$. 然而, 在反向的指数剪切流 ($\varepsilon > 0$) 的情况下, 波高可能大于纯波的极限值, 如表 15.7

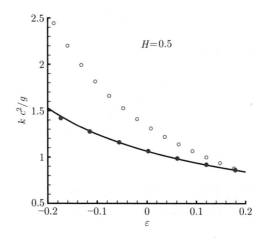

图 15.7 当 $H = 0.5$ 时, 涡量参数 ε 对深水波速 $k\,c^2/g$ 的影响. 实线: $[2,2]$ 阶同伦–帕德近似; 实心圆: $[8,8]$ 阶同伦–帕德近似; 空心圆: 5 阶摄动解 [366]

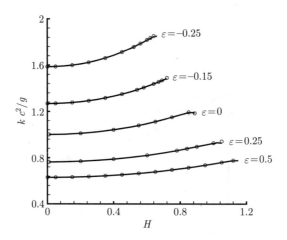

图 15.8 当 $\varepsilon = -0.25, -0.15, 0, 0.25, 0.5$ 时, 波浪在具有呈指数分布涡量 $\Omega = \varepsilon \exp(-\psi)$ 的来流上的色散关系. 实线: $[12, 12]$ 阶同伦–帕德近似; 空心圆: $[8, 8]$ 阶同伦–帕德近似

和图 15.9 所示. 例如, 对于 $0.25 \leqslant \varepsilon \leqslant 0.5$ 的反向指数来流, 当 $H = 0.92$ 时, 我们可以得到收敛的级数解, 这甚至高于 Stokes 理论给出的在静水上纯波的 $H = 0.886$ (见 [357]). 这里需要强调的是, 根据 Stokes 理论, 对于静水上的纯波 ($\varepsilon = 0$), $H = 0.92$ 的行进波不存在. 然而, 对于 $\varepsilon = 0.25$ 时的反向剪切流, 即使 $H = 1.02$, 我们也可以获得收敛的级数解, 如图 15.10 所示. 因此, 基于同伦近似, 在反向剪切流 ($\varepsilon > 0$) 上的行进波的最大波高可能大于在静水中的纯波的最大波高. 注意, 当 $H = 1.02$ 和 $\varepsilon = 0.25$ 时的波浪形状甚至比静水

中 $H = 0.8825$ 的波浪更陡, 如图 15.10 所示. 对于更多的物理讨论, 请参考成钧等人 [329].

表 15.7 具有呈指数分布涡量 $\Omega = \varepsilon \exp(-\psi)$ 的来流上行进波的最大波高

ε	最大波高
-0.25	0.6700
0	0.8825
0.25	1.0400
0.50	1.1700

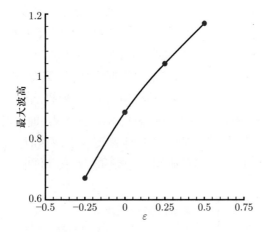

图 15.9 具有呈指数分布涡量 $\Omega = \varepsilon \exp(-\psi)$ 的剪切流的最大波高和 ε 的关系

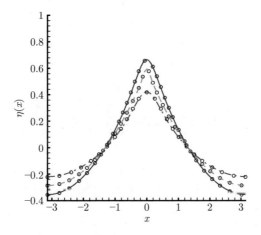

图 15.10 选取 $c_0 = -1/2$ 时, 波浪超高的近似值. 实线: 当 $H = 1.02$ 和 $\varepsilon = 0.25$ 时的 25 阶近似; 虚线: 当 $H = 0.8825$ 和 $\varepsilon = 0$ 时的 45 阶近似; 点划线: 当 $H = 0.64$ 和 $\varepsilon = -0.25$ 时的 25 阶近似; 空心圆: 20 阶同伦近似

正如 Stokes [360,361] 所建议的那样，当波峰处的流体速度等于波速时，静水中的行进波会破碎. 为了检验剪切流上波的这个标准，我们还计算了以速度 c 运动的坐标系中流体微团在波峰顶点处的动能，定义为

$$KE_c = \frac{1}{2}\mathbf{u}^2 = \frac{1}{\mu}[\gamma - y(0,0)]. \tag{15.57}$$

对于助流 ($\varepsilon = -0.25$), 静水 ($\varepsilon = 0$) 和逆流 ($\varepsilon = 0.25$), KE_c 相对于波高 H 的曲线，如图 15.11 所示. 注意，对于无流情况下静水中的纯波 ($\varepsilon = 0$), 当 $H \approx 0.8825$ 时, $KE_c \approx 0$, 正好对应于 Schwartz [360] 和 Longuet-Higgins [356] 给出的最大波高. 然而，当 $H \approx 0.67$ 时，对于助流上的波浪 ($\varepsilon = -0.25$) 和当 $H \approx 1.04$ 时，对于逆流上的波浪 ($\varepsilon = 0.25$), 我们有 $KE_c \approx 0$. 因此，波高最大值对应于 $KE_c = 0$, 即波峰处的流体速度等于波速. 注意，在 $\varepsilon = \pm 0.25$ 和 $\varepsilon = 0$ 的情况下, $KE_c \sim H$ 的三条曲线近似平行. 因此，定性地说，同向剪切流趋向于减小最大波高，而反向剪切流具有相反的效果. 因此，根据同伦分析方法给出的近似值, Stokes 波浪破碎问题具有一般意义，即使对于非均匀来流上的波浪仍然是正确的，即当波峰处的流体速度等于波速时，在剪切流上的一列行进波会发生破碎. 详情请参考成钧、仓杰和廖世俊 [329].

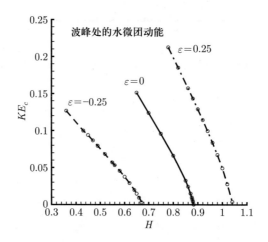

图 15.11 各种来流情况下的深水波波峰处的水微团动能 KE_c 与波高 H 的关系. 实心圆: $[11,11]$ 阶同伦–帕德近似; 实线: $[12,12]$ 阶同伦–帕德近似; 空心圆: $[15,15]$ 阶同伦–帕德近似; 虚线: $[20,20]$ 阶同伦–帕德近似

这个例子很好地表明了同伦分析方法可以用来求解一些复杂非线性偏微分方程，从而加深和丰富我们对一些复杂非线性现象的物理理解.

15.5 本章小结

本章应用同伦分析方法研究了周期性行进波与具有指数性质分布涡量的非均匀来流的非线性相互作用. 通过适当地选择一个简单的辅助线性算子, 将原来的变系数强非线性偏微分方程转化为无穷多个更简单的线性偏微分方程, 这些线性偏微分方程很容易求解. 如果通过最优收敛控制参数 c_0 和同伦-帕德方法得到收敛的级数解, 则对于给定的非均匀来流我们可以得到波高最大值. 这些解析近似揭示了周期性行进波与非均匀来流之间非线性相互作用的基本特征. 这说明了同伦分析方法可作为有用的解析工具来加深我们对某些复杂非线性现象的物理理解.

从数学角度, 这里值得强调的是, 两个辅助线性算子 (15.37) 和 (15.38) 与原始的偏微分方程 (15.9) 和非线性边界条件 (15.10) 几乎没有任何联系. 注意, 其他解析方法, 尤其是摄动方法, 强烈依赖于原始控制方程中的线性算子. 与其他解析方法不同, 同伦分析方法为我们选择辅助线性算子提供了极大的自由, 这种自由使我们可以根据给定非线性问题的物理背景来选择辅助线性算子. 所以, 在同伦分析方法的框架下, 我们不需要在给定非线性问题的控制方程和边界条件的细节上花太多时间, 而应该更多地关注它的物理背景, 从而找到合适的解表达. 这在本质上与其他解析方法完全不同. 注意, 正是同伦分析方法为我们提供的这种极大的自由, 可以极大地简化求解一些非线性问题的过程, 如本章和前几章所述. 此外, 如前几章所述, 这种极大的自由是建立在通过最优收敛控制参数 c_0 保证级数解收敛的基础上的: 如果相应的级数解是发散的, 那么辅助线性算子选择的自由没有任何意义.

从物理角度, 我们发现, 与静水中的波相比, 对于给定幅值的波在同向剪切流上传播得更快, 但在反向剪切流上传播得更慢. 此外, 与均匀来流不同的是, 同向剪切流倾向于使波峰变尖而使波谷变平滑, 而反向剪切流则具有相反的效果. 因此, 影响波形的不是来流的大小, 而是来流的不均匀性. 特别地, 我们发现 Stokes 波浪破碎问题即使对于非均匀来流的行进波仍然是正确的, 并且反向剪切流上的最高波甚至比静水上的最高波更高更陡.

所有这些都验证了同伦分析方法对于一些复杂的非线性偏微分方程的有效性. 这个例子表明同伦分析方法可以作为一种有用的工具来加深我们对一些复杂非线性现象的物理理解.

参考文献

[329] Cheng, J., Cang, J., Liao, S.J.: On the interaction of deep water waves and exponential shear currents. ZAMP. **60**, 450–478 (2009).

[330] Clamond, D.: Steady finite-amplitude waves on a horizontal seabed of arbitrary depth. J. Fluid Mech. **398**, 45–60 (1999).

[331] Clamond, D.: Cnoidal-type surface waves in deep water. J. Fluid Mech. **489**, 101–120 (2003).

[332] Clamond, D., Fructus, D.: Accurate simple approximation for the solitary wave. C. R. Mecanique. **331**, 727–732 (2003).

[333] Dalrymple, R.A.: A numerical model for periodic finite amplitude waves on a rotational fluid. J. Comput. Phys. **24**, 29–42 (1977).

[334] Drennan, W.M, Hui, W.H., Tenti, G.: Accurate calculations of Stokes water waves of large amplitude. J. App. Math. Phys. (ZAMP). **43**, 367–384 (1992).

[335] Dubreil-Jacotin, M.L.: Sur la détermination rigoureuse des ondes permanentes périodiques d'ampleur finie. J. Math. Pures et Appl. **13**, 217–291 (1934).

[336] Eastwood, J.W., Watson, C.J.H.: Implication of wave-current interaction for offshore design. E. and P. Forum, Paris (1989).

[337] Fenton, J.D.: A fifth-order Stokes theory for steady waves. J. Waterw. Port Coastal and Ocean Eng. **111**, 216–234 (1985).

[338] Kishida, N., Sobey, R.J.: Stokes theory for waves on linear shear current. J. Waterw. Port Coastal and Ocean Eng. **114**, 1317–1334 (1988).

[339] Li, Y.J., Nohara, B.T., Liao, S.J.: Series solutions of coupled van der Pol equation by means of homotopy analysis method. J. Mathematical Physics **51**, 063517 (2010). doi:10.1063/1.3445770.

[340] Liang, S.X., Jeffrey, D.J.: Comparison of homotopy analysis method and homotopy perturbation method through an evolution equation. Commun. Nonlinear Sci. Numer. Simulat. **14**, 4057–4064 (2009).

[341] Liao, S.J.: The Proposed Homotopy Analysis Technique for the Solution of Nonlinear Problems. PhD dissertation, Shanghai Jiao Tong University (1992).

[342] Liao, S.J.: A kind of approximate solution technique which does not depend upon small parameters—(II) An application in fluid mechanics. Int. J. Nonlin. Mech. **32**, 815–822 (1997).

[343] Liao, S.J.: An explicit, totally analytic approximation of Blasius viscous flow problems. Int. J. Nonlin. Mech. **34**, 759–778 (1999a).

[344] Liao, S.J.: A uniformly valid analytic solution of 2D viscous flow past a semi-infinite flat plate. J. Fluid Mech. **385**, 101–128 (1999b).

[345] Liao, S.J.: On the analytic solution of magnetohydrodynamic flows of non-Newtonian fluids over a stretching sheet. J. Fluid Mech. **488**, 189–212 (2003a).

[346] Liao, S.J.: Beyond Perturbation - Introduction to the Homotopy Analysis Method. Chapman & Hall/ CRC Press, Boca Raton (2003b).

[347] Liao, S.J.: On the homotopy analysis method for nonlinear problems. Appl. Math. Comput. **147**, 499–513 (2004).

[348] Liao, S.J.: A new branch of solutions of boundary layer flows over an impermeable stretched plate. Int. J. Heat Mass Tran. **48**, 2529–2539 (2005).

[349] Liao, S.J.: Series solutions of unsteady boundary-layer flows over a stretching flat plate. Stud. Appl. Math. **117**, 2529–2539 (2006).

[350] Liao, S.J.: Notes on the homotopy analysis method - Some definitions and theorems. Commun. Nonlinear Sci. Numer. Simulat. **14**, 983–997 (2009).

[351] Liao, S.J.: On the relationship between the homotopy analysis method and Euler transform. Commun. Nonlinear Sci. Numer. Simulat. **15**, 1421–1431 (2010a). doi:10.1016/j.cnsns.2009.06.008.

[352] Liao, S.J.: An optimal homotopy-analysis approach for strongly nonlinear differential equations. Commun. Nonlinear Sci. Numer. Simulat. **15**, 2003–2016 (2010b).

[353] Liao, S.J., Campo, A.: Analytic solutions of the temperature distribution in Blasius viscous flow problems. J. Fluid Mech. **453**, 411–425 (2002).

[354] Liao, S.J., Cheung, K.F.: Homotopy analysis of nonlinear progressive waves in deep water. Journal of Engineering Mathematics. **45**, No.2, 105–116 (2003).

[355] Liao, S.J., Tan, Y.: A general approach to obtain series solutions of nonlinear differential equations. Stud. Appl. Math. **119**, 297–355 (2007).

[356] Longuet-Higgins, M., Tanaka, M.: On the crest instabilities of steep surface waves. J. Fluid Mech. **336**, 51–68 (1997).

[357] Tulin, M. P., Li, J. J.: On the breaking of energetic waves. Int. J. Offshore Polar Eng. **2**, No.1, 46–53 (1992).

[358] Peregrine, D.H.: Interaction of water waves and current. Adv. Appl. Mech. **16**, 9–117 (1977).

[359] Phillips, O.M.: The dynamics of the upper ocean. Cambridge Press, Cambridge (1977).

[360] Schwartz, L.W.: Computer extension and analytic continuation of Stokes' expansion for gravity waves. J. Fluid Mech. **62**, 553–578 (1974).

[361] Stokes, G.G.: Supplement to a paper on the theory of oscillatory waves. Math. Phys. Papers. **1**, 314–326 (1880).

[362] Stokes, G.G.: On the theory of oscillation waves. Trans. Cambridge Phil. Soc. **8**, 441–455 (1894).

[363] Thomas, G.P.: Wave-current interactions: an experimental and numerical study. Part2. Nonlinear waves. J. Fluid Mech. **216**, 505–536 (1990).

[364] Thompson, P.D.: The propagation of small surface disturbances through rotational flow. Ann. N. Y. Acad. Sci. **51**, 463–474 (1949).

[365] Toumazis, A.D., Ahilan, R.V.: Review of recent analytical and approximate solutions of wave-current interaction. Environmental forces on offshore structure and their prediction. **26**, 61–79 (1990).

[366] Wang, T. et al.: Effect of nonlinear wave-current interaction on flow fields and hydrodynamic forces. Science in China - A. **40**, No. 6, 622–632 (1997).

[367] Wang, T. et al.: On wave-current interaction (in Chinese). Advances in Mechanics **29**, No. 3, 331–343 (1999).

[368] Xu, H., Lin, Z.L., Liao, S.J., Wu, J.Z., Majdalani, J.: Homotopy-based solutions of the Navier-Stokes equations for a porous channel with orthogonally moving walls. Physics of Fluids. **22**, 053601 (2010). doi:10.1063/1.3392770.

附录 15.1 波流相互作用的 Mathematica 程序

应用同伦分析方法求解二维非线性行进波和非均匀来流的相互作用.

波流相互作用的 Mathematica 程序

成钧和廖世俊编写
上海交通大学
2010 年 8 月

```
(***********************************************************)
(*       Interaction of wave & nonuniform currents      *)
(*  Governing equation:                                 *)
(*       yxx(yz)^2-2yxyxz(yz)+(1+yx^2)yzz==omega,for z > 0 *)
(*  Boundary condition                                  *)
(*       mu(1+yx^2)+2y(yz)^2-2kk(yz)^2 == 0 , for z = 0  *)
(*       y(x,z)->0 , as z->Infinity                     *)
(*       H =  y(0,0)-y(pi,0)                            *)
(***********************************************************)


(***********************************************************)
(* Define  chi_[k]                                       *)
```

```
(***********************************************************)
chi[k_] := If[k<=1,0,1];


(***********************************************************)
(* Define initial guess                                    *)
(***********************************************************)
y[0] = z + H/2 Exp[-z] * (Exp[I*x]+Exp[-I*x])/2;


(***********************************************************)
(* Define GetfAll[k]                                       *)
(***********************************************************)
GetAll[k_]:=Module[{},
yz[k]      = Expand[D[y[k],z]];
yzz[k]     = Expand[D[yz[k],z]];
yx[k]      = Expand[D[y[k],x]];
yxx[k]     = Expand[D[yx[k],x]];
ygamma[k]  = 2(y[k]-gamma[k])//Expand;
yxz[k]     = Expand[D[yx[k],z]];
yxyz[k]    = Expand[Sum[yx[i]*yz[k-i],{i,0,k}]];
yzyzyz[k]  = Expand[Sum[yzyz[i]*yz[k-i],{i,0,k}]];
yzyz[k]    = Expand[Sum[yz[i]*yz[k-i],{i,0,k}]];
yxyxz[k]   = Expand[Sum[yx[i]*yxz[k-i],{i,0,k}]];
yxyx[k]    = Expand[Sum[yx[i]*yx[k-i],{i,0,k}]];
];




(***********************************************************)
(* Define GetRHS[k]                                        *)
(***********************************************************)
GetRHS[k_]:=Module[{temp,i,j,n},
temp[1]   = Sum[yxx[k-1-i]*yzyz[i],{i,0,k-1}];
temp[2]   = Sum[yxyxz[k-1-i]*yz[i],{i,0,k-1}];
temp[3]   = Sum[yzz[k-1-i]*yxyx[i],{i,0,k-1}];
temp[4]   = yzyzyz[k-1]*Omega;
delta[k]  = Expand[temp[1]-2*temp[2]+yzz[k-1]+temp[3]-temp[4]];
RHS[k]    = Expand[c0*delta[k]];
];
```

```
(***************************************************************)
(* Define GetRHSb[k]                                          *)
(***************************************************************)
GetRHSb[k_]:=Module[{temp,i,j,n},
temp[1]   = Sum[mu[k-1-i]*yxyx[i],{i,0,k-1}];
temp[2]   = Sum[(ygamma[k-1-i])*yzyz[i],{i,0,k-1}];
temp[3]   = mu[k-1];
deltaB[k] = Expand[temp[1] + temp[2] + temp[3]];
RHSb[k]   = Expand[c0*deltaB[k]];
];


(***************************************************************)
(* Define the the auxiliary linear operator L and Lb          *)
(***************************************************************)
L[f_]  := Expand[D[f,{z,2}]  + D[f,{x,2}]];
Lb[f_] := Expand[D[f,z]  + f];


(***************************************************************)
(* Define inverse operator of auxiliary linear operator Linv  *)
(***************************************************************)
Linv[Exp[n_.*z]*Exp[m_.*x]]:= 1/(n^2+m^2)*Exp[n*z]*Exp[m*x];
Linv[Exp[m_.*x]]:= Exp[m*x]/m^2;
Linv[Exp[n_.*z]]:= Exp[n*z]/n^2;


(***************************************************************)
(* The property of the inverse operator Linv                  *)
(***************************************************************)
Linv[p_Plus]    := Map[Linv,p];
Linv[c_*f_]     := c*Linv[f]   /;   FreeQ[c,z] && FreeQ[c,x];


(***************************************************************)
(* Define GetySpecial[k]                                      *)
(***************************************************************)
GetySpecial[k_]:=Module[{temp},
temp[0]   = Expand[RHS[k]];
ySpecial  = Linv[temp[0]] //Expand;
];
```

```
(**************************************************)
(* Define two functions for common solution        *)
(**************************************************)
therest[f_,k_]:=(-Lb[f] + RHSb[k])/.z->0//ComplexExpand;
ccc[f_] := Select[f, FreeQ[#, x] &];


(**************************************************)
(* Define Gety[k]                                  *)
(**************************************************)
Gety[k_] := Module[
{temp,sol,p0,p1,plist,clist,dlist,c1,property},
temp[0]     = therest[ySpecial,k];
temp[1]     = temp[0]//Expand;
p0          = ccc[temp[1]];
p1          = Coefficient[temp[1],Cos[x]];
temp[2]     = temp[1] - p0 - p1*Cos[x];
sol         = Solve[{p0 == 0,p1 == 0},{gamma[k-1],mu[k-1]}];
gamma[k-1]  = gamma[k-1]/.sol[[1]]//Expand;
mu[k-1]     = mu[k-1]/.sol[[1]]//Expand;
plist       = Table[Coefficient[temp[2],Cos[i*x]],{i,2,2k+1}]
                //Simplify;
clist       = Table[plist[[i-1]]/(1-i),{i,2,2k+1}];
dlist       = Table[clist[[i-1]]*(1-(-1)^i),{i,2,2k+1}];
temp[3]     = Apply[Plus,dlist];
temp[4]     = ySpecial /.{z->0,x->0 };
temp[5]     = ySpecial /.{z->0,x->Pi};
temp[6]     = Table[Exp[-i*z]*(Exp[i*I*x]+Exp[-i*I*x])/2,
                {i,2,2k+1}];
c1          = -1/2*(temp[4]-temp[5]+temp[3]);
property    = Dot[temp[6],clist]
                + c1*Exp[-z]*(Exp[I*x] + Exp[-I*x])/2;
y[k]        = Expand[ySpecial + chi[k]*y[k-1]  + property];
];


(**************************************************)
(* Define GetYs[k]                                 *)
(**************************************************)
GetYs[k_]:=Module[{temp},
temp[0] = Y[k] /. z->0;
```

```
temp[1] = Integrate[temp[0], {x,-Pi,Pi}]/2/Pi;
Ys[k]   = temp[0] - temp[1]//Expand;
];


(***************************************************************)
(* Define GetErrB[k]                                           *)
(*     Gain squared residual of boundary condition             *)
(***************************************************************)
GetErrB[k_]:=Module[{temp,i,Yx,Yz,Nx,Np,dx,xx},
Yx = D[Y[k],x];
Yz = D[Y[k],z];
temp[1] = MU[k]*(1+Yx^2) + 2*(Y[k]-GAMMA[k])*Yz^2 /. z->0;
temp[2] = temp[1]^2;
Nx   = 20;
dx   = N[Pi/Nx,100];
sum  = 0;
Np   = 0;
For[i = 0, i <= Nx, i++,
    xx = i*dx;
    sum = sum + temp[2] /. x->xx;
    Np  = Np + 1;
   ];
ErrB[k] = sum/Np;
If[NumberQ[ErrB[k]],
   ErrB[k] = Re[ErrB[k]];
   Print["k = ",k," Squared Residual of B.C. = ",ErrB[k]//N]
   ];
];


(***************************************************************)
(* Define GetErr[k]                                            *)
(*     Gain squared residual of governing equation             *)
(***************************************************************)
GetErr[k_]:=Module[
{temp,i,j,sum,xx,zz,Yx,Yz,Yxx,Yxz,Yzz,Nx,Nz,Np},
Yx = D[Y[k],x];
Yz = D[Y[k],z];
Yxx = D[Yx,x];
Yxz = D[Yx,z];
```

```
Yzz = D[Yz,z];
temp[1] = Yxx*Yz^2 - 2*Yx*Yz*Yxz + (1+Yx^2)*Yzz - Yz^3*Omega;
temp[2] = temp[1]^2;
Nx = 10;
Nz = 10;
dx = N[ Pi/Nx,100];
dz = N[2*Pi/Nz,100];
sum = 0;
Np  = 0;
For[i = 0, i <= Nx, i++,
    For[j = 0, j <= Nz,j++,
        xx  =   i*dx;
        zz  =   j*dz;
        sum = sum + temp[2]/.{x->xx,z->zz};
        Np  = Np + 1;
        ];
    ];
Err[k] = sum/Np;
If[NumberQ[Err[k]],
   Err[k] = Re[Err[k]];
   Print["k = ",k," Squared Residual of G.E. = ",Err[k]//N]
   ];
];


(************************************************************)
(* Define hp[f_,m_,n_]                                      *)
(*    Gain [m,n] Homotopy-Pade approximation                *)
(************************************************************)
hp[f_,m_,n_]:=Block[{k,i,df,res,q},
df[0] = f[0];
For[k = 1, k <= m+n, k++, df[k] = f[k] - f[k-1]//Expand ];
res = df[0] + Sum[df[i]*q^i,{i,1,m+n}];
Pade[res,{q,0,m,n}]/.q->1
];


(************************************************************)
(*                     Main Code                            *)
(************************************************************)
ham[m0_,m1_]:=Module[{temp,k,n},
```

```
For[k=Max[1,m0],k<=m1,k=k+1,Print[" k   =   ",k];
    GetAll[k-1];
    GetRHS[k];
    GetRHSb[k];
    GetySpecial[k];
    Gety[k];
    Y[k]    = Y[k-1] + y[k]//ComplexExpand;
    If[k>1,
       MU[k-1]     = MU[k-2] + mu[k-1]//Expand;
       GAMMA[k-1] = GAMMA[k-2] + gamma[k-1]//Expand;
      ];
    Print[" mu       = ",MU[k-1]//N,"  variation = ",mu[k-1]//N];
    Print[" gamma  = ",GAMMA[k-1]//N,"  variation = ",
           gamma[k-1]//N];
   ];
Print[" Sucessful ! "];
];


(*************************************************************)
(* Define initial guess u[0] and related functions            *)
(*************************************************************)
Y[0]       = y[0];
GAMMA[0]   = gamma[0];
MU[0]      = mu[0];
ERR[0]     = ComplexExpand[(Y[0] - GAMMA[0]) /.x->0/.z->0];
Omega      = epsilon* Exp[-z];

(* Physical and control parameters *)
epsilon =  1/5;
H       =  3/10;
c0      = -11/20;

(* Print input data *)
Print["epsilon =    ",epsilon];
Print["    H    =    ",H];
Print["    c0   =    ",c0];

(* Gain the 10th-order approximation *)
ham[1,11];
```

附录 15.1 波流相互作用的 Mathematica 程序

```
(* Gain squared residual of governing equation *)
For[k=2, k<=10, k=k+2, GetErr[k]];

(* Gain squared residual of boundary condition *)
For[k=2, k<=10, k=k+2, GetErrB[k]];
```

第十六章 任意数量周期性行进波之间的共振

本章验证了同伦分析方法对描述任意数量行进波之间非线性相互作用的复杂非线性偏微分方程的有效性. 在同伦分析方法的框架下, 我们首次获得了任意数量波分量之间的共振准则, 其在逻辑上包含了 Phillips 提出的小幅值条件下四波共振条件. 此外, 还首次发现, 当满足共振条件且波系充分发展时, 存在多个定常共振波系. 其中, 共振波分量的幅值可能远小于基波, 从而只占波系总能量很小的一部分. 这个例子说明了同伦分析方法可以用来加深和丰富我们对一些相当复杂的非线性现象的理解.

16.1 绪论

本章将进一步说明同伦分析方法 [375-389] 可用于求解一些相当复杂的非线性偏微分方程, 从而加深和丰富我们对一些有趣的非线性现象的理解.

这里, 我们考虑深水中任意数量的周期性行进重力波之间的非线性相互作用 [387]. 令 z 表示垂直坐标, x, y 表示水平坐标, t 表示时间, $z = \zeta(x, y, t)$ 表示自由表面. x, y, z 三个轴相互垂直, 其单位向量分别为 $\mathbf{i}, \mathbf{j}, \mathbf{k}$, 即 $\mathbf{i} \cdot \mathbf{j} = \mathbf{i} \cdot \mathbf{k} = \mathbf{j} \cdot \mathbf{k} = 0$, 其中 · 为点乘. 假设涡量忽略不计, 并且存在速度势 $\varphi(x, y, z, t)$, 则 $\mathbf{u} = \nabla \varphi$ 以及

$$\nabla^2 \varphi = 0, \quad z \leqslant \zeta(x, y, t), \tag{16.1}$$

其中
$$\nabla = \mathbf{i}\frac{\partial}{\partial x} + \mathbf{j}\frac{\partial}{\partial y} + \mathbf{k}\frac{\partial}{\partial z}.$$

在自由表面 $z = \zeta(x,y,t)$ 上, 压力是恒定的, 这由伯努利方程给出了动力学边界条件

$$g\zeta + \frac{\partial \varphi}{\partial t} + \frac{1}{2}\mathbf{u}^2 = 0, \quad z = \zeta(x,y,t), \tag{16.2}$$

其中 g 为重力加速度. 此外, $z - \zeta$ 随流体微团运动而消失, 这给出了运动学边界条件

$$\frac{\partial \zeta}{\partial t} - \frac{\partial \varphi}{\partial z} + \left(\frac{\partial \varphi}{\partial x}\frac{\partial \zeta}{\partial x} + \frac{\partial \varphi}{\partial y}\frac{\partial \zeta}{\partial y}\right) = 0, \quad z = \zeta(x,y,t). \tag{16.3}$$

结合以上两个方程可以得到组合边界条件

$$\frac{\partial^2 \varphi}{\partial t^2} + g\frac{\partial \varphi}{\partial z} + \frac{\partial(\mathbf{u}^2)}{\partial t} + \mathbf{u} \cdot \nabla\left(\frac{1}{2}\mathbf{u}^2\right) = 0, \quad z = \zeta(x,y,t), \tag{16.4}$$

其中 $\mathbf{u} = \nabla\varphi$ 以及 $\mathbf{u}^2 = \nabla\varphi \cdot \nabla\varphi$. 在液体底面, 其满足

$$\frac{\partial \varphi}{\partial z} = 0, \quad z \to -\infty. \tag{16.5}$$

详情请参考 Phillips [393] 和 Longuet-Higgins [390].

Phillips [393] 在其关于深水中四个重力波分量之间非线性相互作用的开创性工作中, 提出了波浪共振条件

$$\mathbf{k}_1 \pm \mathbf{k}_2 \pm \mathbf{k}_3 \pm \mathbf{k}_4 = 0, \quad \sigma_1 \pm \sigma_2 \pm \sigma_3 \pm \sigma_4 = 0, \tag{16.6}$$

其中 $\sigma_i = \sqrt{gk_i}$ 表示波的角频率, $k_i = |\mathbf{k}_i|$ $(i = 1,2,3,4)$, $\mathbf{k}_i$ 表示波数. 需要强调的是, Phillips 共振条件 (16.6) 仅适用于小幅值的弱非线性波, 因为 $\sigma_i = \sqrt{gk_i}$ 为线性理论中单个小幅值重力波的角频率.

在 $\mathbf{k}_4 = \mathbf{k}_1$ 的特殊情况下, Phillips [393] 基于摄动方法表明, 如果 $\mathbf{k}_3 = 2\mathbf{k}_1 - \mathbf{k}_2$ 以及 $\sigma_3 = 2\sigma_1 - \sigma_2$, 此时定常解是不存在的, 如果初始时刻第三个波分量的幅值为零, 则其幅值会随着时间呈线性增长. 这一结论得到了 Longuet-Higgins [390] 基于摄动方法的证实, 并得到了一些实验 [391,392] 的支持. 此外, Benney [371] 求解了控制共振模式时间依赖性的方程, 并研究了所涉及的能量转移机制.

尽管 Phillips 的开创性工作 [393] 已经过去了半个世纪, 但关于重力波之间的非线性相互作用仍然存在一些悬而未决的问题. 首先, Bretherton [372] 指

出了 Phillips 使用的摄动方法对于长时间失效, 然后, 通过研究一维色散波模型提出每个波分量的幅度应该有界. 此外, Phillips 的波浪共振条件 (16.6) 仅适用于具有小波幅的四个波分量. 对于具有大幅值的任意数量波分量之间的共振条件是什么? 似乎很难应用 Phillips [393] 和 Longuet-Higgins [390] 使用的摄动方法来回答这个问题, 正如 Phillips [394] 所提到的那样, 相关的代数是令人生畏和 "非常乏味的".

2011 年, 廖世俊 [387] 通过所谓的同伦多变量方法, 成功地应用同伦分析方法给出了任意数量行进波的波共振条件, 该方法保留了摄动技术多尺度的明确物理意义, 但完全放弃了物理小参数. 本章将以周期性行进波之间的非线性相互作用为例, 说明同伦多变量方法的基本思想. 通过该方法, 得到了任意数量大幅值行进波之间的共振条件, 其在逻辑上包含了 Phillips 提出的小幅值条件下四波共振条件. 高阶形变方程的详细数学推导见附录 16.1. 更多详情请参考廖世俊 [387].

16.2 两个小幅值基波的共振条件

16.2.1 简明数学公式

不失一般性, 我们先考虑一个充分发展的重力波系统, 其由深水中的两列行进基波组成. 波数分别为 $\mathbf{k}_1, \mathbf{k}_2$, 相应的角频率为 σ_1, σ_2, 其中 $\mathbf{k}_1 \times \mathbf{k}_2 \neq 0$ (即两列行进基波不共线). 由于非线性相互作用, 该波系包含无穷多个波分量, 其对应波数为 $m\mathbf{k}_1 + n\mathbf{k}_2$, 其中 m, n 为整数. 由于假设波系充分发展, 因此每个波分量的幅值与时间无关. 令 α_1, α_2 分别表示 x 轴正方向 $\mathbf{i}$ 与波数向量 $\mathbf{k}_1$ 和 $\mathbf{k}_2$ 之间的夹角, 其中 $\mathbf{k}_1 \cdot \mathbf{k} = \mathbf{k}_2 \cdot \mathbf{k} = 0$, 即 z 轴垂直于波数 $\mathbf{k}_1, \mathbf{k}_2$. 然后,

$$\mathbf{k}_1 = k_1 \left(\cos\alpha_1 \, \mathbf{i} + \sin\alpha_1 \, \mathbf{j} \right), \quad \mathbf{k}_2 = k_2 \left(\cos\alpha_2 \, \mathbf{i} + \sin\alpha_2 \, \mathbf{j} \right), \quad (16.7)$$

其中 $k_1 = |\mathbf{k}_1|, k_2 = |\mathbf{k}_2|$.

写作 $\mathbf{r} = x\mathbf{i} + y\mathbf{j}$. 根据线性重力波理论, 以波数 $\mathbf{k}_1$ 和 $\mathbf{k}_2$ 传播的两列基波分别由 $a_1 \cos\xi_1, a_2 \cos\xi_2$ 给出, 其中

$$\xi_1 = \mathbf{k}_1 \cdot \mathbf{r} - \sigma_1 t, \quad \xi_2 = \mathbf{k}_2 \cdot \mathbf{r} - \sigma_2 t. \quad (16.8)$$

所以, 以上两个变量的物理意义非常明确: 波形一定是关于 ξ_1 和 ξ_2 的周期函数. 从数学上讲, 基于这两个变量, 对于充分发展的波系, 时间 t 不应显式出现. 换句话说, 对于由两列行进基波组成的充分发展波系的非线性相互作用,

可以分别表示其速度势函数为 $\varphi(x,y,z,t) = \phi(\xi_1,\xi_2,z)$ 和波面为 $\zeta(x,y,t) = \eta(\xi_1,\xi_2)$.

那么

$$\mathbf{u} = \nabla\varphi = \mathbf{i}\frac{\partial\varphi}{\partial x} + \mathbf{j}\frac{\partial\varphi}{\partial y} + \mathbf{k}\frac{\partial\varphi}{\partial z} = \mathbf{k}_1\frac{\partial\phi}{\partial\xi_1} + \mathbf{k}_2\frac{\partial\phi}{\partial\xi_2} + \mathbf{k}\frac{\partial\phi}{\partial z} = \hat{\nabla}\phi, \quad (16.9)$$

其中

$$\hat{\nabla} = \mathbf{k}_1\frac{\partial}{\partial\xi_1} + \mathbf{k}_2\frac{\partial}{\partial\xi_2} + \mathbf{k}\frac{\partial}{\partial z}. \quad (16.10)$$

因此

$$\mathbf{u}^2 = \nabla\varphi \cdot \nabla\varphi = \hat{\nabla}\phi \cdot \hat{\nabla}\phi$$
$$= k_1^2\left(\frac{\partial\phi}{\partial\xi_1}\right)^2 + 2\mathbf{k}_1\cdot\mathbf{k}_2\frac{\partial\phi}{\partial\xi_1}\frac{\partial\phi}{\partial\xi_2} + k_2^2\left(\frac{\partial\phi}{\partial\xi_2}\right)^2 + \left(\frac{\partial\phi}{\partial z}\right)^2, \quad (16.11)$$

其中使用 $\mathbf{k}_1\cdot\mathbf{k} = \mathbf{k}_2\cdot\mathbf{k} = 0$, 并且 $\mathbf{k}_1\cdot\mathbf{k}_2 = k_1k_2\cos(\alpha_1-\alpha_2)$. 一般来说, 对于任意函数 $\phi(\xi_1,\xi_2,z)$ 和 $\psi(\xi_1,\xi_2,z)$, 其满足

$$\hat{\nabla}\phi \cdot \hat{\nabla}\psi = k_1^2\frac{\partial\phi}{\partial\xi_1}\frac{\partial\psi}{\partial\xi_1} + \mathbf{k}_1\cdot\mathbf{k}_2\left(\frac{\partial\phi}{\partial\xi_1}\frac{\partial\psi}{\partial\xi_2} + \frac{\partial\psi}{\partial\xi_1}\frac{\partial\phi}{\partial\xi_2}\right)$$
$$+ k_2^2\frac{\partial\phi}{\partial\xi_2}\frac{\partial\psi}{\partial\xi_2} + \frac{\partial\phi}{\partial z}\frac{\partial\psi}{\partial z}. \quad (16.12)$$

类似地, 我们有

$$\nabla^2\varphi = \hat{\nabla}^2\phi = k_1^2\frac{\partial^2\phi}{\partial\xi_1^2} + 2\mathbf{k}_1\cdot\mathbf{k}_2\frac{\partial^2\phi}{\partial\xi_1\partial\xi_2} + k_2^2\frac{\partial^2\phi}{\partial\xi_2^2} + \frac{\partial^2\phi}{\partial z^2}, \quad (16.13)$$

则原始的控制方程为

$$\hat{\nabla}^2\phi = k_1^2\frac{\partial^2\phi}{\partial\xi_1^2} + 2\mathbf{k}_1\cdot\mathbf{k}_2\frac{\partial^2\phi}{\partial\xi_1\partial\xi_2} + k_2^2\frac{\partial^2\phi}{\partial\xi_2^2} + \frac{\partial^2\phi}{\partial z^2} = 0, \quad (16.14)$$

其通解为

$$\phi = [A\cos(m\xi_1+n\xi_2) + B\sin(m\xi_1+n\xi_2)]e^{|m\mathbf{k}_1+n\mathbf{k}_2|z}, \quad (16.15)$$

其中 m, n 为整数, A, B 为积分常数, $-\infty < z \leqslant \eta(\xi_1,\xi_2)$.

为了简单起见, 定义

$$f = \frac{1}{2}\hat{\nabla}\phi\cdot\hat{\nabla}\phi = \frac{\mathbf{u}^2}{2}$$

$$= \frac{1}{2}\left[k_1^2\left(\frac{\partial\phi}{\partial\xi_1}\right)^2 + 2\mathbf{k}_1\cdot\mathbf{k}_2\frac{\partial\phi}{\partial\xi_1}\frac{\partial\phi}{\partial\xi_2} + k_2^2\left(\frac{\partial\phi}{\partial\xi_2}\right)^2 + \left(\frac{\partial\phi}{\partial z}\right)^2\right]. \quad (16.16)$$

使用新变量 ξ_1 和 ξ_2, 动力学边界条件 (16.2) 变为

$$\eta = \frac{1}{g}\left(\sigma_1\frac{\partial\phi}{\partial\xi_1} + \sigma_2\frac{\partial\phi}{\partial\xi_2} - f\right), \quad z = \eta(\xi_1,\xi_2). \quad (16.17)$$

在自由表面 $z = \eta(\xi_1, \xi_2)$ 上, 运动学边界条件 (16.4) 为

$$\sigma_1^2\frac{\partial^2\phi}{\partial\xi_1^2} + 2\sigma_1\sigma_2\frac{\partial^2\phi}{\partial\xi_1\partial\xi_2} + \sigma_2^2\frac{\partial^2\phi}{\partial\xi_2^2} + g\frac{\partial\phi}{\partial z}$$
$$- 2\left(\sigma_1\frac{\partial f}{\partial\xi_1} + \sigma_2\frac{\partial f}{\partial\xi_2}\right) + \hat{\nabla}\phi\cdot\hat{\nabla}f = 0, \quad (16.18)$$

其中

$$\frac{\partial f}{\partial\xi_1} = \hat{\nabla}\phi\cdot\hat{\nabla}\left(\frac{\partial\phi}{\partial\xi_1}\right), \quad \frac{\partial f}{\partial\xi_2} = \hat{\nabla}\phi\cdot\hat{\nabla}\left(\frac{\partial\phi}{\partial\xi_2}\right)$$

以及

$$\hat{\nabla}\phi\cdot\hat{\nabla}f = k_1^2\frac{\partial\phi}{\partial\xi_1}\frac{\partial f}{\partial\xi_1} + \mathbf{k}_1\cdot\mathbf{k}_2\left(\frac{\partial\phi}{\partial\xi_1}\frac{\partial f}{\partial\xi_2} + \frac{\partial f}{\partial\xi_1}\frac{\partial\phi}{\partial\xi_2}\right) + k_2^2\frac{\partial\phi}{\partial\xi_2}\frac{\partial f}{\partial\xi_2} + \frac{\partial\phi}{\partial z}\frac{\partial f}{\partial z}.$$

在液体底面, 有

$$\frac{\partial\phi}{\partial z} = 0, \quad z \to -\infty. \quad (16.19)$$

给定两个物理上合理的角频率 σ_1 和 σ_2, 我们的目标是找出对应的未知速度势函数 $\phi(\xi_1,\xi_2,z)$ 和未知的自由表面 $\eta(\xi_1,\xi_2)$, 其由线性偏微分方程 (16.14) 控制, 在未知自由表面 $z = \eta(\xi_1,\xi_2)$ 上满足两个非线性边界条件 (16.17) 和 (16.18), 以及在液体底面满足一个线性边界条件 (16.19). 这里需要强调的是, 通过两个自变量 ξ_1 和 ξ_2, 时间 t 在未知的势函数和波面上不会显式出现. 这大大简化了问题的求解过程, 如下所述.

如前所述, ξ_1 和 ξ_2 这两个变量具有明确的物理意义, 该问题的解应该是关于 ξ_1 和 ξ_2 的周期函数. 从物理角度看, 波面为

$$\eta(\xi_1,\xi_2) = \sum_{m=0}^{+\infty}\sum_{n=-\infty}^{+\infty} a_{m,n}\cos(m\xi_1 + n\xi_2), \quad (16.20)$$

其中 $a_{m,n}$ 为波分量 $\cos(m\xi_1 + n\xi_2)$ 的幅值. 注意, (16.15) 是控制方程 (16.14) 的通解. 因此, 相应的势函数为

$$\phi(\xi_1,\xi_2,z) = \sum_{m=0}^{+\infty}\sum_{n=-\infty}^{+\infty} b_{m,n}\,\Psi_{m,n}(\xi_1,\xi_2,z), \quad (16.21)$$

其中
$$\Psi_{m,n}(\xi_1,\xi_2,z) = \sin(m\xi_1+n\xi_2)\,e^{|m\mathbf{k}_1+n\mathbf{k}_2|z}, \quad (16.22)$$

$b_{m,n}$ 是独立于 ξ_1,ξ_2,z 的未知系数. 注意, 由 (16.21) 定义的速度势函数 $\phi(\xi_1,\xi_2,z)$ 自动满足控制方程 (16.14) 和液体底面条件 (16.19). 上述表达式 (16.20) 和 (16.21) 分别称为 η 和 ϕ 的解表达, 且在同伦分析方法的框架中起着重要作用, 如下所述.

为了简单起见, 定义非线性算子

$$\mathcal{N}[\phi(\xi_1,\xi_2,z)] = \sigma_1^2 \frac{\partial^2 \phi}{\partial \xi_1^2} + 2\sigma_1\sigma_2 \frac{\partial^2 \phi}{\partial \xi_1 \partial \xi_2} + \sigma_2^2 \frac{\partial^2 \phi}{\partial \xi_2^2} + g\frac{\partial \phi}{\partial z}$$
$$- 2\left(\sigma_1\frac{\partial f}{\partial \xi_1} + \sigma_2\frac{\partial f}{\partial \xi_2}\right) + \hat{\nabla}\phi \cdot \hat{\nabla} f, \quad (16.23)$$

对应于运动学边界条件 (16.18), 其中角频率 σ_1,σ_2 给定. 注意, 其包含一个线性算子

$$\mathcal{L}_0(\phi) = \sigma_1^2 \frac{\partial^2 \phi}{\partial \xi_1^2} + 2\sigma_1\sigma_2 \frac{\partial^2 \phi}{\partial \xi_1 \partial \xi_2} + \sigma_2^2 \frac{\partial^2 \phi}{\partial \xi_2^2} + g\frac{\partial \phi}{\partial z}. \quad (16.24)$$

令 $\mathcal{L}$ 表示具有性质 $\mathcal{L}(0)=0$ 的辅助线性算子. 如前几章所述, 同伦分析方法为我们选择辅助线性算子提供了极大的自由. 基于线性波理论的结果, 即

$$\sigma_1 \approx \sqrt{g\,k_1} = \bar{\sigma}_1, \quad \sigma_2 \approx \sqrt{g\,k_2} = \bar{\sigma}_2, \quad (16.25)$$

我们选取如下辅助线性算子

$$\mathcal{L}(\phi) = \bar{\sigma}_1^2 \frac{\partial^2 \phi}{\partial \xi_1^2} + 2\bar{\sigma}_1\bar{\sigma}_2 \frac{\partial^2 \phi}{\partial \xi_1 \partial \xi_2} + \bar{\sigma}_2^2 \frac{\partial^2 \phi}{\partial \xi_2^2} + g\frac{\partial \phi}{\partial z}. \quad (16.26)$$

令 $q \in [0,1]$ 表示嵌入变量, $c_0 \neq 0$ 表示收敛控制参数, $\phi_0(\xi_1,\xi_2,z)$ 为速度势函数的初始猜测解, $\phi_0(\xi_1,\xi_2,z)$ 分别满足 $\hat{\nabla}^2 \phi_0 = 0$ 以及液体底面条件 (16.19). 构造零阶形变方程

$$\hat{\nabla}^2 \check{\phi}(\xi_1,\xi_2,z;q) = 0, \quad -\infty < z \leqslant \check{\eta}(\xi_1,\xi_2;q), \quad (16.27)$$

满足 $z = \check{\eta}(\xi_1,\xi_2;q)$ 上的两个边界条件

$$(1-q)\,\mathcal{L}\left[\check{\phi}(\xi_1,\xi_2,z;q) - \phi_0(\xi_1,\xi_2,z)\right] = q\,c_0\,\mathcal{N}\left[\check{\phi}(\xi_1,\xi_2,z;q)\right] \quad (16.28)$$

和

$$(1-q)\check{\eta}(\xi_1,\xi_2;q)$$

$$= q\, c_0 \left\{ \check{\eta}(\xi_1,\xi_2;q) - \frac{1}{g}\left[\sigma_1 \frac{\partial \check{\phi}(\xi_1,\xi_2,z;q)}{\partial \xi_1} + \sigma_2 \frac{\partial \check{\phi}(\xi_1,\xi_2,z;q)}{\partial \xi_2} - \check{f}\right] \right\}, \tag{16.29}$$

其中
$$\check{f} = \frac{1}{2}\hat{\nabla}\check{\phi}(\xi_1,\xi_2,z;q) \cdot \hat{\nabla}\check{\phi}(\xi_1,\xi_2,z;q).$$

此外, 在液体底面有
$$\frac{\partial \check{\phi}(\xi_1,\xi_2,z;q)}{\partial z} = 0, \quad z \to -\infty. \tag{16.30}$$

当 $q = 0$ 时, 由于初始猜测解 $\phi_0(\xi_1,\xi_2,z)$ 满足控制方程 (16.14) 和液体底面条件 (16.19), 以及辅助线性算子 (16.26) 的性质 $\mathcal{L}(0) = 0$, 我们有
$$\check{\phi}(\xi_1,\xi_2,z;0) = \phi_0(\xi_1,\xi_2,z) \tag{16.31}$$

和
$$\check{\eta}(\xi_1,\xi_2;0) = 0, \tag{16.32}$$

这为我们提供了速度势函数 $\phi(\xi_1,\xi_2,z)$ 和自由表面 $\eta(\xi_1,\xi_2)$ 的初始近似值. 当 $q = 1$ 时, 由于 $c_0 \neq 0$, (16.27) 至 (16.30) 等价于原方程 (16.14), (16.17), (16.18) 和 (16.19), 有
$$\check{\phi}(\xi_1,\xi_2,z;1) = \phi(\xi_1,\xi_2,z), \quad \check{\eta}(\xi_1,\xi_2;1) = \eta(\xi_1,\xi_2). \tag{16.33}$$

所以, 随着嵌入变量 q 从 0 增加到 1, $\check{\phi}(\xi_1,\xi_2,z;q)$ 从初始猜测解 $\phi_0(\xi_1,\xi_2,z)$ 连续变化到未知速度势函数 $\phi(\xi_1,\xi_2,z)$, 与此同时, $\check{\eta}(\xi_1,\xi_2;q)$ 从初始猜测解 0 连续形变 (或变化) 到未知波面 $\eta(\xi_1,\xi_2)$. 从数学上讲, (16.27) 至 (16.30) 定义了两个同伦
$$\check{\phi}(\xi_1,\xi_2,z;q) : \phi_0(\xi_1,\xi_2,z) \sim \phi(\xi_1,\xi_2,z),$$
$$\check{\eta}(\xi_1,\xi_2;q) : 0 \sim \eta(\xi_1,\xi_2).$$

需要注意的是, 我们可以自由地选择收敛控制参数 c_0. 假设 c_0 选择恰当, 则同伦-Maclaurin 级数
$$\check{\phi}(\xi_1,\xi_2,z;q) = \phi_0(\xi_1,\xi_2,z) + \sum_{n=1}^{+\infty} \phi_n(\xi_1,\xi_2,z)\, q^n, \tag{16.34}$$

$$\check{\eta}(\xi_1,\xi_2;q) = \sum_{n=1}^{+\infty} \eta_n(\xi_1,\xi_2)\, q^n \qquad (16.35)$$

存在且在 $q=1$ 时绝对收敛，根据 (16.33) 我们有同伦级数解

$$\phi(\xi_1,\xi_2,z) = \phi_0(\xi_1,\xi_2,z) + \sum_{n=1}^{+\infty} \phi_n(\xi_1,\xi_2,z), \qquad (16.36)$$

$$\eta(\xi_1,\xi_2) = \sum_{n=1}^{+\infty} \eta_n(\xi_1,\xi_2), \qquad (16.37)$$

其中

$$\phi_n(\xi_1,\xi_2,z) = \frac{1}{n!} \left.\frac{\partial^n \check{\phi}(\xi_1,\xi_2,z;q)}{\partial q^n}\right|_{q=0} = \mathcal{D}_n\left[\check{\phi}(\xi_1,\xi_2,z;q)\right],$$

$$\eta_n(\xi_1,\xi_2) = \frac{1}{n!} \left.\frac{\partial^n \check{\eta}(\xi_1,\xi_2;q)}{\partial q^n}\right|_{q=0} = \mathcal{D}_n\left[\check{\eta}(\xi_1,\xi_2;q)\right]$$

是 n 阶同伦导数，$\mathcal{D}_n$ 是 n 阶同伦导数算子. m 阶同伦近似为

$$\phi(\xi_1,\xi_2,z) \approx \phi_0(\xi_1,\xi_2,z) + \sum_{n=1}^{m} \phi_n(\xi_1,\xi_2,z),$$

$$\eta(\xi_1,\xi_2) \approx \sum_{n=1}^{m} \eta_n(\xi_1,\xi_2).$$

关于未知量 $\phi_n(\xi_1,\xi_2,z)$ 和 $\eta_n(\xi_1,\xi_2)$ 的偏微分方程可以直接从零阶形变方程 (16.27) 至 (16.30) 导出. 将同伦-Maclaurin 级数 (16.34) 代入控制方程 (16.27) 和液体底面边界条件 (16.30)，然后将嵌入变量 q 的相同次幂取等式，有

$$\hat{\nabla}^2 \phi_m(\xi_1,\xi_2,z) = 0, \qquad (16.38)$$

满足底面边界条件

$$\frac{\partial \phi_m(\xi_1,\xi_2,z)}{\partial z} = 0, \quad z \to -\infty, \qquad (16.39)$$

其中 $m \geqslant 1$. 需要强调的是，在未知边界 $z = \check{\eta}(\xi_1,\xi_2;q)$ 上满足的两个边界条件 (16.28) 和 (16.29) 也依赖于嵌入变量 q. 因此，要推导出相应的方程会复杂一些. 简单来说，将级数 (16.34) 和 (16.35) 代入边界条件 (16.28) 和 (16.29)，其中 $z = \check{\eta}(\xi_1,\xi_2;q)$，然后将 q 的相同次幂取等式，我们在 $z = 0$ 上有两个线性边界条件

$$\bar{\mathcal{L}}(\phi_m) = c_0\, \Delta^{\phi}_{m-1} + \chi_m\, S_{m-1} - \bar{S}_m, \quad m \geqslant 1 \qquad (16.40)$$

和
$$\eta_m(\xi_1, \xi_2) = c_0 \Delta_{m-1}^{\eta} + \chi_m \eta_{m-1}, \qquad (16.41)$$

其中
$$\Delta_{m-1}^{\eta} = \eta_{m-1} - \frac{1}{g} \left[\left(\sigma_1 \bar{\phi}_{m-1}^{1,0} + \sigma_2 \bar{\phi}_{m-1}^{0,1} \right) - \Gamma_{m-1,0} \right] \qquad (16.42)$$

以及
$$\bar{\mathcal{L}}(\phi_m) = \left(\bar{\sigma}_1^2 \frac{\partial^2 \phi_m}{\partial \xi_1^2} + 2 \bar{\sigma}_1 \bar{\sigma}_2 \frac{\partial^2 \phi_m}{\partial \xi_1 \partial \xi_2} + \bar{\sigma}_2^2 \frac{\partial^2 \phi_m}{\partial \xi_2^2} + g \frac{\partial \phi_m}{\partial z} \right) \Big|_{z=0}. \qquad (16.43)$$

上述方程的详细推导及 $\Delta_{m-1}^{\phi}, S_{m-1}, \bar{S}_m, \chi_m, \Gamma_{m-1,0}, \bar{\phi}_{m-1}^{1,0}, \bar{\phi}_{m-1}^{0,1}$ 的定义在附录 16.1 中给出. 注意, ϕ_m 和 η_m 的子问题不仅是线性的而且是解耦的: 给定 ϕ_{m-1} 和 η_{m-1}, 可以直接得到 η_m, 然后通过求解线性拉普拉斯方程 (16.38) 和两个线性边界条件 (16.39) 和 (16.40) 可以得到 ϕ_m. 因此, 很容易求解高阶形变方程, 尤其是借助 Mathematica 等科学计算软件.

我们按照第二章类似的方式证明以下定理:

收敛定理 同伦级数解 (16.36) 和 (16.37) 满足原始的控制方程 (16.14) 和边界条件 (16.17), (16.18) 和 (16.19), 前提是

$$\sum_{m=0}^{+\infty} \Delta_m^{\phi} = 0, \quad \sum_{m=0}^{+\infty} \Delta_m^{\eta} = 0, \qquad (16.44)$$

其中 $\Delta_m^{\phi}, \Delta_m^{\eta}$ 分别由 (16.118) 和 (16.42) 定义.

因为 Δ_m^{ϕ} 和 Δ_m^{η} 是求解高阶形变方程的中间项, 上述定理为我们提供了一种方便的方法来检查同伦级数解的收敛性和精确性. 为此, 我们分别为 ϕ 和 η 的 m 阶近似值定义平方残差

$$E_m^{\phi} = \frac{1}{\pi^2} \int_0^{\pi} \int_0^{\pi} \left(\sum_{n=0}^{m} \Delta_n^{\phi} \right)^2 d\xi_1 \, d\xi_2, \qquad (16.45)$$

$$E_m^{\eta} = \frac{1}{\pi^2} \int_0^{\pi} \int_0^{\pi} \left(\sum_{n=0}^{m} \Delta_n^{\eta} \right)^2 d\xi_1 \, d\xi_2. \qquad (16.46)$$

根据上述收敛定理, 如果当 $m \to +\infty$ 时, $E_m^{\phi} \to 0$ 和 $E_m^{\eta} \to 0$, 则同伦级数解 (16.36) 和 (16.37) 满足原始控制方程和所有边界条件. 此外, E_m^{ϕ} 和 E_m^{η} 的值也分别表明了 ϕ 和 η 的 m 阶同伦近似值的精确性.

需要注意的是，辅助线性算子 (16.26) 具有性质

$$\mathcal{L}(\Psi_{m,n}) = \left[g|m\mathbf{k}_1 + n\mathbf{k}_2| - (m\bar{\sigma}_1 + n\bar{\sigma}_2)^2\right]\Psi_{m,n}, \qquad (16.47)$$

其中 $\Psi_{m,n}$ 由 (16.22) 给出，控制方程 (16.38) 和底面条件 (16.39) 自动满足. 因此，从数学上讲，m 阶线性形变方程具有无穷多个特征函数 $\Psi_{m,n}$ 和特征值

$$\lambda_{m,n} = g|m\mathbf{k}_1 + n\mathbf{k}_2| - (m\bar{\sigma}_1 + n\bar{\sigma}_2)^2 \qquad (16.48)$$

且具有性质

$$\mathcal{L}(\Psi_{m,n}) = \lambda_{m,n}\,\Psi_{m,n}.$$

因此，其逆算子 $\mathcal{L}^{-1}$ 的定义如下

$$\mathcal{L}^{-1}(\Psi_{m,n}) = \frac{\Psi_{m,n}}{\lambda_{m,n}}, \quad \lambda_{m,n} \neq 0. \qquad (16.49)$$

还要注意的是，逆算子 $\mathcal{L}^{-1}$ 仅对非零特征值 $\lambda_{m,n} \neq 0$ 有定义. 当 $\lambda_{m,n} = 0$ 时，我们有

$$g|m\mathbf{k}_1 + n\mathbf{k}_2| = (m\bar{\sigma}_1 + n\bar{\sigma}_2)^2, \qquad (16.50)$$

这正是 Phillips 和 Longuet-Higgins 给出的所谓 "波浪共振" 的条件. 令 N_λ 表示特征值为零的特征函数的数量. 如下所述，N_λ 的值是这个问题的关键.

当 (16.50) 中 $n = 0$ 时，有

$$\lambda_{m,0} = g\,k_1\left(|m| - m^2\right),$$

其仅当 $|m| = 1$ 时才等于 0 (注意，$m = n = 0$ 对应于 $\phi = 0$, 因此这里不考虑). 同理，当 $m = 0$ 时，特征值 $\lambda_{0,n}$ 仅当 $|n| = 1$ 时才为零. 所以，至少存在两个特征函数 $\Psi_{1,0} = e^{k_1 z}\sin\xi_1$ 和 $\Psi_{0,1} = e^{k_2 z}\sin\xi_2$ 的特征值为零，即对于任意常数 C_1 和 C_2, 有

$$\mathcal{L}\left(C_1\,e^{k_1 z}\sin\xi_1 + C_2\,e^{k_2 z}\sin\xi_2\right) = 0. \qquad (16.51)$$

因此，对于两个基波的情况，有 $N_\lambda \geqslant 2$. 正如 Phillips [393] 和 Longuet-Higgins [390] 所提到的那样，对于一些特殊的波数和角频率，该问题满足波共振条件 (16.50). 因此，在 $m = m'$ 和 $n = n'$ 的情况下且满足条件 (16.50) 时，其中 m' 和 n' 为满足 $m'^2 + n'^2 \neq 1$ 的整数，存在三个特征函数 $\Psi_{1,0} = e^{k_1 z}\sin\xi_1$, $\Psi_{0,1} = e^{k_2 z}\sin\xi_2$ 和

$$\Psi_{m',n'} = e^{|m'\mathbf{k}_1 + n'\mathbf{k}_2|z}\sin(m'\xi_1 + n'\xi_2),$$

其特征值为零, 即对于任意常数 C_1, C_2 和 C_3, 有

$$\mathcal{L}\left[C_1\,e^{k_1 z}\sin\xi_1 + C_2\,e^{k_2 z}\sin\xi_2 + C_3\,e^{|m'\mathbf{k}_1+n'\mathbf{k}_2|z}\sin(m'\xi_1+n'\xi_2)\right]=0. \tag{16.52}$$

不失一般性, Longuet-Higgins [390] 讨论了一个特殊情况 $m'=2$ 和 $n'=-1$, 对应的特征函数为

$$\Psi_{2,-1}=\exp(|2\mathbf{k}_1-\mathbf{k}_2|z)\sin(2\xi_1-\xi_2).$$

在本节中, 如果没有提到上述情况, 则在两个基波的情况下, 当存在三个零特征值 ($N_\lambda=3$) 时, 就意味着 $m'=2$ 和 $n'=-1$.

根据定义 (16.26) 和 (16.43), 有

$$\bar{\mathcal{L}}(\phi)=(\mathcal{L}(\phi))|_{z=0}.$$

因此

$$\bar{\mathcal{L}}(\Psi_{m,n})=\lambda_{m,n}\,\Psi_{m,n}|_{z=0}=\lambda_{m,n}\sin(m\xi_1+n\xi_2),$$

这给出了线性逆算子的定义

$$\bar{\mathcal{L}}^{-1}[\sin(m\xi_1+n\xi_2)]=\frac{\Psi_{m,n}}{\lambda_{m,n}},\quad \lambda_{m,n}\neq 0. \tag{16.53}$$

基于这个逆算子, 我们很容易求解线性拉普拉斯方程 (16.38) 和两个线性边界条件 (16.39) 和 (16.40), 如下所述. 这里, 我们强调上述的逆算子只对非零特征值 $\lambda_{m,n}\neq 0$ 有定义. 详情请参考廖世俊 [387].

16.2.2 非共振波

先考虑只存在两个特征函数的情况, 即

$$\Psi_{1,0}=\exp(k_1 z)\sin\xi_1,\quad \Psi_{0,1}=\exp(k_2 z)\sin\xi_2,$$

其特征值为零, 即 $\lambda_{1,0}=\lambda_{0,1}=0$. 考虑这两个特征函数并根据线性波理论, 构造势函数的初始猜测解

$$\phi_0(\xi_1,\xi_2,z)=A_0\sqrt{\frac{g}{k_1}}\,\Psi_{1,0}+B_0\sqrt{\frac{g}{k_2}}\,\Psi_{0,1}, \tag{16.54}$$

其中 A_0 和 B_0 为未知常数.

关于势函数 $\phi_1(\xi_1,\xi_2,z)$ 的一阶形变方程为

$$\hat{\nabla}^2\phi_1(\xi_1,\xi_2,z)=0, \tag{16.55}$$

满足 $z=0$ 处的边界条件

$$\begin{aligned}\bar{\mathcal{L}}(\phi_1)=&\,b_1^{1,0}\sin(\xi_1)+b_1^{0,1}\sin(\xi_2)+b_1^{1,1}\sin(\xi_1+\xi_2)+d_1^{1,1}\sin(\xi_1-\xi_2)\\&+b_1^{2,1}\sin(2\xi_1+\xi_2)+d_1^{2,1}\sin(2\xi_1-\xi_2)\\&+b_1^{1,2}\sin(\xi_1+2\xi_2)+d_1^{1,2}\sin(\xi_1-2\xi_2),\end{aligned} \tag{16.56}$$

以及液体底面边界条件

$$\left.\frac{\partial\phi_1}{\partial z}\right|_{z=0}=0, \tag{16.57}$$

其中 $\bar{\mathcal{L}}$ 由 (16.43) 定义,$b_1^{i,j},d_1^{i,j}$ 为常数. 特别地,我们有

$$b_1^{1,0}=c_0A_0\sqrt{\frac{g}{k_1}}\left[gk_1-\sigma_1^2+gk_1(A_0k_1)^2+2gk_1(B_0k_2)^2+\frac{gB_0^2k_1^2k_2}{2}\sin^2(\alpha_1-\alpha_2)\right],$$

$$b_1^{0,1}=c_0B_0\sqrt{\frac{g}{k_2}}\left[gk_2-\sigma_2^2+gk_2(B_0k_2)^2+2gk_2(A_0k_1)^2+\frac{gA_0^2k_2^2k_1}{2}\sin^2(\alpha_1-\alpha_2)\right].$$

因为 $\lambda_{1,0}=\lambda_{0,1}=0$,根据 (16.53),必须有

$$b_1^{1,0}=b_1^{0,1}=0,$$

以避免所谓的"长期"项 $\xi_1\sin\xi_1$ 和 $\xi_2\sin\xi_2$. 这为我们提供了代数方程

$$(A_0k_1)^2+\left[2+\frac{k_1}{2k_2}\sin^2(\alpha_1-\alpha_2)\right](B_0k_2)^2=\frac{\sigma_1^2}{gk_1}-1, \tag{16.58}$$

$$\left[2+\frac{k_2}{2k_1}\sin^2(\alpha_1-\alpha_2)\right](A_0k_1)^2+(B_0k_2)^2=\frac{\sigma_2^2}{gk_2}-1, \tag{16.59}$$

其解为

$$A_0=\pm\left(\frac{\varepsilon}{k_1}\right)\sqrt{\left(\frac{\sigma_2^2}{gk_2}-1\right)\left[2+\frac{k_1}{2k_2}\sin^2(\alpha_1-\alpha_2)\right]-\left(\frac{\sigma_1^2}{gk_1}-1\right)}, \tag{16.60}$$

$$B_0=\pm\left(\frac{\varepsilon}{k_2}\right)\sqrt{\left(\frac{\sigma_1^2}{gk_1}-1\right)\left[2+\frac{k_2}{2k_1}\sin^2(\alpha_1-\alpha_2)\right]-\left(\frac{\sigma_2^2}{gk_2}-1\right)}, \tag{16.61}$$

其中

$$\varepsilon = \left\{ \left[2 + \frac{k_1}{2k_2} \sin^2(\alpha_1 - \alpha_2) \right] \left[2 + \frac{k_2}{2k_1} \sin^2(\alpha_1 - \alpha_2) \right] - 1 \right\}^{-1/2}.$$

注意, A_0 和 B_0 有多个值: 它们可以是正值或负值.

然后, 通过 (16.53) 定义的线性逆算子 $\bar{\mathcal{L}}^{-1}$, 得到 $\phi_1(\xi_1, \xi_2, z)$ 的通解为

$$\phi_1 = A_1 \sqrt{\frac{g}{k_1}}\, \Psi_{1,0} + B_1 \sqrt{\frac{g}{k_2}}\, \Psi_{0,1} + b_1^{1,1} \left(\frac{\Psi_{1,1}}{\lambda_{1,1}} \right) + d_1^{1,1} \left(\frac{\Psi_{1,-1}}{\lambda_{1,-1}} \right)$$
$$+ b_1^{2,1} \left(\frac{\Psi_{2,1}}{\lambda_{2,1}} \right) + d_1^{2,1} \left(\frac{\Psi_{2,-1}}{\lambda_{2,-1}} \right) + b_1^{1,2} \left(\frac{\Psi_{1,2}}{\lambda_{1,2}} \right) + d_1^{1,2} \left(\frac{\Psi_{1,-2}}{\lambda_{1,-2}} \right), \quad (16.62)$$

其中 A_1 和 B_1 为未知系数, 特征函数 $\Psi_{m,n}$ 和特征值 $\lambda_{m,n}$ 由 (16.22) 和 (16.48) 定义. 换句话说, ϕ_1 是特征函数之和 (或线性组合). 注意, 对于任意常数 A_1 和 B_1, ϕ_1 自动满足拉普拉斯方程 (16.38) 和液体底面条件 (16.39). 另外, 给定初始猜测解 ϕ_0, 可以通过公式 (16.41) 直接计算得到 $\eta_1(\xi_1, \xi_2)$.

上述方法具有一般意义. 类似地, 我们可以按照 $m = 1, 2, 3, \cdots$ 的顺序依次得到 $\eta_m(\xi_1, \xi_2)$ 和 $\phi_m(\xi_1, \xi_2, z)$. 注意, 两个未知系数 A_m 和 B_m ($m \geqslant 1$) 可以像 A_0 和 B_0 一样通过避免"长期"项 $\xi_1 \sin \xi_1$ 和 $\xi_2 \sin \xi_2$ 来精确确定. 注意, 上述方法只需要基本运算, 因此可以方便地通过科学计算软件 (例如 Mathematica) 获得高阶近似值.

不失一般性, 这里我们考虑两个基波的特殊情况, 即

$$\frac{\sigma_1}{\sqrt{gk_1}} = \frac{\sigma_2}{\sqrt{gk_2}} = 1.0003, \quad \alpha_1 = 0, \quad \alpha_2 = \frac{\pi}{36}, \quad k_2 = \frac{\pi}{5}, \quad (16.63)$$

具有不同的 k_1/k_2 比值. 在这里, 选择数字 1.0003, 以使摄动理论具有高精度, 因此我们可以将结果与 Phillips [393] 和 Longuet-Higgins [390] 的结果进行比较. Phillips [393] 和 Longuet-Higgins [390] 指出如果满足共振条件 (16.50), 则波浪共振 (幅值随时间增长) 发生, 即在当前算例中

$$\frac{k_2}{k_1} \approx 0.8925, \quad \frac{\sigma_1}{\sigma_2} = \sqrt{\frac{k_1}{k_2}} \approx 1.0585.$$

如前所述, 当满足上述共振条件时, 存在三个特征值为零的特征函数 ($N_\lambda = 3$). 为了避免这种情况, 在这里我们考虑除 $k_2/k_1 = 0.8925$ 外的具有不同波数 k_1 外的非共振波.

先考虑 $k_2/k_1 = 1$ 的情况. 不失一般性, 我们先选取由 (16.60) 和 (16.61) 给出的 A_0 和 B_0 的负值. 我们发现, 随着近似阶数 m 的增加, 平方残差 E_m^ϕ

在区间 $-1.8 \leqslant c_0 < 0$ 内减小, c_0 的最优值接近 -1, 如图 16.1 所示. 实际上, 当 $c_0 = -1$ 时, 两个边界条件对应的平方残差在 10 阶近似处迅速减小到 10^{-24} 量级, 如表 16.1 所示. 根据上述收敛定理, 对应的同伦级数 (16.36) 和 (16.37) 一定是该问题的解.

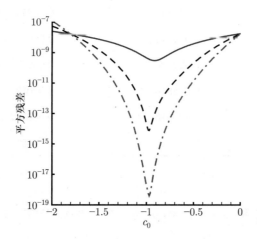

图 16.1 当 (16.63) 和 $k_2/k_1 = 1$ 时, 平方残差 E_m^ϕ 相对于 c_0 的曲线. 实线: 1 阶近似; 虚线: 3 阶近似; 点划线: 5 阶近似

表 16.1 当 (16.63) 和 $k_2/k_1 = 1$ 时, 两个非线性边界条件的平方残差

近似阶数 m	E_m^ϕ	E_m^η
1	1.9×10^{-8}	5.1×10^{-4}
3	3.5×10^{-12}	1.2×10^{-9}
5	2.0×10^{-16}	4.3×10^{-14}
8	2.8×10^{-22}	3.8×10^{-20}
10	4.7×10^{-26}	6.4×10^{-24}

令 $a_{1,0}, a_{0,1}$ 和 $a_{2,-1}$ 分别表示波分量 $\cos\xi_1$, $\cos\xi_2$ 和 $\cos(2\xi_1 - \xi_2)$ 的幅值. 显然, 在 $k_1 = k_2$ 的情况下, $a_{1,0} = a_{0,1}$. 如表 16.2 所示, 每个波分量快速收敛, 这与同伦-帕德方法 [380,389] 得到的结果非常吻合, 如表 16.3 所示. 所有这些都表明了上述基于同伦分析方法的方法的有效性.

表 16.2 当 (16.63) 和 $k_2/k_1 = 1$ 时，一些波分量幅值的解析近似

近似阶数 m	$a_{1,0} = a_{0,1}$	$a_{2,-1}$
1	-0.022502	0
2	-0.022846	0.00059739
3	-0.022814	0.00057198
4	-0.022816	0.00057208
6	-0.022816	0.00057226
8	-0.022816	0.00057226
10	-0.022816	0.00057226

表 16.3 当 (16.63) 和 $k_2/k_1 = 1$ 时，$[m, m]$ 同伦–帕德方法给出的一些波分量幅值的解析近似

m	$a_{1,0} = a_{0,1}$	$a_{2,-1}$
2	-0.022816	0.00057211
3	-0.022816	0.00057226
4	-0.022816	0.00057226
5	-0.022816	0.00057226

同理，我们可以得到 (16.63) 在不同 k_2/k_1 比值情况下收敛的级数解，如表 16.4 所示。注意，随着 k_2/k_1 比值的减小，波分量 $\cos(2\xi_1 - \xi_2)$ 的幅值首先单调增加到 $k_2/k_1 = 0.8925$ 时的最大值 (对应的结果由 §16.2.3 中描述的方法给出)，然后单调减少。注意，$a_{1,0}$ 和 $a_{0,1}$ 的变化不是单调的。

Phillips [393] 指出，当满足共振条件 (16.50) 时，会发生波浪共振，使得波分量 $\cos(2\xi_1 - \xi_2)$ 的幅值随着时间的增加而增加，也就是说，它包含越来越多的波能。然而，如表 16.4所示，对于 k_2/k_1 的所有值，即使满足共振条件 (16.50)，$|a_{1,0}|$ 和 $|a_{0,1}|$ 比 $|a_{2,1}|$ 的值都大得多。从物理上讲，这意味着共振波不需要拥有大部分的波能! 我们稍后将详细讨论这个非常有趣的现象。

表 16.4 在 (16.63) 的情况下, 具有不同 k_2/k_1 值的波分量幅值

k_2/k_1	$a_{1,0}$	$a_{0,1}$	$a_{2,-1}$
1.000	-0.0228	-0.0228	0.00057
0.950	-0.0229	-0.0218	0.00084
0.930	-0.0230	-0.0215	0.00116
0.920	-0.0231	-0.0214	0.00150
0.910	-0.0231	-0.0215	0.00219
0.905	-0.0232	-0.0216	0.00290
0.900	-0.0230	-0.0220	0.00433
0.893	-0.0205	-0.0232	0.00898
0.880	-0.0235	-0.0179	0.00307
0.860	-0.0221	-0.0189	0.00120
0.850	-0.0226	-0.0187	0.00087
0.830	-0.0229	-0.0182	0.00054
0.800	-0.0232	-0.0172	0.00033
0.700	-0.0239	-0.0121	0.00010

16.2.3 共振波

这里, 我们进一步考虑 (16.63) 具有特殊比值 $k_2/k_1 = 0.8925$ 的情况, 即满足共振条件 (16.50). 该条件由 Phillips [393] 首次提出, 然后由 Longuet-Higgins [390] 证实, 当所谓的波浪共振发生时, 如果 $a_{2,-1}$ 最初为零, 那么波分量 $\cos(2\xi_1 - \xi_2)$ 的幅值将随时间的增加而增加.

在这种情况下, 满足共振条件 (16.50), 即

$$g|2\mathbf{k}_1 - \mathbf{k}_2| = (2\bar{\sigma}_1 - \bar{\sigma}_2)^2.$$

根据 (16.63), 我们有

$$k_1 = 0.703998, \quad k_2 = 0.628319, \quad k_3 = |2\mathbf{k}_1 - \mathbf{k}_2| = 0.783981.$$

因此, 根据 (16.48), 我们有一个额外的零特征值 $\lambda_{2,-1} = 0$ 和特征函数

$$\Psi_{2,-1} = e^{|2\mathbf{k}_1 - \mathbf{k}_2|z} \sin(2\xi_1 - \xi_2).$$

所以，我们现在有三个特征函数 $\Psi_{1,0}, \Psi_{0,1}$ 和 $\Psi_{2,-1}$，它们的特征值均为零，即 $\lambda_{1,0} = 0$, $\lambda_{0,1} = 0$ 和 $\lambda_{2,-1} = 0$. 这里需要强调的是，Phillips [393] 提出的波浪共振条件在数学上等价于

$$\lambda_{m,n} = 0, \quad m^2 + n^2 > 1.$$

注意，额外的零特征值 $\lambda_{2,-1}$ 使得 §16.2.2 中提到的方法失效，因为 (16.62) 中的 $\Psi_{2,-1}/\lambda_{2,-1}$ 项会变得无穷大：这正是 Phillips [393] 和 Longuet-Higgins [390] 提出幅值随时间的增加而增加的所谓波浪共振的原因. 但是，由于波的能量不可能是无限的，而且对于足够大的幅值波会发生波破碎，因此，在物理上这种增长仅在很短的时间内发生是可能的.

需要注意的是，§16.2.2 中研究的非共振波的初始猜测解 (16.54) 是特征值 $\lambda_{1,0}$ 和 $\lambda_{0,1}$ 为零的特征函数 $\Psi_{1,0}$ 和 $\Psi_{0,1}$ 两者的线性组合. 对于波浪共振的情况，唯一的区别是我们有一个额外的特征函数 $\Psi_{2,-1}$，其特征值 $\lambda_{2,-1}$ 也为零. 注意，同伦分析方法为我们提供了极大的自由来选择初始猜测解. 有了这样的自由度，为什么不使用这三个特征函数 (特征值为零) 来表达初始猜测解 ϕ_0？换句话说，我们可以通过特征值为零的所有特征函数来表达初始猜测解 ϕ_0，即

$$\phi_0(\xi_1, \xi_2, z) = \bar{A}_0 \sqrt{\frac{g}{k_1}} \Psi_{1,0} + \bar{B}_0 \sqrt{\frac{g}{k_2}} \Psi_{0,1} + \bar{C}_0 \sqrt{\frac{g}{k_3}} \Psi_{2,-1}, \quad (16.64)$$

其中 $\bar{A}_0, \bar{B}_0, \bar{C}_0$ 为未知常数. 类似地，将上述表达式代入形变方程 (16.38) 至 (16.40)，我们有相同的一阶形变方程 (16.55) 以及相同的液体底面边界条件 (16.57)，但在 $z = 0$ 上的边界条件对应更复杂的方程，即

$$\begin{aligned}
\bar{\mathcal{L}}(\phi_1) &= \bar{b}_1^{1,0} \sin(\xi_1) + \bar{b}_1^{0,1} \sin(\xi_2) + \bar{b}_1^{2,0} \sin(2\xi_1) + \bar{b}_1^{3,0} \sin(3\xi_1) \\
&+ \bar{b}_1^{1,1} \sin(\xi_1 + \xi_2) + \bar{d}_1^{1,1} \sin(\xi_1 - \xi_2) \\
&+ \bar{b}_1^{2,1} \sin(2\xi_1 + \xi_2) + \bar{d}_1^{2,1} \sin(2\xi_1 - \xi_2) \\
&+ \bar{b}_1^{1,2} \sin(\xi_1 + 2\xi_2) + \bar{d}_1^{1,2} \sin(\xi_1 - 2\xi_2) \\
&+ \bar{d}_1^{2,2} \sin(2\xi_1 - 2\xi_2) + \bar{d}_1^{2,3} \sin(2\xi_1 - 3\xi_2) \\
&+ \bar{d}_1^{3,1} \sin(3\xi_1 - \xi_2) + \bar{d}_1^{3,2} \sin(3\xi_1 - 2\xi_2) \\
&+ \bar{d}_1^{4,1} \sin(4\xi_1 - \xi_2) + \bar{d}_1^{4,2} \sin(4\xi_1 - 2\xi_2) + \bar{d}_1^{4,3} \sin(4\xi_1 - 3\xi_2) \\
&+ \bar{d}_1^{5,2} \sin(5\xi_1 - 2\xi_2) + \bar{d}_1^{6,3} \sin(6\xi_1 - 3\xi_2), \quad (16.65)
\end{aligned}$$

其中 $\bar{b}_1^{m,n}, \bar{d}_1^{m,n}$ 为常系数，线性算子 $\bar{\mathcal{L}}$ 由 (16.43) 定义. 注意，此时存在三个零特征值，即 $\lambda_{1,0} = 0$, $\lambda_{0,1} = 0$ 和 $\lambda_{2,-1} = 0$. 因此，根据逆算子 $\bar{\mathcal{L}}^{-1}$ 的定义

(16.53), 不仅 $\bar{b}_1^{1,0}$ 和 $\bar{b}_1^{0,1}$ 这两个系数, 还有附加系数 $\bar{d}_1^{2,1}$ 必须为零, 以避免长期项. 令

$$\bar{b}_1^{1,0} = 0, \quad \bar{b}_1^{0,1} = 0, \quad \bar{d}_1^{2,1} = 0,$$

对于上述的特殊情况, 给出了一组非线性代数方程

$$\begin{cases} 12.7576\bar{A}_0^2 + 20.3675\bar{B}_0^2 + 31.6768\bar{C}_0^2 + 25.6718\bar{B}_0\bar{C}_0 = 0.01545, \\ 24.1456\bar{A}_0^2 + 9.6004\bar{B}_0^2 + 30.0398\bar{C}_0^2 + 14.9558\bar{A}_0^2\bar{C}_0/\bar{B}_0 = 0.01459, \\ 26.9621\bar{A}_0^2 + 21.6116\bar{B}_0^2 + 16.6956\bar{C}_0^2 + 10.7158\bar{A}_0^2\bar{B}_0/\bar{C}_0 = 0.01630, \end{cases} \quad (16.66)$$

该非线性代数方程组有四个复数解和十二个实数解. 由于复数解没有物理意义, 我们在表 16.5 中只列出了十二个实数解. 我们发现, 这十二个实数解分为三组, 不同的组给出不同的解, 如下所述. 求解这组非线性代数方程后, 初始猜测解 $\phi_0(\xi_1, \xi_2, z)$ 已知, 因此可以直接通过 (16.41) 得到 $\eta_1(\xi_1, \xi_2)$. 更重要的是, 此时, (16.65) 的右端项 $\sin\xi_1, \sin\xi_2$, 尤其是 $\sin(2\xi_1 - \xi_2)$ 消失了. 然后, 根据逆算子 (16.53), 很容易得到一阶近似的通解

$$\begin{aligned} \phi_1 = & \bar{A}_1 \sqrt{\frac{g}{k_1}} \Psi_{1,0} + \bar{B}_1 \sqrt{\frac{g}{k_2}} \Psi_{0,1} + \bar{C}_1 \sqrt{\frac{g}{k_3}} \Psi_{2,-1} \\ & + \bar{b}_1^{2,0} \left(\frac{\Psi_{2,0}}{\lambda_{2,0}} \right) + \bar{b}_1^{3,0} \left(\frac{\Psi_{3,0}}{\lambda_{3,0}} \right) + \bar{b}_1^{1,1} \left(\frac{\Psi_{1,1}}{\lambda_{1,1}} \right) + \bar{d}_1^{1,1} \left(\frac{\Psi_{1,-1}}{\lambda_{1,-1}} \right) \\ & + \bar{b}_1^{2,1} \left(\frac{\Psi_{2,1}}{\lambda_{2,1}} \right) + \bar{b}_1^{1,2} \left(\frac{\Psi_{1,2}}{\lambda_{1,2}} \right) + \bar{d}_1^{1,2} \left(\frac{\Psi_{1,-2}}{\lambda_{1,-2}} \right) + \bar{d}_1^{2,2} \left(\frac{\Psi_{2,-2}}{\lambda_{2,-2}} \right) \\ & + \bar{d}_1^{2,3} \left(\frac{\Psi_{2,-3}}{\lambda_{2,-3}} \right) + \bar{d}_1^{3,1} \left(\frac{\Psi_{3,-1}}{\lambda_{3,-1}} \right) + \bar{d}_1^{3,2} \left(\frac{\Psi_{3,-2}}{\lambda_{3,-2}} \right) + \bar{d}_1^{4,1} \left(\frac{\Psi_{4,-1}}{\lambda_{4,-1}} \right) \\ & + \bar{d}_1^{4,2} \left(\frac{\Psi_{4,-2}}{\lambda_{4,-2}} \right) + \bar{d}_1^{4,3} \left(\frac{\Psi_{4,-3}}{\lambda_{4,-3}} \right) + \bar{d}_1^{5,2} \left(\frac{\Psi_{5,-2}}{\lambda_{5,-2}} \right) + \bar{d}_1^{6,3} \left(\frac{\Psi_{6,-3}}{\lambda_{6,-3}} \right). \end{aligned}$$
(16.67)

这里需要强调的是, 上述表达式中的所有特征值 $\lambda_{m,n}$ 都是非零的, 因此 $\phi_1(\xi_1, \xi_2, z)$ 是有限的. 此外, 上述表达式中的所有系数都与时间无关, 因此相应的波形不会随时间的增加而增长. 与 (16.64) 定义的初始猜测解 $\phi_0(\xi_1, \xi_2, z)$ 一样, (16.67) 给出的通解 $\phi_1(\xi_1, \xi_2, z)$ 具有三个未知系数 $\bar{A}_1, \bar{B}_1$ 和 $\bar{C}_1$, 类似地, 其可通过避免 $\phi_2(\xi_1, \xi_2, z)$ 中的 "长期" 项来确定. 所以, 上述方法具有普遍意义. 因此, 通过类似的方式, 我们可以按照 $m = 1, 2, 3, \cdots$ 的顺序依次得到 $\eta_m(\xi_1, \xi_2)$ 和 $\phi_m(\xi_1, \xi_2, z)$.

表 16.5 (16.66) 的解

解的序列数 K	$\bar{A}_0$	$\bar{B}_0$	$\bar{C}_0$
(Group-I)			
1	−0.0156112	0.0282054	−0.0084973
2	−0.0156112	−0.0282054	0.0084973
3	0.0156112	0.0282054	−0.0084973
4	0.0156112	−0.0282054	0.0084973
(Group-II)			
5	−0.0155774	−0.0141927	−0.0113800
6	−0.0155774	0.0141927	0.0113800
7	0.0155774	−0.0141927	−0.0113800
8	0.0155774	0.0141927	0.0113800
(Group-III)			
9	−0.0155626	0.0106109	−0.0226353
10	−0.0155626	−0.0106109	0.0226353
11	0.0155626	0.0106109	−0.0226353
12	0.0155626	−0.0106109	0.0226353

需要注意的是，表 16.4 中的波分量幅值 $a_{1,0}$ 和 $a_{0,1}$ 为负值. 为了计算 $k_2/k_1 = 0.8925$ 时 (16.63) 的对应波分量幅值，我们选择 Group-I 中的第 2 个解, 对应于 $K = 2$, 即

$$\bar{A}_0 = -0.0156112, \quad \bar{B}_0 = -0.0282054, \quad \bar{C}_0 = 0.0084973.$$

类似地，图 16.2 给出了对应的平方残差 E_m^ϕ 相对于 c_0 的曲线，可以看出 c_0 的最优值接近 −1. 确实, 选取 $c_0 = -1$, 两个边界条件的平方残差 E_m^ϕ 和 E_m^η 在 20 阶同伦近似处迅速减小到 10^{-19} 量级, 如表 16.6 所示. 根据上述收敛定理, 同伦级数 (16.36) 和 (16.37) 满足原控制方程 (16.14) 和所有边界条件 (16.17), (16.18) 和 (16.19). 此外, 我们发现对应的波分量幅值 $a_{1,0}, a_{0,1}$ 和 $a_{2,-1}$ 分别收敛于 −0.020512, −0.023212, 0.0089752, 如表 16.7 所示. 为了验证其收敛性, 我

16.2 两个小幅值基波的共振条件

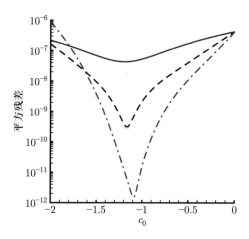

图 16.2 在 (16.63) 和 $k_2/k_1 = 0.8925$ 的情况下, 平方残差 E_m^ϕ 相对于 c_0 的曲线. 实线: 1 阶近似; 虚线: 3 阶近似; 点划线: 5 阶近似

表 16.6 当 $k_2/k_1 = 0.8925$ 时, (16.63) 在两个边界条件处的平方残差

近似阶数 m	E_m^ϕ	E_m^η
1	4.0×10^{-7}	4.8×10^{-4}
3	1.8×10^{-8}	1.0×10^{-6}
5	2.4×10^{-10}	1.2×10^{-8}
8	1.2×10^{-13}	1.5×10^{-11}
10	1.3×10^{-14}	1.2×10^{-12}
15	1.2×10^{-17}	3.4×10^{-16}
20	1.2×10^{-20}	1.9×10^{-19}

们进一步采用同伦 – 帕德技术 [380, 389] 来加速收敛并得到了相同的收敛波分量幅值

$$a_{1,0} = -0.0205119, \quad a_{0,1} = -0.0232118, \quad a_{2,-1} = 0.0089752,$$

如表 16.8 所示. 因此, 即使在完全满足波浪共振条件 (16.50) 的情况下, 我们也能得到具有恒定幅值的共振波系收敛的级数解.

将上述结果与表 16.4 的结果相结合, 我们得到了 (16.63) 对应的无量纲波分量幅值 $k_3(a_{2,-1})$ 相对于 k_2/k_1 的关系, 如图 16.3 所示. 的确, 当 k_2/k_1 的比

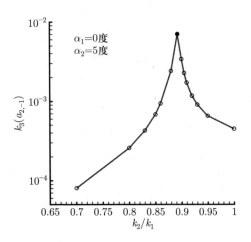

图 16.3 (16.63) 对应的无量纲波分量幅值 $k_3(a_{2,-1})$ 相对于 k_2/k_1 的曲线. 实心圆: $k_2/k_1 = 0.8925$ 时的结果; 空心圆: $k_2/k_1 \neq 0.8925$ 时的结果

表 16.7 当 $k_2/k_1 = 0.8925$ 时, (16.63) 的波分量幅值的同伦近似

近似阶数 m	$a_{1,0}$	$a_{0,1}$	$a_{2,-1}$
1	−0.015616	−0.028214	0.0084999
3	−0.020587	−0.023469	0.0094143
5	−0.020533	−0.023231	0.0090083
7	−0.020504	−0.023220	0.0089755
9	−0.020511	−0.023213	0.0089757
11	−0.020512	−0.023212	0.0089753
13	−0.020512	−0.023212	0.0089752
15	−0.020512	−0.023212	0.0089752
18	−0.020512	−0.023212	0.0089752

值变为 0.8925 时, 对应于波浪共振条件 (16.50), 无量纲波分量幅值 $k_3\,a_{2,-1}$ 增加到最大值. 注意, 即使共振条件 (16.50) 精确满足, 波分量 $\cos(2\xi_1 - \xi_2)$ 的幅值 $|a_{2,-1}|$ 也为有限常数. 此外, 非常有趣的是, 在 $k_2/k_1 = 0.8925$ 的情况下, 共振波分量的幅值 $|a_{2,-1}|$ 甚至要比两个基波的波幅 $|a_{1,0}|$ 和 $|a_{0,1}|$ 小得多!

上述结果是基于表 16.5 中 Group I 的第二个解得到的, 对应于 $K = 2$. 同样, 基于表 16.5 中的不同解, 我们可以得到对应的不同的收敛级数解. 我

表 16.8 当 $k_2/k_1 = 0.8925$ 时，由 $[m,m]$ 阶同伦−帕德方法给出的 (16.63) 的波分量幅值的同伦近似

m	$a_{1,0}$	$a_{0,1}$	$a_{2,-1}$
2	−0.0206010	−0.0232058	0.0089660
3	−0.0204911	−0.0232251	0.0089568
4	−0.0205102	−0.0232172	0.0089780
5	−0.0205122	−0.0232118	0.0089752
6	−0.0205119	−0.0232118	0.0089752
7	−0.0205119	−0.0232118	0.0089752
8	−0.0205119	−0.0232118	0.0089752
9	−0.0205119	−0.0232118	0.0089752

们发现表 16.5 中每组的四个不同的解给出的波分量幅值的绝对值 $|a_{1,0}|, |a_{0,1}|$ 和 $|a_{2,-1}|$ 完全相同，如表 16.9 所示. 考虑到波能谱是由波分量幅值的平方决定的，我们将每组中的四种不同解认为是相同的. 因此，当 $k_2/k_1 = 0.8925$ 时，(16.63) 存在具有不同波能谱的三种不同共振波模态. 这里需要强调的是，共振波的幅值 $|a_{2,-1}|$ 在 Group I 中是最小的，在 Group II 中处于中间，在 Group III 中是最大的. 所以，共振波的幅值 $|a_{2,-1}|$ 一点也不特殊：它和两个基波的幅值分量 $|a_{1,0}|$ 和 $|a_{0,1}|$ 是一样的！非常有趣的是，对于一个完全发展的波系，即使精确满足 Phillips 共振条件，也存在多个定常解，而且共振波幅值可能比基波幅值小得多. 据我们所知，其他数值和解析方法从未报道过这一有趣的结果. 这说明了同伦分析方法可以作为有用的工具来求解复杂非线性偏微分方程，从而加深我们对一些非线性现象的物理理解.

表 16.9 表 16.5 中 (16.66) 的不同根给出的 $k_2/k_1 = 0.8925$ 时的 (16.63) 情况下共振波系的多个幅值

	$|a_{1,0}|$	$|a_{0,1}|$	$|a_{2,-1}|$	$|a_{1,-2}|$
Group I	0.02051186921	0.02321179687	0.00897520547	0.00022907754
Group II	0.01475607438	0.01002488089	0.01464333477	0.00096077706
Group III	0.00971236473	0.01032248128	0.02576462018	0.00059968416

令 Π 表示所有波分量幅值的平方和，写作

$$\Pi_0 = a_{1,0}^2 + a_{0,1}^2 + a_{2,-1}^2.$$

对于 (16.63)，$k_2/k_1 = 0.8925$ 时波浪共振的发生，我们发现此时 Π_0/Π 对应 Group I，II 和 III 的值分别为 $98.82\%, 98.27\%$ 和 99.76%. 这主要是因为 $a_{1,-2}$ 等其他波分量的幅值要小得多，如表 16.9 所示. 所以，这三个波分量几乎包含了整个波系的能量. 注意，给定两个波数为 $\mathbf{k}_1$ 和 $\mathbf{k}_2$ 的基波，存在无穷多个不同的波分量 $a_{m,n} \cos(m\xi_2 + n\xi_2)$ 与波数 $m\mathbf{k}_1 + n\mathbf{k}_2$，其中 m 和 n 为任意整数. 令

$$\mathbf{K} = \{m\mathbf{k}_1 + n\mathbf{k}_2 | m, n \text{ 为整数}\}$$

表示这些波数的集合. $\mathbf{K}$ 中的每个波数均对应于 (16.22) 定义的特征函数和 (16.48) 定义的特征值. 计算表明，对于一个由两个小幅值的行进基波组成的充分发展的波系，波能主要集中在特征值为零（或接近于零）的波分量上. 这为我们提供了对所谓的"波浪共振"的另一种解释. 根据这一解释，共振波分量 $a_{2,-1}\cos(2\xi_1 - \xi_2)$ 与两个基波 $a_{1,0}\cos\xi_1$ 和 $a_{0,1}\cos\xi_2$ 一样重要，因此并不特别. 这个观点可以很好地解释为什么共振波的幅值往往小于非共振波的幅值.

当 $\alpha_1 = 0, \alpha_2 = \pi/36, k_2/k_1 = 0.8925$ 时，$\sigma_1/\sqrt{gk_1} = \sigma_2/\sqrt{gk_2}$ 不同比值对应的由两个行进基波组成的充分发展波系（对应 Group I）的波能分布，如表 16.10 所示. 我们发现两个基波包含了整个波系的大部分波能. 此外，随着 $\sigma_1/\sqrt{gk_1} = \sigma_2/\sqrt{gk_2}$ 比值的增加，共振波分量 $a_{2,-1}\cos(2\xi_1 - \xi_2)$ 包含总波能的百分比越来越小. 特别地，当 $\sigma_1/\sqrt{gk_1} = \sigma_2/\sqrt{gk_2} = 1.0008$ 时，Group I

表 16.10 当 $\alpha_1 = 0, \alpha_2 = \pi/36, k_2/k_1 = 0.8925$ 时，$\sigma_1/\sqrt{gk_1} = \sigma_2/\sqrt{gk_2}$ 不同比值对应的共振波 (Group I) 波能分布

$\sigma_1/\sqrt{gk_1} =$ $\sigma_2/\sqrt{gk_2}$	$a_{1,0}^2/\Pi$	$a_{0,1}^2/\Pi$	$a_{2,-1}^2/\Pi$	Π_0/Π
1.0001	36.97%	53.95%	8.99%	99.91%
1.0002	38.40%	52.74%	8.42%	99.56%
1.0003	39.97%	51.19%	7.65%	98.82%
1.0004	41.59%	49.28%	6.70%	97.57%
1.0005	43.10%	47.09%	5.60%	95.78%
1.0006	44.38%	44.75%	4.43%	93.56%
1.0007	45.65%	42.73%	3.58%	91.95%
1.0008	46.23%	39.07%	2.22%	87.52%

的共振波分量仅包含总波能的 2.22%. 这个结果非常有趣但令人惊讶, 因为根据传统理论, 共振波应该具有很大的波幅, 因此包含整个波系的大部分能量. 类似地, 表 16.11 中给出了充分发展波系 (即 Group II) 的波能分布, 发现 Group II 的共振波具有与其中一个基波相当的幅值以及相当的总波能百分比. 在三组中, 只有一组 (即 Group III) 共振波具有最大的幅值, 因此包含了整个波系的大部分能量, 如图 16.4 和表 16.12 所示.

表 16.11 当 $\alpha_1 = 0, \alpha_2 = \pi/36, k_2/k_1 = 0.8925$ 时, $\sigma_1/\sqrt{gk_1} = \sigma_2/\sqrt{gk_2}$ 不同比值对应的共振波 (Group II) 波能分布

$\sigma_1/\sqrt{gk_1} =$ $\sigma_2/\sqrt{gk_2}$	$a_{1,0}^2/\Pi$	$a_{0,1}^2/\Pi$	$a_{2,-1}^2/\Pi$	Π_0/Π
1.0001	42.30%	20.43%	37.07%	99.79%
1.0002	41.26%	19.50%	38.44%	99.20%
1.0003	40.17%	18.54%	39.56%	98.27%
1.0004	39.05%	17.56%	40.44%	97.06%
1.0005	37.93%	16.56%	41.12%	95.61%
1.0006	36.79%	15.54%	41.64%	93.97%
1.0007	35.65%	14.51%	42.02%	92.18%
1.0008	34.49%	13.46%	42.29%	90.24%

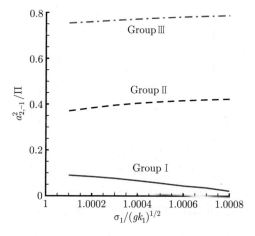

图 16.4 当 $k_2/k_1 = 0.8925$ 以及 $\sigma_1/\sqrt{gk_1} = \sigma_2/\sqrt{gk_2}$ 时, 对于不同的解组, 在 (16.63) 的情况下, $a_{2,-1}^2/\Pi$ 相对于 $\sigma_1/\sqrt{gk_1}$ 的变化曲线. 实线: Group I; 虚线: Group II; 点划线: Group III

表 16.12 当 $\alpha_1 = 0, \alpha_2 = \pi/36, k_2/k_1 = 0.8925$ 时，$\sigma_1/\sqrt{gk_1} = \sigma_2/\sqrt{gk_2}$ 不同比值对应的共振波 (Group III) 波能分布

$\sigma_1/\sqrt{gk_1} = \sigma_2/\sqrt{gk_2}$	$a_{1,0}^2/\Pi$	$a_{0,1}^2/\Pi$	$a_{2,-1}^2/\Pi$	Π_0/Π
1.0001	12.13%	12.47%	75.37%	99.97%
1.0002	11.53%	12.38%	75.99%	99.90%
1.0003	10.88%	12.29%	76.58%	99.76%
1.0004	10.20%	12.20%	77.15%	99.55%
1.0005	9.47%	12.11%	77.68%	99.26%
1.0006	8.72%	12.01%	78.15%	98.88%
1.0007	7.94%	11.90%	78.53%	98.38%
1.0008	7.15%	11.78%	78.82%	97.75%

需要注意的是，基波和共振波包含了整个波系的大部分能量，特别是当所有波分量幅值都很小时；然而，随着波幅的增加，它们包含总波能的百分比越来越小，如图 16.5 所示. 这也表明 Phillips 波浪共振条件 (16.6) 仅适用于小幅值的行进波.

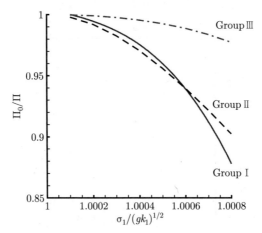

图 16.5 当 $k_2/k_1 = 0.8925$ 以及 $\sigma_1/\sqrt{gk_1} = \sigma_2/\sqrt{gk_2}$ 时，不同的解组，在 (16.63) 的情况下，Π_0/Π 相对于 $\sigma_1/\sqrt{gk_1}$ 的变化曲线. 实线: Group I; 虚线: Group II; 点划线: Group III

Phillips [393] 研究了两个小幅值基波 $a_{1,0}\cos\xi_1$ 和 $a_{0,1}\cos\xi_2$ 在 $t = 0$ 时的非线性相互作用. Phillips [393] 揭示了当波分量 $a_{m,n}\cos(m\xi_1 + n\xi_2)$ 初始时

刻为零, 即 $a_{m,n} = 0$ 时, 如果满足共振条件 (16.50), 则波分量幅值 $a_{m,n}$ 随时间线性增长. 因此, Phillips 共振条件适用于短时间情况下的波系评估, 例如 $t \ll 1$. 由于在物理学中波幅不可能增加到无穷大, 因此尚不清楚当 $t \to +\infty$ 时共振波分量 $a_{m,n} \cos(m\xi_1 + n\xi_2)$ 是否包含整个波系的主要能量. 与 Phillips [393] 不同, 我们在这里考虑两个小幅值基波的"定常"非线性相互作用, 即 $t \to +\infty$ 时的波系评估, 也可以说是, 一个充分发展的波系. 我们的计算表明, 如果所有波分量都充分发展, 使它们彼此间处于平衡状态, 并且满足 Phillips 共振条件 (16.50), 则存在多个共振波分量 $a_{m,n} \cos(m\xi_1 + n\xi_2)$. 此外, 共振波分量的幅值 $a_{m,n}$ 可能比基波幅值 $a_{1,0}$ 和 $a_{0,1}$ 要小得多, 即共振波分量不是必须具有整个波系的大部分能量. 因此, Phillips 共振条件不是大共振波分量的充分条件: 即使满足 Phillips 共振条件, 共振波的幅值也可能远小于两个基波幅值.

尽管上述结果是由 (16.63) 的特殊情况得到的, 但仍具有一般意义. 这强烈地表明, 对于由两个行进基波组成的充分发展的波系, 即使 Phillips 共振条件精确满足, 也存在多个定常"共振波系", 其中共振波分量幅值可能远小于基波的幅值. 对于任意数量的行进波, 该结论仍然成立, 如下所述. 详情请参考廖世俊 [387].

目前, 基于对具有连续谱的非线性随机波浪场演化的 DNS (直接数值模拟), Annenkov 等人 [370] 研究了精确共振、近似共振和非共振波相互作用, 结果表明, 波包的幅值趋于常数. 他们的结果虽然是针对连续波谱得到的, 但也支撑了我们上面提到的一些结论.

16.3 任意数量基波的共振条件

本章的后续部分将对 §16.2 中的波浪共振条件进行推广, 使其适用于任意数量的行进基波.

16.3.1 小幅值波的共振条件

基于同伦分析方法, 我们在 §16.2 中验证了对于由两个行进基波组成的充分发展的波系, 小幅值波的 Phillips 共振条件 (16.50) 在数学上等价于零特征值, 即

$$\lambda_{m,n} = g|m\mathbf{k}_1 + n\mathbf{k}_2| - (m\bar{\sigma}_1 + n\bar{\sigma}_2)^2 = 0, \qquad (16.68)$$

其中 m, n 为整数, $\bar{\sigma}_i = \sqrt{g|\mathbf{k}_i|}$ $(i = 1, 2)$. 这个结论具有普遍意义, 并且很容易在同伦分析方法框架中扩展, 以给出任意数量的小幅值行进波的共振条件. 关

键是在任意数量的行进波的情况下给出特征值的显式表达式.

让我们考虑 κ 个小幅值周期行进波的非线性相互作用, 其中 $2 \leqslant \kappa < +\infty$ 是一个任意整数. 假设波系充分发展, 即每个波分量的幅值与时间无关. 令 $\mathbf{k}_n$ 和 σ_n ($1 \leqslant n \leqslant \kappa$) 分别表示深水中第 n 个周期性行进波的波数和角频率. 定义变量

$$\xi_n = \mathbf{k}_n \cdot \mathbf{r} - \sigma_n t.$$

然后, 我们有速度势和波浪高程

$$\varphi(x,y,z,t) = \phi(\xi_1, \xi_2, \cdots, \xi_\kappa, z), \quad \zeta(x,y,t) = \eta(\xi_1, \xi_2, \cdots, \xi_\kappa).$$

类似地, 我们有

$$\frac{\partial^2 \varphi}{\partial t^2} = \sum_{m=1}^{\kappa} \sum_{n=1}^{\kappa} \sigma_m \sigma_n \frac{\partial^2 \phi}{\partial \xi_m \partial \xi_n},$$

$$\nabla \varphi = \left(\sum_{m=1}^{\kappa} \mathbf{k}_m \frac{\partial \phi}{\partial \xi_m} \right) + \mathbf{k} \frac{\partial \phi}{\partial z} = \mathbf{u} = \hat{\nabla} \phi,$$

$$\nabla^2 \varphi = \left(\sum_{m=1}^{\kappa} \sum_{n=1}^{\kappa} \mathbf{k}_m \cdot \mathbf{k}_n \frac{\partial^2 \phi}{\partial \xi_m \partial \xi_n} \right) + \frac{\partial^2 \phi}{\partial z^2} = \hat{\nabla}^2 \phi,$$

$$\hat{\nabla} \phi \cdot \hat{\nabla} \phi = \left(\sum_{m=1}^{\kappa} \sum_{n=1}^{\kappa} \mathbf{k}_m \cdot \mathbf{k}_n \frac{\partial \phi}{\partial \xi_m} \frac{\partial \phi}{\partial \xi_n} \right) + \left(\frac{\partial \phi}{\partial z} \right)^2 = \mathbf{u}^2,$$

$$\hat{\nabla} \phi \cdot \hat{\nabla} \psi = \left(\sum_{m=1}^{\kappa} \sum_{n=1}^{\kappa} \mathbf{k}_m \cdot \mathbf{k}_n \frac{\partial \phi}{\partial \xi_m} \frac{\partial \psi}{\partial \xi_n} \right) + \frac{\partial \phi}{\partial z} \frac{\partial \psi}{\partial z}.$$

注意,

$$\Psi_{m_1, m_2, \cdots, m_\kappa} = \exp\left(\left| \sum_{n=1}^{\kappa} m_n \mathbf{k}_n \right| z \right) \sin\left(\sum_{n=1}^{\kappa} m_n \xi_n \right) \tag{16.69}$$

满足 Laplace 方程 $\hat{\nabla} \phi = 0$, 即

$$\hat{\nabla}^2 \Psi_{m_1, m_2, \cdots, m_\kappa} = 0$$

以及液体底面边界条件 (16.5). 自由表面处的两个非线性边界条件可以通过上述定义的算子以类似的方式得到. 当然, 相应的偏微分方程和边界条件比 §16.2 中的要复杂得多. 即便如此, 我们也可以通过与上述类似的方式构造零阶形变方程并推导出相应的高阶形变方程.

同样，我们选择如下辅助线性算子

$$\mathcal{L}(\phi) = \left(\sum_{m=1}^{\kappa} \sum_{n=1}^{\kappa} \bar{\sigma}_m \, \bar{\sigma}_n \frac{\partial^2 \phi}{\partial \xi_m \partial \xi_n} \right) + g \frac{\partial^2 \phi}{\partial z^2}, \tag{16.70}$$

其中 $\bar{\sigma}_m = \sqrt{g\,|\mathbf{k}_m|}$ 是基于小幅值波的线性理论得到的结果. 根据 (16.69) 和 (16.70), 有

$$\mathcal{L}\left(\Psi_{m_1, m_2, \cdots, m_\kappa}\right) = \lambda_{m_1, m_2, \cdots, m_\kappa} \Psi_{m_1, m_2, \cdots, m_\kappa}, \tag{16.71}$$

其中

$$\lambda_{m_1, m_2, \cdots, m_\kappa} = g \left| \sum_{n=1}^{\kappa} m_n \mathbf{k}_n \right| - \left(\sum_{n=1}^{\kappa} m_n \bar{\sigma}_n \right)^2 \tag{16.72}$$

为特征值, 由 (16.69) 定义的 $\Psi_{m_1, m_2, \cdots, m_\kappa}$ 是线性高阶形变方程的特征函数. 类似地, (16.70) 的逆算子满足

$$\mathcal{L}^{-1}\left(\Psi_{m_1, m_2, \cdots, m_\kappa}\right) = \frac{\Psi_{m_1, m_2, \cdots, m_\kappa}}{\lambda_{m_1, m_2, \cdots, m_\kappa}}, \quad \lambda_{m_1, m_2, \cdots, m_\kappa} \neq 0. \tag{16.73}$$

需要注意的是, 逆算子 (16.73) 仅对非零特征值 $\lambda_{m_1, m_2, \cdots, m_\kappa} \neq 0$ 有效. 此外, 每个行进基波的特征值为零, 即

$$\lambda_{m_1, m_2, \cdots, m_\kappa} = 0, \quad \sum_{n=1}^{\kappa} m_n^2 = 1.$$

所以, κ 个基波至少存在 κ 个零特征值. 因此, 当零特征值多于 κ 个时, 就会发生所谓的波浪共振, 即

$$\lambda_{m_1, m_2, \cdots, m_\kappa} = 0, \quad \sum_{n=1}^{\kappa} m_n^2 > 1.$$

然后, 根据 (16.72), 我们有广义波浪共振条件

$$g \left| \sum_{n=1}^{\kappa} m_n \mathbf{k}_n \right| = \left(\sum_{n=1}^{\kappa} m_n \bar{\sigma}_n \right)^2, \quad \sum_{n=1}^{\kappa} m_n^2 > 1, \tag{16.74}$$

其中 $\bar{\sigma}_n = \sqrt{g\,|\mathbf{k}_n|}$ 是基于小幅值波的线性理论得到的结果.

还要注意的是, (16.50) 是上述广义波浪共振条件的一个特例. 此外, 它还包含 Phillips [393] 给出的共振条件. 因此, 其更具普遍性. 将 $\bar{\sigma}_n = \sqrt{g\,|\mathbf{k}_n|}$ 代入 (16.74) 给出小幅值波的广义波浪共振条件

$$\left| \sum_{n=1}^{\kappa} m_n \mathbf{k}_n \right| = \left(\sum_{n=1}^{\kappa} m_n \sqrt{|\mathbf{k}_n|} \right)^2, \quad \sum_{n=1}^{\kappa} m_n^2 > 1, \tag{16.75}$$

其与重力加速度无关.

假设对于给定的 κ 个行进基波, 有 $N_\lambda \geq \kappa$ 个其特征值为零的特征函数. 当 $N_\lambda = \kappa$ 时, 不发生波浪共振. 但是, 当 $N_\lambda > \kappa$ 时, 存在所谓的共振波. 为了简单起见, 令 Ψ_m^* ($1 \leq m \leq N_\lambda$) 表示特征值为零的第 m 个特征函数. 根据 (16.71), 对于任意常数 A_m 有

$$\mathcal{L}\left(\sum_{m=1}^{N_\lambda} A_m \Psi_m^*\right) = 0, \quad N_\lambda \geq \kappa. \tag{16.76}$$

所以, 我们可以选择这样一个初始猜测解

$$\phi_0 = \sum_{m=1}^{N_\lambda} B_{0,m} \Psi_m^*, \tag{16.77}$$

其中 $B_{0,m}$ 为未知数. 类似地, N_λ 个未知常数 $B_{0,m}$ ($1 \leq m \leq N_\lambda$) 可以通过避免 ϕ_1 中的 "长期" 项来确定. 此外, 根据 (16.76), ϕ_1 的通解中也包含 N_λ 个未知常数 $B_{1,m}$, 类似地, 可以通过避免 ϕ_2 中的 "长期" 项来确定. 通过这种方式, 可以依次求解相应的高阶形变方程. 类似地, 可以找到一个最优收敛控制参数 c_0 来保证同伦级数的收敛. 上述方法是通用的, 适用于任意数量的小幅值的周期性行进基波. 这为我们研究由任意数量的小幅值的周期性行进基波组成的充分发展波系提供了一种方法.

16.3.2 大幅值波的共振条件

需要注意的是, 广义波浪共振条件 (16.74) 仅在 $\bar{\sigma}_n = \sqrt{g|\mathbf{k}_n|}$ 时成立, 即小幅值重力波. 那么任意数量的大幅值行进重力波的共振条件是什么呢?

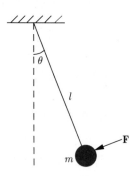

图 16.6 单摆的共振

为了回答这个问题, 我们考虑 (16.74) 的物理意义. 一般来说, 当外力 (或

扰动) 的频率等于动力系统的 "固有" 频率时, 就会发生动力系统所谓的共振现象. 比如我们考虑一个单摆的共振, 如图 16.6 所示, 其中 $\mathbf{F} = A\cos(\omega t + \alpha)$ 为外力, 频率为 ω, 相位差为 α. 令 ω_0 表示单摆的固有频率. 当最大振动角 $\theta_{\max}$ 较小时, $\sin\theta \approx \theta$, $\omega_0 \approx \sqrt{g/l}$ 近似良好. 所以, 如果外力 $\mathbf{F}$ 的频率 ω 等于单摆的固有频率 ω_0, 即 $\omega = \sqrt{g/l}$, 则单摆的总能量 (因此 $\theta_{\max}$) 在相位差 $\alpha = 0$ 的情况下迅速增加 (或在 $\alpha = \pi$ 的情况下迅速减小): 发生所谓的共振. 然而, $\omega_0 \approx \sqrt{g/l}$ 只对小的 $\theta_{\max}$ 有效: 固有频率 ω_0 随着 $\theta_{\max}$ 的变大而增加. 所以, 随着 $\theta_{\max}$ 越来越大, 固有频率 ω_0 与外力 $\mathbf{F}$ 的频率 $\omega = \sqrt{g/l}$ 的偏差越来越大, 那么单摆从外力获得的能量越来越少: 当单摆在一个振动周期内不能再从 $\mathbf{F}$ 中获得能量时, 最大振动角 $\theta_{\max}$ 停止增加.

重力波共振现象在物理上与其本质上是相似的. 对于波数为 $\mathbf{k}'$ 和 "固有" 角频率为 σ'_0 的单个行进波, 当存在具有相同的角频率 σ' 的 "外部" 周期扰动时, 即 $\sigma' = \sigma'_0$, 就会发生共振. 如上所述, 即使对于大幅值波, 这种共振机制在物理上也是合理的.

让我们考虑具有波数 $\mathbf{k}_n$ 和角频率 σ_n 的 κ 个行进基波, 其中 $1 \leqslant n \leqslant \kappa$. 由于非线性相互作用, 存在无穷多个波分量

$$\cos\left(\sum_{n=1}^{\kappa} m_n\, \xi_n\right),$$

其中 m_n 为整数, 可以是负整数、零或正整数. 注意,

$$\sum_{n=1}^{\kappa} m_n\, \xi_n = \left(\sum_{n=1}^{\kappa} m_n \mathbf{k}_n\right) \cdot \mathbf{r} - \left(\sum_{n=1}^{\kappa} m_n\, \sigma_n\right) t.$$

所以

$$\mathbf{k}' = \sum_{n=1}^{\kappa} m_n\, \mathbf{k}_n \tag{16.78}$$

为非线性相互作用波的波数以及

$$\sigma' = \left|\sum_{n=1}^{\kappa} m_n\, \sigma_n\right| \tag{16.79}$$

为对应的角频率. 为了简单起见, 我们称 $\mathbf{k}'$ 为非线性相互作用波数以及 σ' 为非线性相互作用角频率, 其中 $\sum_{n=1}^{\kappa} m_n^2 > 1$.

对于小幅值波, 我们有 $\sigma_n \approx \sqrt{g\,k_n} = \bar{\sigma}_n$, 因此有

$$\left|\sum_{n=1}^{\kappa} m_n\,\bar{\sigma}_n\right| \approx \left|\sum_{n=1}^{\kappa} m_n\,\sigma_n\right| = \sigma', \tag{16.80}$$

将上述表达式和 (16.78) 代入 (16.74), 我们得到共振条件 (对于 κ 个小幅值基波) 如下

$$g|\mathbf{k}'| = \sigma'^2,$$

即

$$\sigma' = \sqrt{g|\mathbf{k}'|}. \tag{16.81}$$

上述共振条件清楚地揭示了非线性相互作用波数 $\mathbf{k}' = \sum\limits_{n=1}^{\kappa} m_n\,\mathbf{k}_n$ 与非线性相互作用角频率 $\sigma' = \left|\sum\limits_{n=1}^{\kappa} m_n\,\sigma_n\right|$ 之间的物理关系.

令 σ'_0 表示具有波数 $\mathbf{k}'$ 和幅值 a' 的单个行进波的 "固有" 角频率. 在小幅值波的情况下, 根据线性理论, 我们有 "固有" 角频率 $\sigma'_0 \approx \sqrt{g\,|\mathbf{k}'|}$. 那么, 上述波浪共振条件变为

$$\sigma' = \sigma'_0, \tag{16.82}$$

即

$$\left|\sum_{n=1}^{\kappa} m_n\,\sigma_n\right| = \sigma'_0, \quad \sum_{n=1}^{\kappa} m_n^2 > 1. \tag{16.83}$$

从物理上讲, 当对应非线性相互作用波数为 $\mathbf{k}' = \sum\limits_{n=1}^{\kappa} m_n\,\mathbf{k}_n$ 的非线性相互作用角频率 $\sigma' = \left|\sum\limits_{n=1}^{\kappa} m_n\,\sigma_n\right|$ 等于其 "固有" 角频率 σ'_0 时, 会发生波浪共振. 注意, 与基波角频率叠加的非线性相互作用角频率 σ' 不同, 对应波数 $\mathbf{k}'$ 的 "固有" 角频率 σ'_0 只取决于波数 $\mathbf{k}'$ 及其幅值 a', 而与基波的角频率无关. 因此, 一般来说, 非线性相互作用角频率 σ' 不等于波数为 $\mathbf{k}'$ 的非线性相互作用波的 "固有" 角频率 σ'_0. 所以, 波浪共振条件确实比较特殊. 这种物理解释与传统的共振理论非常吻合. 因此, (16.82) 和 (16.83) 揭示了重力波共振的物理本质.

虽然 (16.83) 是从小幅值波的共振条件 (16.74) 推导出来的, 但重力波共振的这种物理机制具有普遍意义, 且对于大幅值波仍然适用, 即使 $\sigma_n \approx \sqrt{g\,k_n}$ 不

是一个很好的近似值. 因此, (16.83) 是任意数量的具有大幅值周期性行进基波的广义波浪共振条件. 这里需要强调的是, (16.83) 在逻辑上包含任意数量的小幅值条件下的波共振条件 (16.74), 以及 Phillips 提出的小幅值条件下四波共振条件 (16.6). 因此, 该波浪共振条件是相当普遍的.

当满足波浪共振条件 (16.83) 时, 波能从基波转移到共振波, 基波的幅值减小, 共振波的幅值增加. 然后, 每个基波的角频率 σ_n 减小, 但共振波的 "固有" 角频率增加, 因此将不再满足共振条件 (16.83). 结果表明, "固有" 角频率 σ'_0 越来越偏离非线性相互作用角频率 σ', 因此非线性相互作用波从基波中获得的能量越来越少, 直到整个波系处于平衡状态, 即充分发展. 这解释了为什么共振重力波具有有限的幅值.

如上所述, 如图 16.6 所示, 受相位差为 π 的外力作用的共振单摆失去其能量, 因此最大振动角 $\theta_{\max}$ 减小. 类似地, 当满足波浪共振条件 (16.83) 时, 波能也可能从共振波转移到基波, 使得共振波幅值减小, 基波幅值增加: 这很好地解释了为什么共振波的幅值可能比基波小得多, 如表 16.10 所示.

例如, 考虑由波数 $\mathbf{k}_1$ 和 $\mathbf{k}_2$ 表示的两个基波, 以及一个共振波 $2\mathbf{k}_1 - \mathbf{k}_2$, 其幅值分别用 $a_{1,0}, a_{0,1}$ 和 $a_{2,-1}$ 表示. 假设满足 Phillips 波浪共振条件. 如果 $a_{2,-1}$ 最初为 0, 那么 $a_{2,-1}$ 在小时间 $t \ll 1$ 内随时间呈线性增加. 此为 Phillips [393] 和 Longuet-Higgins [390] 研究的情况. 但是, 如果 $a_{2,-1}$ 最初不等于 0, 即当 $t = 0$ 时 $a_{2,-1} \neq 0$, 那么 $a_{2,-1}$ 可能增加, 即波能从基波转移到共振波; 也可能降低, 即波能从共振波转移到基波, 这取决于 $t = 0$ 时的初始条件. 从这个观点来看, 很容易理解为什么共振波幅值可能比充分发展波系中的基波幅值小得多.

更多结论, 请参考廖世俊 [387].

16.4 本章小结

本章验证了同伦分析方法对于任意数量行进波的非线性相互作用的复杂非线性偏微分方程的有效性. 在同伦分析方法框架中, 我们首次获得了任意数量小幅值波的波浪共振条件, 其在逻辑上包含了 Phillips 提出的小幅值条件下四波共振条件. 此外, 我们首次发现, 当满足波浪共振条件且波系充分发展时, 存在多个定常共振波系, 其幅值可能比基波小得多, 所以共振波可能包含总波系很小百分比的能量. 因此, Phillips 共振条件不足以保证较大的共振波幅值. 此外, 上述波浪共振条件被进一步推广到了任意数量的大幅值波, 从物理角度来说, 这为研究任意数量的具有大幅值的行进重力基波之间的强非线性相互作用开辟了一条新途径. 这个例子说明了同伦分析方法可以用来加深和丰富我们对一些相当复杂的非线性现象的理解.

需要注意的是, 任意数量的大幅值行进波的波浪共振条件 (16.83) 仅从共振的物理角度得出. 尽管其很好地解释了为什么共振波的幅值是有限的, 以及为什么它可能比基波幅值小得多, 但仍需要物理实验和 (或) 解析 (数值) 结果来支撑它. 此外, 研究任意数量的远离平衡态的行进波系的演化, 特别是不同波分量之间的波能传递也十分有趣.

本章使用的同伦多变量方法 [387] 与 Phillips [393] 和 Longuet-Higgins [390] 使用的摄动方法不同, 它不依赖于任何物理小参数, 并且为我们提供了一种便捷的方法来保证级数解的收敛. 与多尺度摄动方法一样, 其解法具有明确的物理意义. 例如, 使用同伦多变量方法 [388], 对于一个充分发展的波系, 时间 t 将不会显式出现: 这不仅在数学上极大地简化了问题的求解过程, 而且更有助于清晰地揭示相关物理意义[①], 如本章所述.

同伦多变量方法 [387] 比著名的多尺度摄动方法更通用. 通过多尺摄动方法, 人们经常将未知解 $u(t)$ 改写为

$$u(t) = u_0(T_0, T_1, T_2) + u_1(T_0, T_1, T_2)\,\varepsilon + u_2(T_0, T_1, T_2)\,\varepsilon^2 + \cdots,$$

其中

$$T_0 = t, \quad T_1 = \varepsilon\, t, \quad T_2 = \varepsilon^2\, t$$

为用物理小参数 ε 表示的不同时间尺度, 然后基于物理小参数 ε 将非线性问题转化为一系列线性摄动问题. 使用同伦多变量方法 [387], 我们也可以将 $u(t)$ 重写为 $\breve{u}(\xi_0, \xi_1, \xi_2)$, 且有定义

$$\xi_n = \varepsilon^n\, t.$$

然而, 与多尺度摄动方法不同, 我们不需要依赖任何物理小参数. 此外, 其更容易获得高阶近似值, 特别是如果多个变量被正确定义并具有明确的物理意义, 则会给出具有重要物理意义的结果, 如本章所述. 因此, 同伦多变量方法具有一般意义. 尽管未来仍需要进一步修正, 考虑到多尺度摄动方法已经得到广泛应用, 同伦多变量方法也可以应用于求解科学和工程中的多种非线性问题, 例如著名的关于畸形波的自然现象 [369, 373, 374], 任意数量远离平衡态的行进波之间的非线性相互作用等.

本书的第三部分通过求解一些非线性偏微分方程验证了同伦分析方法的有效性. 第十三章应用同伦分析方法求解金融界著名的美式认沽期权问题. 最优

[①] 一些研究人员将解展开成关于时间 t 的泰勒级数来求解非线性波的偏微分方程. 不幸的是, 这通常会导致非常复杂的解表达式, 且几乎没有任何物理意义.

行权边界 $B(\tau)$ 的显式解析近似以 $\sqrt{\tau}$ 级数展开到 $o(\tau^M)$ 的有效期可能是几十年，而其他渐近 (摄动) 公式的有效期只有几天或几周. 这种长期有效的公式从未被报道过，其对商人很有帮助. 第十四章通过将二阶二维 (或三维) 非线性 Gelfand 方程转化为无穷多个 4 阶 (或 6 阶) 线性偏微分方程以一种相当简单的方式求解该方程，这种变换从未被其他数值和解析方法采用过. 这还表明，人类求解非线性问题的自由度可能比我们传统上认为和相信的要大得多. 这些都体现了同伦分析方法的原创性和灵活性. 第十五章应用同伦分析方法成功地求解了水波与指数剪切流之间非线性相互作用的复杂非线性偏微分方程. 还首次发现，即使存在指数剪切流，传统的波浪破碎准则仍然有效. 本章基于同伦分析方法首次给出任意数量的行进波的波浪共振条件. 这些都说明了同伦分析方法可以用来求解一些复杂的非线性偏微分方程，从而加深和丰富我们对一些有趣的非线性现象的物理理解.

这里必须强调的是，同伦分析方法并非对所有非线性问题都有效，因为我们的目标是开发一种对尽可能多的非线性问题有效的解析方法. 是否存在对所有非线性问题都有效的解析近似方法，这甚至是一个悬而未决的问题. 从本质上讲，理解复杂的非线性现象是相当困难的，尤其是那些与混沌和湍流有关的现象. 正如 Rabindranth Tagore (1861—1941) 所指出的那样，"浅理字字可言明，深道讱讱默无声". 然而，尽管非线性常微分方程和偏微分方程比线性微分方程更难求解，但同伦分析方法仍然为我们研究这些方程提供了一个有用的工具.

参考文献

[369] Adcock, T.A.A., Taylor, P.H.: Focusing of unidirectional wave groups on deep water: an approximate nonlinear Schrödinger equation-based model. Proceedings of the Royal Society: A **465**, 3083–3102 (2009).

[370] Annenkov, S.Y., Shrira, V.I.: Role of non-resonant interactions in the evolution of nonlinear random water wave fields. J. Fluid Mech. **561**, 181–207 (2006).

[371] Benney, D.T.: Non-linear gravity wave interactions. J. Fluid Mech. **14**, 577–584 (1962).

[372] Bretherton, F.P.: Resonant interactions between waves: the case of discrete oscillations. J. Fluid Mech. **20**, 457–479 (1964).

[373] Gibbs, R.H., Taylor, P.H.: Formation of walls of water in "fully" nonlinear simulations. Applied Ocean Research **27**, 142–257 (2005).

[374] Kharif, C., Pelinovsky, E.: Physical mechanisms of the rogue wave phenomenon. Eur. J. Mech. B - Fluids. **22**, 603–634 (2003).

[375] Liao, S.J.: The Proposed Homotopy Analysis Technique for the Solution of Nonlinear Problems. PhD dissertation, Shanghai Jiao Tong University (1992).

[376] Liao, S.J.: A kind of approximate solution technique which does not depend upon small parameters—(II) An application in fluid mechanics. Int. J. Nonlin. Mech. **32**, 815–822 (1997).

[377] Liao, S.J.: An explicit, totally analytic approximation of Blasius viscous flow problems. Int. J. Nonlin. Mech. **34**, 759–778 (1999a).

[378] Liao, S.J.: A uniformly valid analytic solution of 2D viscous flow past a semi-infinite flat plate. J. Fluid Mech. **385**, 101–128 (1999b).

[379] Liao, S.J.: On the analytic solution of magnetohydrodynamic flows of non-Newtonian fluids over a stretching sheet. J. Fluid Mech. **488**, 189–212 (2003a).

[380] Liao, S.J.: Beyond Perturbation - Introduction to the Homotopy Analysis Method. Chapman & Hall/ CRC Press, Boca Raton (2003b).

[381] Liao, S.J.: On the homotopy analysis method for nonlinear problems. Appl. Math. Comput. **147**, 499–513 (2004).

[382] Liao, S.J.: A new branch of solutions of boundary-layer flows over an impermeable stretched plate. Int. J. Heat Mass Tran. **48**, 2529–2539 (2005).

[383] Liao, S.J.: Series solutions of unsteady boundary-layer flows over a stretching flat plate. Stud. Appl. Math. **117**, 2529–2539 (2006).

[384] Liao, S.J.: Notes on the homotopy analysis method - Some definitions and theorems. Commun. Nonlinear Sci. Numer. Simulat. **14**, 983–997 (2009).

[385] Liao, S.J.: On the relationship between the homotopy analysis method and Euler transform. Commun. Nonlinear Sci. Numer. Simulat. **15**, 1421–1431 (2010a). doi:10.1016/j.cnsns.2009.06.008.

[386] Liao, S.J.: An optimal homotopy-analysis approach for strongly nonlinear differential equations. Commun. Nonlinear Sci. Numer. Simulat. **15**, 2003–2016 (2010b).

[387] Liao, S.J.: On the homotopy multiple-variable method and its applications in the interactions of nonlinear gravity waves. Commun. Nonlinear Sci. Numer. Simulat. **16**, 1274–1303 (2011).

[388] Liao, S.J., Campo, A.: Analytic solutions of the temperature distribution in Blasius viscous flow problems. J. Fluid Mech. **453**, 411–425 (2002).

[389] Liao, S.J., Tan, Y.: A general approach to obtain series solutions of nonlinear differential equations. Stud. Appl. Math. **119**, 297–355 (2007).

[390] Longuet-Higgins, M.S.: Resonant interactions between two trains of gravity waves. J. Fluid Mech. **12**, 321–332 (1962).

[391] Longuet-Higgins, M.S., Smith, N.D.: An experiment on third order resonant wave interactions. J. Fluid Mech. **25**, 417–435 (1966).

[392] McGoldrick, L.F., Phillips, O.M., Huang, N., Hodgson, T.: Measurements on resonant wave interactions. J. Fluid Mech. **25**, 437–456 (1966).

[393] Phillips, O.M.: On the dynamics of unsteady gravity waves of finite amplitude. Part 1. The elementary interactions. J. Fluid Mech. **9**, 193–217 (1960).

[394] Phillips, O.M.:Wave interactions – the evolution of an idea. J. Fluid Mech. **106**, 215–227 (1981).

附录 16.1　高阶形变方程的详细推导

写作

$$\left(\sum_{i=1}^{+\infty} \eta_i\, q^i\right)^m = \sum_{n=m}^{+\infty} \mu_{m,n}\, q^n, \tag{16.84}$$

包含定义

$$\mu_{1,n}(\xi_1,\xi_2) = \eta_n(\xi_1,\xi_2), \quad n \geqslant 1. \tag{16.85}$$

然后

$$\left(\sum_{i=1}^{+\infty} \eta_i\, q^i\right)^{m+1} = \left(\sum_{n=m}^{+\infty} \mu_{m,n}\, q^n\right)\left(\sum_{i=1}^{+\infty} \eta_i\, q^i\right) = \sum_{n=m+1}^{+\infty} \mu_{m+1,n}\, q^n, \tag{16.86}$$

有

$$\mu_{m,n}(\xi_1,\xi_2) = \sum_{i=m-1}^{n-1} \mu_{m-1,i}(\xi_1,\xi_2)\, \eta_{n-i}(\xi_1,\xi_2), \quad m \geqslant 2,\, n \geqslant m. \tag{16.87}$$

定义

$$\psi_{i,j}^{n,m}(\xi_1,\xi_2) = \frac{\partial^{i+j}}{\partial \xi_1^i \partial \xi_2^j}\left(\frac{1}{m!}\left.\frac{\partial^m \phi_n}{\partial z^m}\right|_{z=0}\right).$$

根据泰勒级数，对任意的 z 我们有

$$\phi_n(\xi_1,\xi_2,z) = \sum_{m=0}^{+\infty}\left(\frac{1}{m!}\left.\frac{\partial^m \phi_n}{\partial z^m}\right|_{z=0}\right) z^m = \sum_{m=0}^{+\infty} \psi_{0,0}^{n,m}\, z^m \tag{16.88}$$

和

$$\frac{\partial^{i+j}\phi_n}{\partial \xi_1^i \partial \xi_2^j} = \sum_{m=0}^{+\infty} \frac{\partial^{i+j}}{\partial \xi_1^i \partial \xi_2^j}\left(\frac{1}{m!}\left.\frac{\partial^m \phi_n}{\partial z^m}\right|_{z=0}\right) z^m = \sum_{m=0}^{+\infty} \psi_{i,j}^{n,m}\, z^m. \tag{16.89}$$

然后，在 $z = \check{\eta}(\xi_1, \xi_2; q)$ 时，基于 (16.84) 我们有

$$\frac{\partial^{i+j}\phi_n}{\partial \xi_1^i \partial \xi_2^j} = \sum_{m=0}^{+\infty} \psi_{i,j}^{n,m} \left(\sum_{s=1}^{+\infty} \eta_s\, q^s\right)^m = \psi_{i,j}^{n,0} + \sum_{m=1}^{+\infty} \psi_{i,j}^{n,m} \left(\sum_{s=m}^{+\infty} \mu_{m,s}\, q^s\right)$$

$$= \sum_{m=0}^{+\infty} \beta_{i,j}^{n,m}(\xi_1, \xi_2)\, q^m, \qquad (16.90)$$

其中

$$\beta_{i,j}^{n,0} = \psi_{i,j}^{n,0}, \qquad (16.91)$$

$$\beta_{i,j}^{n,m} = \sum_{s=1}^{m} \psi_{i,j}^{n,s}\, \mu_{s,m}, \quad m \geqslant 1. \qquad (16.92)$$

类似地，在 $z = \check{\eta}(\xi_1, \xi_2; q)$ 时，有

$$\frac{\partial^{i+j}}{\partial \xi_1^i \partial \xi_2^j}\left(\frac{\partial \phi_n}{\partial z}\right) = \sum_{m=0}^{+\infty} \gamma_{i,j}^{n,m}(\xi_1, \xi_2)\, q^m, \qquad (16.93)$$

$$\frac{\partial^{i+j}}{\partial \xi_1^i \partial \xi_2^j}\left(\frac{\partial^2 \phi_n}{\partial z^2}\right) = \sum_{m=0}^{+\infty} \delta_{i,j}^{n,m}(\xi_1, \xi_2)\, q^m, \qquad (16.94)$$

其中

$$\gamma_{i,j}^{n,0} = \psi_{i,j}^{n,1}, \qquad (16.95)$$

$$\gamma_{i,j}^{n,m} = \sum_{s=1}^{m} (s+1)\psi_{i,j}^{n,s+1}\, \mu_{s,m}, \quad m \geqslant 1, \qquad (16.96)$$

$$\delta_{i,j}^{n,0} = 2\psi_{i,j}^{n,2}, \qquad (16.97)$$

$$\delta_{i,j}^{n,m} = \sum_{s=1}^{m} (s+1)(s+2)\psi_{i,j}^{n,s+2}\, \mu_{s,m}, \quad m \geqslant 1. \qquad (16.98)$$

然后，在 $z = \check{\eta}(\xi_1, \xi_2; q)$ 时，基于 (16.90) 我们有

$$\check{\phi}(\xi_1, \xi_2, \check{\eta}; q) = \sum_{n=0}^{+\infty} \phi_n(\xi_1, \xi_2, \check{\eta})\, q^n = \sum_{n=0}^{+\infty} q^n \left[\sum_{m=0}^{+\infty} \beta_{0,0}^{n,m}(\xi_1, \xi_2)\, q^m\right]$$

$$= \sum_{n=0}^{+\infty} \sum_{m=0}^{+\infty} \beta_{0,0}^{n,m}(\xi_1, \xi_2)\, q^{m+n} = \sum_{s=0}^{+\infty} q^s \left[\sum_{m=0}^{s} \beta_{0,0}^{s-m,m}(\xi_1, \xi_2)\right]$$

$$= \sum_{n=0}^{+\infty} \bar{\phi}_n^{0,0}(\xi_1, \xi_2)\, q^n, \qquad (16.99)$$

其中
$$\bar{\phi}_n^{0,0}(\xi_1,\xi_2) = \sum_{m=0}^{n} \beta_{0,0}^{n-m,m}. \tag{16.100}$$

类似地，我们有
$$\frac{\partial^{i+j}\check{\phi}}{\partial \xi_1^i \partial \xi_2^j} = \sum_{n=0}^{+\infty} \bar{\phi}_n^{i,j}(\xi_1,\xi_2)\, q^n, \tag{16.101}$$

$$\frac{\partial^{i+j}}{\partial \xi_1^i \partial \xi_2^j}\left(\frac{\partial \check{\phi}}{\partial z}\right) = \sum_{n=0}^{+\infty} \bar{\phi}_{z,n}^{i,j}(\xi_1,\xi_2)\, q^n, \tag{16.102}$$

$$\frac{\partial^{i+j}}{\partial \xi_1^i \partial \xi_2^j}\left(\frac{\partial^2 \check{\phi}}{\partial z^2}\right) = \sum_{n=0}^{+\infty} \bar{\phi}_{zz,n}^{i,j}(\xi_1,\xi_2)\, q^n, \tag{16.103}$$

其中
$$\bar{\phi}_n^{i,j}(\xi_1,\xi_2) = \sum_{m=0}^{n} \beta_{i,j}^{n-m,m}, \tag{16.104}$$

$$\bar{\phi}_{z,n}^{i,j}(\xi_1,\xi_2) = \sum_{m=0}^{n} \gamma_{i,j}^{n-m,m}, \tag{16.105}$$

$$\bar{\phi}_{zz,n}^{i,j}(\xi_1,\xi_2) = \sum_{m=0}^{n} \delta_{i,j}^{n-m,m}. \tag{16.106}$$

然后，在 $z = \check{\eta}(\xi_1,\xi_2;q)$ 时，基于 (16.101) 和 (16.102) 我们有

$$\begin{aligned}\check{f} &= \frac{1}{2}\hat{\nabla}\check{\phi}\cdot\hat{\nabla}\check{\phi} \\ &= \frac{k_1^2}{2}\left(\frac{\partial \check{\phi}}{\partial \xi_1}\right)^2 + \mathbf{k}_1\cdot\mathbf{k}_2\frac{\partial \check{\phi}}{\partial \xi_1}\frac{\partial \check{\phi}}{\partial \xi_2} + \frac{k_2^2}{2}\left(\frac{\partial \check{\phi}}{\partial \xi_2}\right)^2 + \frac{1}{2}\left(\frac{\partial \check{\phi}}{\partial z}\right)^2 \\ &= \sum_{m=0}^{+\infty}\Gamma_{m,0}(\xi_1,\xi_2)\, q^m, \end{aligned} \tag{16.107}$$

其中
$$\begin{aligned}\Gamma_{m,0}(\xi_1,\xi_2) = &\frac{k_1^2}{2}\sum_{n=0}^{m}\bar{\phi}_n^{1,0}\bar{\phi}_{m-n}^{1,0} + \mathbf{k}_1\cdot\mathbf{k}_2\sum_{n=0}^{m}\bar{\phi}_n^{1,0}\bar{\phi}_{m-n}^{0,1} \\ &+ \frac{k_2^2}{2}\sum_{n=0}^{m}\bar{\phi}_n^{0,1}\bar{\phi}_{m-n}^{0,1} + \frac{1}{2}\sum_{n=0}^{m}\bar{\phi}_{z,n}^{0,0}\bar{\phi}_{z,m-n}^{0,0}.\end{aligned} \tag{16.108}$$

类似地，在 $z = \check{\eta}(\xi_1, \xi_2; q)$ 时，有

$$\frac{\partial \check{f}}{\partial \xi_1} = \hat{\nabla}\check{\phi} \cdot \hat{\nabla}\left(\frac{\partial \check{\phi}}{\partial \xi_1}\right)$$

$$= k_1^2 \frac{\partial \check{\phi}}{\partial \xi_1} \frac{\partial^2 \check{\phi}}{\partial \xi_1^2} + k_2^2 \frac{\partial \check{\phi}}{\partial \xi_2} \frac{\partial^2 \check{\phi}}{\partial \xi_1 \partial \xi_2} + \frac{\partial \check{\phi}}{\partial z} \frac{\partial}{\partial \xi_1}\left(\frac{\partial \check{\phi}}{\partial z}\right)$$

$$+ \mathbf{k}_1 \cdot \mathbf{k}_2 \left(\frac{\partial \check{\phi}}{\partial \xi_1} \frac{\partial^2 \check{\phi}}{\partial \xi_1 \partial \xi_2} + \frac{\partial \check{\phi}}{\partial \xi_2} \frac{\partial^2 \check{\phi}}{\partial \xi_1^2}\right)$$

$$= \sum_{m=0}^{+\infty} \Gamma_{m,1}(\xi_1, \xi_2)\, q^m, \qquad (16.109)$$

$$\frac{\partial \check{f}}{\partial \xi_2} = \hat{\nabla}\check{\phi} \cdot \hat{\nabla}\left(\frac{\partial \check{\phi}}{\partial \xi_2}\right)$$

$$= k_1^2 \frac{\partial \check{\phi}}{\partial \xi_1} \frac{\partial^2 \check{\phi}}{\partial \xi_1 \partial \xi_2} + k_2^2 \frac{\partial \check{\phi}}{\partial \xi_2} \frac{\partial^2 \check{\phi}}{\partial \xi_2^2} + \frac{\partial \check{\phi}}{\partial z} \frac{\partial}{\partial \xi_2}\left(\frac{\partial \check{\phi}}{\partial z}\right)$$

$$+ \mathbf{k}_1 \cdot \mathbf{k}_2 \left(\frac{\partial \check{\phi}}{\partial \xi_1} \frac{\partial^2 \check{\phi}}{\partial \xi_2^2} + \frac{\partial \check{\phi}}{\partial \xi_2} \frac{\partial^2 \check{\phi}}{\partial \xi_1 \partial \xi_2}\right)$$

$$= \sum_{m=0}^{+\infty} \Gamma_{m,2}(\xi_1, \xi_2)\, q^m, \qquad (16.110)$$

其中

$$\Gamma_{m,1}(\xi_1, \xi_2) = \sum_{n=0}^{m} \left(k_1^2\, \bar{\phi}_n^{1,0}\, \bar{\phi}_{m-n}^{2,0} + k_2^2\, \bar{\phi}_n^{0,1}\, \bar{\phi}_{m-n}^{1,1} + \bar{\phi}_{z,n}^{0,0}\, \bar{\phi}_{z,m-n}^{1,0}\right)$$

$$+ \mathbf{k}_1 \cdot \mathbf{k}_2 \sum_{n=0}^{m} \left(\bar{\phi}_n^{1,0}\, \bar{\phi}_{m-n}^{1,1} + \bar{\phi}_n^{2,0}\, \bar{\phi}_{m-n}^{0,1}\right), \qquad (16.111)$$

$$\Gamma_{m,2}(\xi_1, \xi_2) = \sum_{n=0}^{m} \left(k_1^2\, \bar{\phi}_n^{1,0}\, \bar{\phi}_{m-n}^{1,1} + k_2^2\, \bar{\phi}_n^{0,1}\, \bar{\phi}_{m-n}^{0,2} + \bar{\phi}_{z,n}^{0,0}\, \bar{\phi}_{z,m-n}^{0,1}\right)$$

$$+ \mathbf{k}_1 \cdot \mathbf{k}_2 \sum_{n=0}^{m} \left(\bar{\phi}_n^{1,0}\, \bar{\phi}_{m-n}^{0,2} + \bar{\phi}_n^{0,1}\, \bar{\phi}_{m-n}^{1,1}\right). \qquad (16.112)$$

此外，在 $z = \check{\eta}(\xi_1, \xi_2; q)$ 时，基于 (16.101), (16.102) 和 (16.103)，我们有

$$\frac{\partial \check{f}}{\partial z} = \hat{\nabla}\check{\phi} \cdot \hat{\nabla}\left(\frac{\partial \check{\phi}}{\partial z}\right)$$

$$= k_1^2 \frac{\partial \check{\phi}}{\partial \xi_1} \frac{\partial}{\partial \xi_1}\left(\frac{\partial \check{\phi}}{\partial z}\right) + k_2^2 \frac{\partial \check{\phi}}{\partial \xi_2} \frac{\partial}{\partial \xi_2}\left(\frac{\partial \check{\phi}}{\partial z}\right) + \frac{\partial \check{\phi}}{\partial z} \frac{\partial^2 \check{\phi}}{\partial z^2}$$

$$+ \mathbf{k}_1 \cdot \mathbf{k}_2 \left[\frac{\partial \check{\phi}}{\partial \xi_1} \frac{\partial}{\partial \xi_2}\left(\frac{\partial \check{\phi}}{\partial z}\right) + \frac{\partial \check{\phi}}{\partial \xi_2} \frac{\partial}{\partial \xi_1}\left(\frac{\partial \check{\phi}}{\partial z}\right)\right]$$

$$= \sum_{m=0}^{+\infty} \Gamma_{m,3}(\xi_1,\xi_2)\, q^m, \tag{16.113}$$

其中

$$\Gamma_{m,3}(\xi_1,\xi_2) = \sum_{n=0}^{m} \left(k_1^2\, \bar{\phi}_n^{1,0}\, \bar{\phi}_{z,m-n}^{1,0} + k_2^2\, \bar{\phi}_n^{0,1}\, \bar{\phi}_{z,m-n}^{0,1} + \bar{\phi}_{z,n}^{0,0}\, \bar{\phi}_{zz,m-n}^{0,0} \right)$$

$$+ \mathbf{k}_1 \cdot \mathbf{k}_2 \sum_{n=0}^{m} \left(\bar{\phi}_n^{1,0}\, \bar{\phi}_{z,m-n}^{0,1} + \bar{\phi}_n^{0,1}\, \bar{\phi}_{z,m-n}^{1,0} \right). \tag{16.114}$$

进一步，根据 (16.101), (16.109), (16.110) 和 (16.113)，我们有

$$\hat{\nabla}\check{\phi}\cdot\hat{\nabla}\check{f} = k_1^2 \frac{\partial\check{\phi}}{\partial\xi_1}\frac{\partial\check{f}}{\partial\xi_1} + k_2^2 \frac{\partial\check{\phi}}{\partial\xi_2}\frac{\partial\check{f}}{\partial\xi_2} + \frac{\partial\check{\phi}}{\partial z}\frac{\partial\check{f}}{\partial z}$$

$$+ \mathbf{k}_1 \cdot \mathbf{k}_2 \left(\frac{\partial\check{\phi}}{\partial\xi_1}\frac{\partial\check{f}}{\partial\xi_2} + \frac{\partial\check{\phi}}{\partial\xi_2}\frac{\partial\check{f}}{\partial\xi_1} \right)$$

$$= \sum_{m=0}^{+\infty} \Lambda_m(\xi_1,\xi_2)\, q^m, \tag{16.115}$$

其中

$$\Lambda_m(\xi_1,\xi_2) = \sum_{n=0}^{m} \left(k_1^2\, \bar{\phi}_n^{1,0}\, \Gamma_{m-n,1} + k_2^2\, \bar{\phi}_n^{0,1}\, \Gamma_{m-n,2} + \bar{\phi}_{z,n}^{0,0}\, \Gamma_{m-n,3} \right)$$

$$+ \mathbf{k}_1 \cdot \mathbf{k}_2 \sum_{n=0}^{m} \left(\bar{\phi}_n^{1,0}\, \Gamma_{m-n,2} + \bar{\phi}_n^{0,1}\, \Gamma_{m-n,1} \right). \tag{16.116}$$

然后，根据 (16.101), (16.102), (16.109), (16.110) 和 (16.115)，在 $z = \check{\eta}(\xi_1,\xi_2;q)$ 时，我们有

$$\mathcal{N}\left[\check{\phi}(\xi_1,\xi_2,z;q)\right]$$

$$= \sigma_1^2 \frac{\partial^2\check{\phi}}{\partial\xi_1^2} + 2\sigma_1\sigma_2 \frac{\partial^2\check{\phi}}{\partial\xi_1\partial\xi_2} + \sigma_2^2 \frac{\partial^2\check{\phi}}{\partial\xi_2^2} + g\frac{\partial\check{\phi}}{\partial z}$$

$$- 2\left(\sigma_1 \frac{\partial\check{f}}{\partial\xi_1} + \sigma_2 \frac{\partial\check{f}}{\partial\xi_2} \right) + \hat{\nabla}\check{\phi}\cdot\hat{\nabla}\check{f}$$

$$= \sum_{m=0}^{+\infty} \Delta_m^{\phi}(\xi_1,\xi_2)\, q^m, \tag{16.117}$$

其中对于 $m \geqslant 0$，有

$$\Delta_m^{\phi}(\xi_1,\xi_2) = \sigma_1^2\, \bar{\phi}_m^{2,0} + 2\sigma_1\sigma_2\, \bar{\phi}_m^{1,1} + \sigma_2^2\, \bar{\phi}_m^{0,2} + g\bar{\phi}_{z,m}^{0,0}$$

$$-2\left(\sigma_1\,\Gamma_{m,1}+\sigma_2\,\Gamma_{m,2}\right)+\Lambda_m. \tag{16.118}$$

根据 (16.34) 和 (16.90), 在 $z=\check{\eta}(\xi_1,\xi_2;q)$ 时, 我们有

$$\check{\phi}-\phi_0=\sum_{n=1}^{+\infty}\phi_n(\xi_1,\xi_2,\check{\eta})\,q^n=\sum_{n=1}^{+\infty}q^n\left(\sum_{m=0}^{+\infty}\beta_{0,0}^{n,m}\,q^m\right)$$
$$=\sum_{n=1}^{+\infty}q^n\left(\sum_{m=0}^{n-1}\beta_{0,0}^{n-m,m}\right) \tag{16.119}$$

以及类似的

$$\frac{\partial}{\partial z}\left(\check{\phi}-\phi_0\right)=\sum_{n=1}^{+\infty}\frac{\partial\phi_n}{\partial z}\,q^n=\sum_{n=1}^{+\infty}q^n\left(\sum_{m=0}^{+\infty}\gamma_{0,0}^{n,m}\,q^m\right)$$
$$=\sum_{n=1}^{+\infty}q^n\left(\sum_{m=0}^{n-1}\gamma_{0,0}^{n-m,m}\right). \tag{16.120}$$

然后, 在 $z=\check{\eta}(\xi_1,\xi_2;q)$ 时, 由算子 (16.26) 的线性性质, 有

$$\mathcal{L}\left(\check{\phi}-\phi_0\right)=\sum_{n=1}^{+\infty}S_n(\xi_1,\xi_2)\,q^n, \tag{16.121}$$

其中

$$S_n(\xi_1,\xi_2)$$
$$=\sum_{m=0}^{n-1}\left(\bar{\sigma}_1^2\,\beta_{2,0}^{n-m,m}+2\bar{\sigma}_1\bar{\sigma}_2\,\beta_{1,1}^{n-m,m}+\bar{\sigma}_2^2\,\beta_{0,2}^{n-m,m}+g\,\gamma_{0,0}^{n-m,m}\right). \tag{16.122}$$

然后, 在 $z=\check{\eta}(\xi_1,\xi_2;q)$ 时, 有

$$(1-q)\mathcal{L}\left(\check{\phi}-\phi_0\right)=(1-q)\sum_{n=1}^{+\infty}S_n\,q^n=\sum_{n=1}^{+\infty}(S_n-\chi_n\,S_{n-1})\,q^n, \tag{16.123}$$

其中

$$\chi_n=\begin{cases}0, & n\leqslant 1,\\ 1, & n>1.\end{cases} \tag{16.124}$$

将 (16.123), (16.117) 代入 (16.27), 取 q 的相同次幂为等式, 我们有边界条件

$$S_m(\xi_1,\xi_2)-\chi_m\,S_{m-1}(\xi_1,\xi_2)=c_0\,\Delta_{m-1}^{\phi}(\xi_1,\xi_2),\quad m\geqslant 1. \tag{16.125}$$

定义

$$\bar{S}_n(\xi_1, \xi_2) = \sum_{m=1}^{n-1} \left(\bar{\sigma}_1^2 \, \beta_{2,0}^{n-m,m} + 2\bar{\sigma}_1\bar{\sigma}_2 \, \beta_{1,1}^{n-m,m} + \bar{\sigma}_2^2 \, \beta_{0,2}^{n-m,m} + g \, \gamma_{0,0}^{n-m,m} \right). \quad (16.126)$$

然后

$$\begin{aligned} S_n &= \left(\bar{\sigma}_1^2 \, \beta_{2,0}^{n,0} + 2\bar{\sigma}_1\bar{\sigma}_2 \, \beta_{1,1}^{n,0} + \bar{\sigma}_2^2 \, \beta_{0,2}^{n,0} + g \, \gamma_{0,0}^{n,0} \right) + \bar{S}_n \\ &= \left(\bar{\sigma}_1^2 \, \frac{\partial^2 \phi_n}{\partial \xi_1^2} + 2\bar{\sigma}_1\bar{\sigma}_2 \, \frac{\partial^2 \phi_n}{\partial \xi_1 \partial \xi_2} + \bar{\sigma}_2^2 \, \frac{\partial^2 \phi_n}{\partial \xi_2^2} + g \, \frac{\partial \phi_n}{\partial z} \right) \bigg|_{z=0} + \bar{S}_n. \quad (16.127) \end{aligned}$$

将上述表达式代入 (16.125), 在 $z=0$ 处的边界条件如下

$$\bar{\mathcal{L}}(\phi_m) = c_0 \, \Delta_{m-1}^\phi + \chi_m \, S_{m-1} - \bar{S}_m, \quad m \geqslant 1, \quad (16.128)$$

其中 $\bar{\mathcal{L}}$ 由 (16.43) 定义.

将 (16.35), (16.101) 和 (16.107) 代入 (16.29), 取 q 的相同次幂为等式, 我们有

$$\eta_m(\xi_1, \xi_2) = c_0 \, \Delta_{m-1}^\eta + \chi_m \, \eta_{m-1}, \quad m \geqslant 1, \quad (16.129)$$

其中

$$\Delta_m^\eta = \eta_m - \frac{1}{g} \left(\sigma_1 \, \bar{\phi}_m^{1,0} + \sigma_2 \, \bar{\phi}_m^{0,1} - \Gamma_{m,0} \right).$$

现代数学基础图书清单

序号	书号	书名	作者
1	9787040217179	代数和编码（第三版）	万哲先 编著
2	9787040221749	应用偏微分方程讲义	姜礼尚、孔德兴、陈志浩
3	9787040235975	实分析（第二版）	程民德、邓东皋、龙瑞麟 编著
4	9787040226171	高等概率论及其应用	胡迪鹤 著
5	9787040243079	线性代数与矩阵论（第二版）	许以超 编著
6	9787040244656	矩阵论	詹兴致
7	9787040244618	可靠性统计	茆诗松、汤银才、王玲玲 编著
8	9787040247503	泛函分析第二教程（第二版）	夏道行 等编著
9	9787040253177	无限维空间上的测度和积分——抽象调和分析（第二版）	夏道行 著
10	9787040257724	奇异摄动问题中的渐近理论	倪明康、林武忠
11	9787040272611	整体微分几何初步（第三版）	沈一兵 编著
12	9787040263602	数论 I——Fermat 的梦想和类域论	[日]加藤和也、黑川信重、斋藤毅 著
13	9787040263619	数论 II——岩泽理论和自守形式	[日]黑川信重、栗原将人、斋藤毅 著
14	9787040380408	微分方程与数学物理问题（中文校订版）	[瑞典]纳伊尔·伊布拉基莫夫 著
15	9787040274868	有限群表示论（第二版）	曹锡华、时俭益
16	9787040274318	实变函数论与泛函分析（上册，第二版修订本）	夏道行 等编著
17	9787040272482	实变函数论与泛函分析（下册，第二版修订本）	夏道行 等编著
18	9787040287073	现代极限理论及其在随机结构中的应用	苏淳、冯群强、刘杰 著
19	9787040304480	偏微分方程	孔德兴
20	9787040310696	几何与拓扑的概念导引	古志鸣 编著
21	9787040316117	控制论中的矩阵计算	徐树方 著
22	9787040316988	多项式代数	王东明 等编著
23	9787040319668	矩阵计算六讲	徐树方、钱江 著
24	9787040319583	变分学讲义	张恭庆 编著
25	9787040322811	现代极小曲面讲义	[巴西] F. Xavier、潮小李 编著
26	9787040327113	群表示论	丘维声 编著
27	9787040346756	可靠性数学引论（修订版）	曹晋华、程侃 著
28	9787040343113	复变函数专题选讲	余家荣、路见可 主编

续表

序号	书号	书名	作者
29	9787040357387	次正常算子解析理论	夏道行
30	9787040348347	数论——从同余的观点出发	蔡天新
31	9787040362688	多复变函数论	萧荫堂、陈志华、钟家庆
32	9787040361681	工程数学的新方法	蒋耀林
33	9787040345254	现代芬斯勒几何初步	沈一兵、沈忠民
34	9787040364729	数论基础	潘承洞 著
35	9787040369502	Toeplitz 系统预处理方法	金小庆 著
36	9787040370379	索伯列夫空间	王明新
37	9787040372526	伽罗瓦理论——天才的激情	章璞 著
38	9787040372663	李代数(第二版)	万哲先 编著
39	9787040386516	实分析中的反例	汪林
40	9787040388909	泛函分析中的反例	汪林
41	9787040373783	拓扑线性空间与算子谱理论	刘培德
42	9787040318456	旋量代数与李群、李代数	戴建生 著
43	9787040332605	格论导引	方捷
44	9787040395037	李群讲义	项武义、侯自新、孟道骥
45	9787040395020	古典几何学	项武义、王申怀、潘养廉
46	9787040404586	黎曼几何初步	伍鸿熙、沈纯理、虞言林
47	9787040410570	高等线性代数学	黎景辉、白正简、周国晖
48	9787040413052	实分析与泛函分析(续论)(上册)	匡继昌
49	9787040412857	实分析与泛函分析(续论)(下册)	匡继昌
50	9787040412239	微分动力系统	文兰
51	9787040413502	阶的估计基础	潘承洞、于秀源
52	9787040415131	非线性泛函分析(第三版)	郭大钧
53	9787040414080	代数学(上)(第二版)	莫宗坚、蓝以中、赵春来
54	9787040414202	代数学(下)(修订版)	莫宗坚、蓝以中、赵春来
55	9787040418736	代数编码与密码	许以超、马松雅 编著
56	9787040439137	数学分析中的问题和反例	汪林
57	9787040440485	椭圆型偏微分方程	刘宪高

续表

序号	书号	书名	作者
58	9787040464832	代数数论	黎景辉
59	9787040456134	调和分析	林钦诚
60	9787040468625	紧黎曼曲面引论	伍鸿熙、吕以辇、陈志华
61	9787040476743	拟线性椭圆型方程的现代变分方法	沈尧天、王友军、李周欣
62	9787040479263	非线性泛函分析	袁荣
63	9787040496369	现代调和分析及其应用讲义	苗长兴
64	9787040497595	拓扑空间与线性拓扑空间中的反例	汪林
65	9787040505498	Hilbert 空间上的广义逆算子与 Fredholm 算子	海国君、阿拉坦仓
66	9787040507249	基础代数学讲义	章璞、吴泉水
67.1	9787040507256	代数学方法（第一卷）基础架构	李文威
67.2	9787040627541	代数学方法（第二卷）线性代数	李文威
68	9787040522631	科学计算中的偏微分方程数值解法	张文生
69	9787040534597	非线性分析方法	张恭庆
70	9787040544893	旋量代数与李群、李代数（修订版）	戴建生
71	9787040548846	黎曼几何选讲	伍鸿熙、陈维桓
72	9787040550726	从三角形内角和谈起	虞言林
73	9787040563665	流形上的几何与分析	张伟平、冯惠涛
74	9787040562101	代数几何讲义	胥鸣伟
75	9787040580457	分形和现代分析引论	马力
76	9787040583915	微分动力系统（修订版）	文兰
77	9787040586534	无穷维 Hamilton 算子谱分析	阿拉坦仓、吴德玉、黄俊杰、侯国林
78	9787040587456	p 进数	冯克勤
79	9787040592269	调和映照讲义	丘成桐、孙理察
80	9787040603392	有限域上的代数曲线：理论和通信应用	冯克勤、刘凤梅、廖群英
81	9787040603568	代数几何（英文版，第二版）	扶磊
82	9787040621068	代数基础：模、范畴、同调代数与层（修订版）	陈志杰
83	9787040621761	微分方程和代数	黎景辉
84.1	9787040627602	非线性微分方程的同伦分析方法（上卷）	廖世俊著

续表

序号	书号	书名	作者
84.2	9787040629032	非线性微分方程的同伦分析方法（下卷）	廖世俊著
85		常微分方程（第二版）	Stephen Salaff、丘成桐
86	9787040619010	分析学（第二版）	Elliott H. Lieb, Michael Loss

购书网站：高教书城（www.hepmall.com.cn），高教天猫（gdjycbs.tmall.com），京东，当当，微店

其他订购办法：

各使用单位可向高等教育出版社电子商务部汇款订购。书款通过银行转账，支付成功后请将购买信息发邮件或传真，以便及时发货。购书免邮费，发票随书寄出（大批量订购图书，发票随后寄出）。

通过银行转账：

户　　名：高等教育出版社有限公司
开　户　行：交通银行北京马甸支行
银行账号：110060437018010037603

单位地址：北京西城区德外大街4号
电　　话：010-58581118
传　　真：010-58581113
电子邮箱：gjdzfwb@pub.hep.cn

郑重声明

高等教育出版社依法对本书享有专有出版权。任何未经许可的复制、销售行为均违反《中华人民共和国著作权法》，其行为人将承担相应的民事责任和行政责任；构成犯罪的，将被依法追究刑事责任。为了维护市场秩序，保护读者的合法权益，避免读者误用盗版书造成不良后果，我社将配合行政执法部门和司法机关对违法犯罪的单位和个人进行严厉打击。社会各界人士如发现上述侵权行为，希望及时举报，我社将奖励举报有功人员。

反盗版举报电话　（010）58581999　58582371
反盗版举报邮箱　dd@hep.com.cn
通信地址　北京市西城区德外大街4号　高等教育出版社知识产权与法律事务部
邮政编码　100120